HORTICULTURE BASED INTEGRATED FARMING SYSTEMS

About the Editors

Dr. Anil Kumar Shukla worked as Scientist and Sr. Scientist at CIAH, Bikaner, Associate Professor at Maharana Pratap University of Agriculture and Technology, Udaipur and as Principal Scientist (Horticulture) at Directorate of Research on Women in Agriculture, Bhubaneswar, Odisha. Presently, he is head, ICAR-CAZRI, RRS, Pali (Rajasthan). He is the recipient of N.E. Borlaug Fellowship for year 2008; Fellowship Award 2012, from Indian Society of Horticultural Research and Development, Uttarakhand (Society); Fellowship Award 2012, from Hi-Tech Horticultural Society, Meerut (Society); Fellowship Award 2012, from Confederation of Horticulture Association of India, New Delhi (Society); Fellowship Award 2018, from Indian Society of Arid Horticulture, Bikaner; Fellowship Award 2019, ISNS, Chennai; Fellowship Award 2019, IASH, New Delhi; Eminent Scientist Award 2016, from SVWS, Lucknow; Himadri Young Scientist Award 2010 from GBPUAT, Pant Nagar, Uttarakhand and Indian Society of Horticulture Research and Development, Uttarakhand.

Dr. Anil Kumar Shukla immensely contributed in developing two varieties of Ber (Thar Sevika and Thar Bhubraj) and two varieties of guava MPUAT-S-1 and MPUAT-S-2. He has developed several Agro techniques in ber and guava crops. He has guided 02 Ph.D. and 01 M.Sc. (Ag) students as Major Advisor and 05 Ph.D. and 04 M Sc. (Ag.) students as Co-Advisor. He has published 70, research papers, 38 book chapters, 76 research abstracts, 68 popular articles, 11 bulletins and 03 books.

Dr. Dipak Kumar Gupta is presently working as Scientist (Environmental science) at ICAR- Central Arid Zone Research Institute, Regional Research Station, Pali-Marwar. He obtained his B.Sc. (Ag) from Birsa Agricultural University, Ranchi, Jharkhand and M.Sc and Ph.D. (Environmental Science) from IARI, Pusa, New Delhi. He has received Junior and Senior research fellowship of ICAR and CSIR as well as NET of ICAR, UGC and CSIR. He has received Jawaharlal Nehru Award 2016 by ICAR, New Delhi and Young Scientist Award by S&T SIRI, Thorrur, Telangana. He is working in the field of remediation of contaminated water and soil; mitigation of greenhouse gas emission; biomass and soil carbon sequestration and impact of climate change on crops. He has published about

28 research papers, 01 book, 16 book chapters, 20 research abstracts, and 12 popular articles in reputed international and national publications.

 **Dr. B L Jangid** is presently working as Principal Scientist (Agricultural Extension) at ICAR- Agricultural Technology Application Research Institute, Zone-II, Jodhpur. He obtained his B.Sc. (Ag), M.Sc. (Extension Education) from Rajasthan College of Agriculture, Udaipur, Rajasthan Agricultural University, Bikaner and Ph.D. (Extension Education) from MPUA&T, Udaipur. He has received University Gold Medal for his M.Sc. and also received NET of ICAR and joined ARS in 1997. He is working in the field of agriculture technology transfer, impact assessment, livelihood security, economics of crops and livestock production, and new extension methodologies and approaches. He has published about 40 research papers, 17 book chapters, 85 research abstracts, 10 extension literature, 17 training manual/technical bulletins and 48 popular articles in reputed international and national publications.

 **Mis Keerthika A.** is working as Scientist (Agroforestry) at ICAR- Central Arid Zone Research Institute, Regional Research Station, Pali-Marwar. She obtained her B.Sc. M.Sc. and Ph.D. (continuing) in Forestry from Forest College and Research Institute, TNAU, Mettupalayam (Tamil Nadu). She has also qualified ICAR-NET. She is actively involved in germplasm collection, evaluation and improvement in trees and fruit crops. Her working area includes tree improvement, germplasm collection, tree born oil seed, carbon sequestration and development of horti-based models. She has published about 30 research papers, 16 book chapters, 25 research abstracts, and 15 popular articles in reputed international and national publications.

 **Dr. M.B. Noor mohamed,** is presently working as Scientist (Agroforestry) at ICAR- Central Arid Zone Research Institute, RRS, Pali Marwar (Rajasthan) with more than 5 years of experience in tree improvement, agroforestry, carbon sequestration and breeding works in shrubs and trees. He completed his UG, PG and Ph.D. in Forestry at Forest College and Research Institute, TNAU, Mettupalayam (Tamil Nadu). He has been awarded with Gold medal in Ph.D. (Forestry). He is actively involved in germplasm collection, evaluation and improvement in trees and fruit crops. He has published 20 research papers in international as well as national journals, 2 popular articles, 8 book chapters, 2 books, 1 training compendium and more than 20 abstracts. He has been

honoured with Young scientist award, Innovative scientist award, scientists of the year, 2019 and Best oral presentation award by various societies.

Dr. R.S Mehta is presently working as Principal Scientist in Agronomy at ICAR-Central Arid Zone Research Institute- Regional Research Station, Pali-Marwar (Rajasthan). He has graduated and post-graduated from Rajasthan Agricultural University Bikaner (Rajasthan) and obtained Ph.D (Agronomy) from S.D. Agricultural University, Sardarkrushinagar(Gujarat).

He worked for 9 year at ICAR-Central Sheep and Wool Research Institute Avikanagr (Raj) on feed and fodder resource development for small ruminant,15 year on development production technology for seed spices and 2 and half year on natural resource management at ICAR- CAZRI- Regional Research Station-Pali- Marwar. During service he has also been awarded Senior Research Fellowship by ICAR for doing Ph.D. He received best scientist award, best oral presentation awards and best poster presentation award by various organization. He has published 110 research papers in different national and international journals, and 190 publications in proceedings of seminar, symposiums, book chapters, bulletin, books and popular articles.

HORTICULTURE BASED INTEGRATED FARMING SYSTEMS

Editors

A.K. Shukla
D.K. Gupta
B.L. Jangid
Keerthika A.
M.B. Noor mohamed
R.S. Mehta

CRC Press
Taylor & Francis Group
Boca Raton London New York

CRC Press is an imprint of the
Taylor & Francis Group, an **informa** business

NEW INDIA PUBLISHING AGENCY
New Delhi – 110 034

First published 2022
by CRC Press
4 Park Square, Milton Park, Abingdon, Oxon, OX14 4RN

and by CRC Press
6000 Broken Sound Parkway NW, Suite 300, Boca Raton, FL 33487-2742

© 2022 selection and editorial matter, NIPA; individual chapters, the contributors
CRC Press is an imprint of Taylor & Francis Group, an Informa business

British Library Cataloguing-in-Publication Data
A catalogue record for this book is available from the British Library

Library of Congress Cataloging-in-Publication Data
A catalog record has been requested

ISBN: 978-1-032-15822-8 (hbk)
ISBN: 978-1-003-24581-0 (ebk)

DOI: 10.1201/9781003245810

Preface

The production and life support systems in the hot arid regions are constrained by low and erratic precipitation, high evapotranspiration, poor soil physical and fertility conditions. In addition to these adverse conditions, farmers are finding it increasingly difficult to earn a sustainable livelihood from crops alone. Farmers of this region are in need of continuous, reliable and balanced supply of food, fodder, as well as ensured income to satisfy basic needs and farm expenditure. Horticulture based integrated farming system is considered as effective strategy for obtaining continuous high income, enhanced productivity and enhancing economic, employment, fodder and nutritional security. Therefore, this book has been written with aim to educate farmers, extension workers, stakeholders and planners about the different aspects of horti-based system of Arid zone. This book may be helpful in enhancing knowledge about horti-based integrated farming system to the readers. The content of book has been synthesised and organised in such a way so as to provide information on advanced knowledge in the sphere of importance and scope of arid horticulture in India, horticulture based integrated farming systems for arid region, integration of livestock in horticulture based farming systems, emerging issues, natural resource management, disease and pest management, organic farming and certification, post-harvest measures and value addition in arid fruits and vegetables, marketing aspects, status and export promotion measures and procedures. All the book chapters have been written by highly specialized scientists from institutes of national importance. We are grateful to the authors for providing chapters of the book. Their invaluable contribution has been major driving force in completing the assignment.

Editors

Contents

1

Prospects of Horticulture in Arid Region

S.K. Sharma[1], B.D. Sharma[1] and A.K. Shukla[2]

[1]ICAR-Central Institute for Arid Horticulture, Bikaner, Rajasthan-334006
[2]ICAR-Central Arid Zone Research Institute, Regional Research Station
Pali, Rajasthan-306401

The Indian arid zone is characterized by high temperature and low and variable precipitation which limit the scope for high crops productivity. However, these conditions greatly favour the development of high quality in a number of fruits such as date palm, ber, aonla, bael, pomegranate, kinnow, lasoda and in vegetables cucurbits, legumes and solanaceous crops, spices, medicinal and aromatic plants. The existing low productivity could be increased by following improved technologies and inputs. It is now realized that there is a limited scope for quantum jump in fruit and vegetable production in the traditional production areas. The amelioration of the extreme conditions is also considered vital for life support to the inhabitants of this area. The recent awareness regarding the potential of these ecologically fragile lands for production of quality produce has not only opened up scope for providing sustainability for the people of this region, but also for bringing in new areas to increase horticultural production. The area expansion and yield potential of arid horticultural crops has increased many folds because of development of new varieties and advancement in production techniques in arid region.

Constraints in arid horticulture

The soils of arid region are very poor in fertility. The soils of the north-western arid region described as 'desert soils' and 'grey brown soils' of the Order Aridisols are light textured. Most of arid areas (about 64.6%) are duny where the soils often contain only about 3.2-4 per cent clay and 1.4-1.8 per cent silt. Besides this, about 5.9 per cent area is covered by soils having hard pan, 5.6 per cent is under hills and pediments, 6.8 per cent area is alluvial duny and 1.6 per cent is sierozems extending from the soils of Haryana and the Punjab. In the peninsular India, a considerable part of arid region has red sandy soil and some parts have mixed black soils. The soils are poor in organic matter having organic carbon of 0.03 % in bare sand dunes to 0.1 % in the stabilized dunes. The water holding capacity of soil is also poor. Soils are generally rich in total potassium and boron but are low in nitrogen, phosphorus and micronutrients such as copper, zinc and iron. The soils often have high salinity. The ground water resource is not only limited owing to poor surface and sub-surface drainage but is also saline in quality. The irrigation water resources in the region are seasonal rivers and rivulets, surface wells and some runoff water storage devices (e.g., *nadi, tanka, khadins*) and canal irrigation in arid region. Thus, the water resources in arid region are limited and can irrigate hardly 4% of the area.

The annual average rainfall in the Indian arid regions is very low and varies from 100 mm in north-western sector of Jaisalmer to 450 mm in the eastern boundary or arid zone of Rajasthan. Most of the precipitation in north western arid region occurs during July-September in about 19-21 rain spells. Due to low and erratic rainfall pattern in arid region, appropriate technology is needed to increase productivity. Water is precious input in hot arid region of the country therefore, adoption of micro-irrigation system is desirable to save water and enhance productivity. For arid environment, the variety is needed which are resistant to biotic and abiotic stresses for sustainable production. In some parts of arid region, occurrence of frost is also common features during winter season which affects vegetative growth of plants as well as productivity, quality of fruits especially in ber and aonla. There is no heat tolerance variety of arid horticultural crops which should be developed to achieve higher production.

One of the major bottlenecks in development of horticulture scenario in arid parts of the country is lack of sufficient quality seed and planting material. Seed is a precious input to increase quality production of vegetable and flowers as well as some fruit crops. Thus production of quality seed material and their distribution to farmers will boost the arid horticulture production. The post-harvest management is essential to over-come the losses at different stages of grading, packing, storage, transport and finally marketing of both fresh and processed products. The weak processing infrastructure, as it exists today, has been one of the contributing factors for ineffective utilization of the raw materials resulting in huge post-harvest losses. Lack of sufficient processing units for production of quality output is a major bottleneck for the arid fruit crops. Marketing of horticultural produce is a major constraint in the production and disposal system and has a major role to play in making the industry viable. The high capital cost involved in establishing orchards, or rejuvenation of existing old unproductive plantation poses serious constraint in area expansion. The situation becomes all the more difficult in view of the large number of small holdings devoted to these crops which are essentially owned by weaker section, who have no means to invest, nor can afford to stand the burden of credit even if available. Added to this is the long gestation period that the perennial horticultural crops like mango, sapota, citrus and date palm coming to the economic bearing age. High cost of inputs and lack of enough incentives for production of quality varieties /species, product diversification, value addition, etc. also hinder crops development. Lack of proper cold storage facility and knowledge and equipment for grading and packaging of fruits and vegetables is also constraint for the growers of hot arid region.

Prospects in Arid Horticulture

Presently, there is a scope in expansion of horticultural crops in arid region and it has vast potential for changing scenario of horticulture of the country Vast land resource, surplus family labours, increasing canal irrigated area, developing infrastructural facilities, plenty of solar and wind energy, etc., are the strength in arid region for research and development in arid horticulture. Further, minimum pressure of diseases and insects in the region is good scope for production of seed and planting material. Ber is commercially grown in more than 50,000 ha with production of 633000 tonnes in India. It requires more attention for value addition. Pomegranate production (area 246000 ha) and production 2865000 tonnes) is increasing very fast in dry part of the country. Since, it has tremendous scope of export of this crop from semi-arid and arid regions of the country. Presently, the export value from pomegranate is Rs.70 million. The crops like fig, custard apple are also coming very well under dry land conditions. At present, fig is cultivated in more than 1,000 ha area in Maharashtra, Karnataka. Likewise, custard apple is grown in about 50,000 ha area in the state of Maharashtra, A.P., Karnataka, Rajasthan and Tamil Nadu. Aonla is a medicinal fruit plant and cultivated in more than 94,000 ha area and produced 1098 MT fruits/year. Date palm is most suitable fruit tree of hot arid region and it is grown in Rajasthan, Gujarat, Punjab and Haryana in more than 12, 493 ha and producing 85,000 tonnes fresh fruit. However, date is imported to India from Gulf countries due to its meager production. Bael is also an important fruit crop of semi-arid and arid region. Now, attention is being given on its commercial production. India is the second largest producer of vegetables (187.47 million tonnes from 5.77-million-hectare area in 2018-19). In 2018-19 under major spices, the increase in area and production was recorded to be 4086000 ha and 8509000 tonnes. By the improvement in production technology in arid region, many seed spices likes Coriander, Cumin, Fenugreek, Ajowain, Fennel, Dill and Nigella, are being cultivated on large scale and also exported to earn foreign exchange. There is a vast potential of floriculture in some parts of Rajasthan because of low infection of disease and insects. The prospect of floriculture under hot arid condition is also important from seed and plant production point of view. At present, Roses, Marigold, Chrysanthemum and other flowers are being grown nearby cities of Udaipur, Ajmer, Jaipur, Kota, Sri Ganganagar districts of Rajasthan.

Selection of fruit crops and varieties

The environmental conditions of arid region are very harsh for sustainability of plants hence; selection of a plant species for such region is an important factor for growth and production. While selecting the fruit species for dry land horticulture, one of the basic requirements is that those crops, which complete their vegetative growth and reproductive phase during the period of maximum moisture availability, should be selected. The fruit such as ber, guava, pomegranate, custard apple, aonla and sour lime, conform to this prerequisite. The crops must have xeric characters such as deep root system (e.g. aonla, ber), summer dormancy (e.g. ber), high 'bound water' in the tissues (e.g. cactus pear, fig), reduced leaf area (e.g. Indian gooseberry, tamarind), leaf surface having sunken stomata, thick cuticle, wax coating and pubescence (fig, ber, phalsa, tamarind), and ability to adapt to shallow soils, rocky, gravelly, and undulating wastelands (pomegranate, aonla, bael) (Pareek and Sharma,1991). Some of the fruit crops suitable for dryland areas are given in Table -1.

Table 1: Fruit crops for drylands in different rainfall zones of India

Rainfall (mm)	Suitable fruit crops
< 500	Ber, Bordi, Lasora, Karonda, Ker, Khejri, Phalsa, Custard apple, Bael, Pilu, Pomegranate,
500-1000	Ber, Aonla, Jamun, Woodapple, Mahua, Custard apple, Wild date palm, Indian almond, Guava, Sour lime, Lemon, Mango, Palmyrah palm, Tamarind, Bael, Chironji, Karonda, Grape fruit, Pomegranate, Passion fruit.
> 1000	Mango, Litchi, Jackfruit, Persimmon, Mandarin, Avocado, Tamarind, Jamun, Kokum, Palmyrah palm, Guava, Cashew nut, Barbados cherry, Pomegranate.

Crop Improvement

The Central Institute of Arid Horticulture (CIAH), Bikaner has released 16 varieties of arid fruits and vegetable crops which includes, Thar Bhubharaj, Thar Sevika, Goma Kirti of ber, Thar Shobha of khejri, Goma Aishwarya of aonla, Thar Samridhi of bottle guard; AHW-19, AHW-65, Thar Manak of mateera; AHK-119, AHK-200 of Kachri, AHS-10, AHS-82 of snap melon, AHC-2, AHC-13 of kakri and Goma Manjri of cluster bean. There are a number of varieties of arid horticultural crops are at prerelease stage which includes AHRM-1 of round melon; AHG-13 in cluster bean, Indian bean AHDB-3,

AHDB-16 and Sward bean (AHSB-1) besides the some promising lines in ber, mulberry, lasora and ker, has been identified for evaluation and release (More *et al.*, 2008). Further, a number of varieties of arid fruit and vegetable crops have recommended for cultivation in different parts of the country after evaluation of germplasm for high yield and quality and given in Table 2.

Table 2: Promising varieties of fruit and vegetable crops for cultivation in semi-arid and arid regions

Crops	Varieties
Fruits	
Ber	Gola, Seb, Umran, Mundia, Kaithali, Banarasi Kadaka, Thar Bhubharaj, Thar Sevika, Goma Kirti
Bael	NB-5, NB 9, Pant Aparna, Pant Sujata, Pant Shivani, CISH Bael-1, CISH Bael-2
Pomegranate	G-137, GKVK-1, Ganesh, Jalore seedless, Mridula, Bhagwa, Phule Arakta
Aonla	NA 7, Kanchan, Krishna, Balwant, NA-6, NA-10, Laxmi-52
Sweet orange	Blood Red Malta, Mosambi, Pineapple, Valencia
Custard apple	Arka Sahan, Balanagar, Mammoth, Island, Gem, Red Sitaphal, APK(Ca)-1
Guava	Allahabad Safeda, L-49, Kohir Safed, Safed Jam, Chittidar, Lalit, Hisar Surkha
Date palm	Halawy, Barhee, Medjool, Shamran, Khuneizi, Khadrawy, Zahidi
Sapota	Kalipatti, Cricket Ball
Fig	Poona Fig, Dianna, Dinkar, Conadaria, Excel
Mango	Banglora, Kesar, Rajapuri, Bombay Green, Dashehari, Vanraj
Tamarind	PKM 1, Pratisthan, Yogeshwari
Vegetables	
Chilli	Pusa Jwala, Mathania, Pant C-1, Arka Mohani, Arka Gaurav, Arka Basant, Bharat, Indira
Cowpea	Pusa Dofasali, Pusa Phalguni, Pusa Barsati, Pusa Rituraj, Pusa komal
Cluster bean	Pusa Sadabahar, Pusa Mausami, Pusa Navbahar, Durga Bahar, AHG-13
Onion	Patna Red, Nasik Red, N-53, Pusa Red, Pusa Ratnar, Pusa White Round, Pusa White Flat, Punjab Selection, Agrifound Dark Red, Arka Pragati
Tomato	Pusa Ruby, Pusa Early Dwarf, Pusa-120, HS-102, Sweet-72, S-12, Mangla, Punjab Chhuhara
Brinjal	Pusa Purple Long, Pusa Purple Round, Pusa Kranti, Pusa Anmol, Arka Sheet, Arka Shirish, Arka Kusumakar, Arka Navneet
Amaranth	Chhoti Chauali, Badi Chaulai, CO-1, CO-2, CO-3
Okra	Pusa Makhmali, Punjab No. 13, Punjab Padmini, P-7, Parbhani Kranti
Pumpkin	Arka Chandan, CO-1, CO-2
Muskmelon	Pusa Sharbati, Pusa Madhuras, Hara Madhu, Punjab Sunehri, Durgapura Madhu
Watermelon	Sugar Baby, Arka Manik, Arka Jyoti, Durgapura Meetha, Kesar, Mateera (AHW-19 and AHW 65)
Bottle gourd	Pusa Summer Prolific Round, Pusa Summer Prolific Long, Pusa Meghdoot, Pusa Manjari, Pusa Naveen, Thar Samridhi,
Bitter gourd	Pusa Do Mausmi, Arka Harit, Pride of Gujarat
Kachri	AHK-119, AHK-200
Snap melon	AHS-10 , AHS-82

Varietal variation in endurance to drought has also been observed in horticultural crops. Early ripening cultivars seem to escape stress conditions caused by the receding soil moisture sotred in the soil profile during the monsoon. Ber cultivars Gola, Seb and Mundia for extremely dry areas, Banarasi Kadaka, Kaithli, Umran and Maharwali for dry regions, and Sanaur-2, Umran and Mehrun for comparatively humid regions have been recommended. Apart from morphological parameters, plants should also have physiological parameters for endurance to drought for commercial cultivation in this region. Some physiological parameters identified in ber are no mid day depression in photosynthetic rate, low rate of transpiration, maintenance of leaf water balance, growth, canopy development, dry matter allocation, high water use efficiency, etc. It has been demonstrated that plant having capacity for drought endurance are able to maintain turgour, dry matter allocation, leaf and fruit growth even under low soil moisture level.

Biotechnological tools for improvement

Micropropagation has been commerciallized in many ornamental crops and herbacious fruit crop species worldwide. However, its wide spread commercial use is still limited in fruit tree species because of several inherent problems of *in vitro* culture system such as hyperhydricity of cultures, frequent subculturing for shoot proliferation, poor morphogeneic responces of explants from mature tree, contamination in culture either systemic infection or infection during long term culturing process, problem of somaclonal variation due to reapeated subculturing process, poor root formation and low rate of servival during acclimatization of the plantlets ex-vitro. Keeping in view these inherent limitaions of micropropagation of fruit tree species, a new concept of micropropagations of fruit tree species has developed using single or double node explants from the mature trees. Direct morphogenesis of shoot and root formation was achieved in lasoda (*Cordia myxa*), mulberry (*Morus alba*) and lime (*Citrus aurantifolia*) using single or double node explant having physiological active axillary buds. Under this *in vitro* system two type of media were used, one for shoot induction in preexisting axillary buds and another for formation roots at the basal of the original explant.

In another important study with citrus, direct shoot and root formation was achieved in double node explant within 35 days of culture period. These results conform the production of plantlets within a short period eliminating subculturing process completely. Thus, using this technique of micropropagation, the fruit tree species can be multiply *in vitro* with minimizing inherent problems of tissue culture in greater way to obtain a large number of genetical identical, physiological uniform and developmentally normal plantlets preferably with high photosynthetic or phototrophic potential to survive the harsh *ex-vitro* condition.

Attempts have been made for mass multiplication of ker through tissue culture technique since it is a hardy plant and suitable for hot arid environment. Work on date palm tissue culture is being done at various places in the country However; some good results have been achieved through organogenesis and embryogenesis in date palm tissue culture at CIAH, Bikaner.

Orchard establishment

The fruit plants propagated in the nursery are generally used to establish orchards. Such plants invariably lose their tap roots as a result of repeated transplanting. Plants raised in containers develop oiled roots. For success in drylands, plants must have root architecture with a strong tendency to penetrate deep into the soil. *In situ* technique of orchard establishment is found suitable under arid conditions (Vishalnath *et al,* 2000). Rootstock seedling of ber are raised in the nursery in 300 gauge polythene tubes (25 cm length and 10 cm diameter, open at both ends), filled with a mixture of farm yard manure (FYM), sand, and clay in 1:1:1 ratio. The seedlings can be budded when about 90-100 days old. These plants become ready for transplanting, 1-2 months after budding. This technique helps to retain the straight growth of the tap root as the tubes are open at the bottom. Thus, the tubes neither restrict root growth nor induce coiling. Budded plants raised by this technique are also suitable for transportation to distant place (Pareek, 1978). Planting of stem cuttings of pomegranate, phasla, fig and mulberry in such polytubes would also induce straight roots. The pit size 2 x 2 x 2 ft and filling mixtures (FYM + pond silt + soil in 1:1:1 ratio) has standardized for planting of ber and pomegranate in arid region.

The plant density mainly depend upon the plant type, soil fertility status and management practices while planting system to be adopted in dry lands depends largely upon the topography of the land, fruit species and soil type. In the plains, planting, is generally done is square or rectangular system. On slopy lands, fruit trees are planted on contour terraces, half moon terraces, trenches and bunds, and micro-catchments. On marshy and wet areas mounding and ridge-ditch method of planting have been suggested. The trenches and bunds made across the slope are staggered (Saroj *et al.,* 1994). In a micro-catchment, which may be triangular or rectangular, trees are planted at the lowest point where runoff accumulates. The planting distance 6 x 6 m or 8 x 8 m for ber cultivation is optimum for arid region. Date palm, Bael and Aonla is recommended for planting at 8 x 8 m or 10 x 10 m distance under rainfed and irrigated conditions.

Plant architecture and canopy management

The canopy of plant plays a vital role to increase quality production of fruit trees. Canopy management work has been done for high yield and quality of fruits in guava at CISH, Lucknow and in citrus at NRC on Citrus, Nagpur. Training at initial stages of growth gives proper shape and strong frame to the trees. The bushy pomegranate should be trained keeping 3-5 stems from the ground level while in other fruits, single stem training keeping 3-4 main branches is adopted. However, pruning is essential to regulate reproductive phase of plants. Ber is pruned during January in Tamil Nadu, by the end of April in Maharashtra, and by the end of May in North India. The main shoots of the previous season are cut back retaining 15 - 25 nodes, depending upon location, cultivar, and age and vigour of tree. All the secondary shoots are completely removed. As a result of light pruning for several years, long non-flowering shoots develop. To eliminate this, half the number of shoots on the tree should be pruned keeping normal length and remaining half should be pruned keeping one to two nodes to induce new growth for fruiting in the following year. In phalsa, the time of pruning should be regulated according to the flowering period and should result in maximum number of new shoots on which bearing takes place. Established phalsa bushes should be pruned at 150 cm height once a year during January in north India and twice a year (December and June) in south India. Pruning from ground level is done either to rejuvenate old bushes or to train young plants into bush from. Defoliation of leaves in lasoda trees in the month of December-January produces early flowering and fruiting in arid region. High density planting is also beneficial in aonla fruit trees to achieve high yield under semi arid conditions (Singh et al., 2009).

Water management

In arid region, the major constraint in commercial cultivation of arid horticultural crops is water resources. Hence, the need of the hour is to develop technologies, which not only requires low water input but also have high water use efficiency. Water being a rare commodity in arid eco-system, the first and foremost requirement is to conserve the available soil or rain water. For conservation of rain water both in situ and ex-situ technologies have been developed. It has been reported that micro-catchment slopes greater than 5 per cent did not significantly affect run off at Jodhpur and that the highest ber yields were obtained when 0.5 per cent and 5 per cent slopes had 8.5m and 7m length of run, and 72 m^2 and 54 m^2 catchment area per tree, respectively (Sharma et al., 1986). Work done at Aruppukottai (Tamil Nadu) and Anantapur (Andhra Pradesh) has indicated usefulness of in situ water harvesting technique for fruit production.

Arora and Mohan (1988) found V-shaped micro-catchments with run-on surface mulched with grass to enhance the productivity of lemon, sweet orange and plum in Doon valley. At Hyderabad, micro-relief of 3 m width and 25 cm height, spaced 9 m from ridge to ridge, have been used to store extra rain water for fruit trees such as kagzi lime, coorg mandarin, and sweet orange with tomato and okra as intercrops.

Mulching with organic materials (e.g., hay, straw, dry leaves, and local weeds) has been found highly beneficial in reducing evaporation loss. The practice also suppresses weed growth, prevents erosion, and adds organic matter to the soil (Gupta, 1995). Black polythene mulch is very effective in ber orchards in western India, Although, local organic mulch materials are cheaper than polythene mulches but these require proper care to maintain effective cover thickness. Leaf mulch has been used to conserve soil moisture in sapota orchards in Karnataka, Tamil Nadu, and Andhra Pradesh. Sugarcane trash mulch in pomegranate, fig, and custard apple was found effective in Maharashtra.

At CIAH, Bikaner, the work on in-situ water harvesting has been undertaken in Pomegranate, aonla and vegetable. It has been demonstrated that application of black polythene mulch and local weeds helps in conserving soil moisture status in above crops. It has been demonstrated that plant growth and development remains optimum with use of above mulching materials. Mulching studies with respects to soil hydro thermal regimes in brinjal revealed that organic mulches curtailed soil temperature during warm months, while an increase was recorded during the winter month. Significant increase in fruit yield by 66 and 58% could be obtained through lasora *(Cordia* sp.*)* and kheep *(Leptodenia pyrotechnica)* (Awasthi *et al.,* 2006).

Among the *ex-situ* water conservation methods, in arid ecosystem, emphasis has been given mostly on pressurized irrigation system. It has been demonstrated that fruit s and vegetables can be grown economically by use of drip or sprinkler irrigation system. At CIAH, Bikaner and its regional station it has been demonstrated that crops such as pomegranate and ber can be grown successfully under drip irrigation system. It has been proved that water saving to the tune of 25 per cent can be achieved if pressurized irrigation system is used as compared to conventional flooding or bubbler system.

The use of drip alone or in combination with mulching has been demonstrated as a successful technology for cultivation of pomegranate at Anantapur (Anon., 2006). The studies have shown that highest number of 'B' grade pomegranate can be harvested under drip + mulch. FYM mulching is found beneficial for production of brinjal crops than other mulches in arid region.

Application of pitcher irrigation was attempted in cactus pear at CIAH, Bikaner and it was recorded that growth of cactus pear was better under this treatment as compared to control. The use of double ring system to conserve the moisture applied for production of fruit crops was attempted in aonla. It was observed that by this method the water is applied in zone having functional roots and hence, water use efficiency is enhanced (Shukla *et al.*, 2006).

Water loss due to transpiration can be reduced by use of radiation reflectants, stomata closing chemicals, and plastic films. Spraying of 4-6 per cent Kaolin, 0.5-1.0 per cent liquid paraffin, and 1.5 per cent power oil, after occasional rains in low rainfall areas, considerably reduce plant water losses (Pareek and Sharma, 1991). Chemicals such as Phenyl mercuric acetate (PMA), Decinyl succinic acid (DSA), Abscisic acid (ABA), and Cetylalcohal cause stomata clousure and thereby reduce transpiration (Jones and Mansfield, 1991; Chundawat 1990). Shelterbelt and windbreaks can reduce evapo-transpiration by reducing the wing speed and stabilizing microclimate (Muthana *et al.*, 1984).

Control of weeds has special significance in rainfed orchards in reducing soil moisture losses. Timely weeding is essential to improve fruit quality even in high rainfall areas. Application of pre-emergence weedicides such as Diuron, Bromacil, and Atrazine @ 2-3 kg ha[-1] and post emergence weedicides such as Grammaxone (Paraquat) and Glyphosate @ 1 L ha[-1] have proved effective in checking weed growth in the orchards.

Integrated Nutrient Management

The balanced nutrition in plants is required at appropriate time according to the age of plants. The application methods also play important role for availability of nutrients to the plants. In ber orchards, besides 10-15 kg organic manure, annual application of 100 g N, 50 g P_2O_5 and 50 g K_2O per tree is recommended. Fertilizer doses should be raised according to the age of plants and soil fertility of the region. Application of 15-20 kg FYM per tree has been found beneficial in aonla, custard apple, and tamarind. At MPKV, Rahuri, in addition to 50 kg FYM, 625 g N, 225 g P_2O_5 and 225 g K_2O has been recommended for application to 5-year-old pomegranate trees. At Bangalore, application of 500 g N + 250 g P + 125 g K produced six times higher yield than in the control. In 6 to 7 years old fig trees planted at 5m x 5m spacing, fertilization with 900 g N + 250 g K improved fruit production.

The nutritional trials have been undertaken in arid fruits at CIAH, Bikaner and centres of AICRP on Arid Zone Fruits. The studies conducted on Date palm at Abohar showed that application of 300-400 g N/tree/year gave maximum number and weight of bunch. Similarly in pomegranate it has been demonstrated that application of 50 per cent recommended dose of nitrogen at monthly interval gave best performance (Anon., 2006).

In order to conserve the costly input such as fertilizer, attempts were made to supply this along with water under pressurized irrigation system. The studies conducted in pomegranate and ber has demonstrated that fertilizer saving to the tune of 25 per cent can be achieved if plants are feti-irrigated through drip.

Keeping in view the export potential of pomegranate, attempts have been made to assess the organically production of this crop. In this pursuit, substitution of in-organic with organic fertilizers was attempted. The results have demonstrated that a good crop of pomegranate can be harvested by giving 50 per cent RD of NPK through Vermicompost and 50 per cent through inorganic fertilizer. Thus, the use of inorganic fertilizers can be reduced to half through this technology.

Micronutrients are often found deficient in semi-arid and arid soils. Foliar feeding of nutrients such as nitrogen (0.5-2.0% urea), zinc (0.05-1.0% zinc sulphate), and boron (0.05- 1.0% borex) has given beneficial results in these areas (Pareek and Sharma, 1991). In the medium rainfall region of eastern Uttar Pradesh, application of FYM, pond soil, gypsum, and pyrite in sodic soils resulted in better establishment and growth of aonla and bael plants. Foliar spray of micronutrients (Fe 0.50% + Zn 0.50% + Cu 0.25%) improved the yield and fruit quality in kinnow mandarin in arid region. Foliar spray of zinc sulphate 0.5-1.0 per cent improved fruit quality in ber cv. Seb in semi-arid region.

Fruit Based Cropping Systems

Monoculture in arid zone is highly risk prone due to crop failures, hence a suitable tree crop combination is essential for alleviating the risk, generation of income, improvement productivity per unit area/volume as a result of efficient use of natural resources and inputs, and ameliorate and improve adverse agro-climate. Agri-horticultural combinations with legume intercrops such as mung bean, moth bean, cluster bean, and cowpea are beneficial. In the rainfed orchards of guava and ber, cluster bean okra, and cowpea in kharif (rainy season) proved good in the medium rainfall region of Gujarat (Raturi and Hiwale, 1988). Under South Indian conditions of Hyderabad; cowpea, green gram, cluster bean and horse gram in ber orchards and bitter gourd, tomato and okra in acid lime orchards have been grown as intercrops.

In areas with large livestock population, horti-pastoral system would be beneficial. In the arid areas, the system could have combinations such as khejri *(Prosopis cineraria)* +ber+ dhaman *(Cenchrus ciliaris, C. setigerus) or* sewan *(Laisurus sindicus,* or tumba. In semi-arid areas, perennial trees (mango, tamarind, sapota, jackfruit and palmyrah palm) could be grown with fodder crops.

Fruit trees can also be planted in association with forest trees, and they yield wood for packaging and fuel. Multi-storey combinations incorporating large trees, small trees, and ground crops can be used. In low rainfall (300-500 mm) zone, combinations such as khejri or ber + ber or drumstick + vegetables (legumes and cucurbits); in 500-700 mm rainfall zone, combination of mango or ber or aonla or guava + pomegranate or sour lime or lemon or drumstick + solanaceous or leguminous or cucurbitaceous vegetables; and in 700-1000 mm rainfall zone, combination of mango or jackfruit or mahua or palmyra palm or tamarind or guava + sour lime or lemon or pomegranate or aonla + vegetables can be adopted.

In arid ecosystem, attempts have been made to develop models for crop diversification. Keeping in view the traditional over storey crops as ber and new introduction aonla, the cropping models have been developed. It has been demonstrated that in ber based cropping system cultivation of Indian aloe can be taken up as a remunerative model Dhandar *et al.* (2004). Similarly, in aonla based cropping system, it has been demonstrated that model consisting of aonla + ber along with moth bean or fenugreek can be adopted as a sustainable model for nutritional and income security of the inhabitants (Awasthi *et al.,* 2007).

Mono cropping of either fruit or seasonal crops is highly risk prone in arid areas, hence to mitigate the effect of total crop risk failures, fruit based multistory cropping system such as Aonla-ber-brinjal-moth bean, Aonla- drumstick-senna-moth bean-cumin can be profitably adopted by the farmers of arid region for better cash flow, nutritional and environmental security and sustainable livelihood. In areas where frost is a severe Aonla-Khejri-suaeda-moth bean and mustard can be another lucrative option (Awasthi *et al.*, 2007).

Crop diversification studies in ber (*Ziziphus mauritiana*) and aonla (*Emblica officinalis*) based cropping studies led to the recommendations that in pre-establishment phase of ber orchard, Indian aloe (*Aloe barbedensis*) and cluster bean (*Cyamopsis tetragonoloba)* are the low input and high returning crops in arid region. In aonla based multi storey cropping system, the model-4 with crop combination of aonla- drumstick- senna- moth bean- cumin recorded highest net return followed by cropping model-I (aonla - ber - brinjal - mothbean – fenugreek) has been recommended for sustainable and remunerative under arid ecosystem. Under semi arid conditions of Godhra, Gujarat fruit based farming system like aonla / ber + okra / brinjal / cowpea have been recommended to the farming community for sustainable production.

Future strategies for arid horticulture development

Although, great efforts have been made to develop technology compatible for commercial production of arid horticultural crops, yet there is a need to address various issues for further refinement of technology, improvement in socio-economic status of arid region and development of sustainable agro-horti-system.

The major issues are:

- Crop improvement
- Exploitation of Biotechnology in arid horticultural crops
- Protected cultivation and off season production
- Hi-tech crop production
- Efficient utilization of water resource
- Rehabilitation of degraded lands
- Utilization of solar and wind energy
- Organic farming
- Breeding for resistance to abiotic stresses
- Diversified farming systems
- Value addition
- Marketing and export
- Transfer of technology
- Human resource development

References

Anonymous. 2006. Biennial Report (2004-05) of XIII Group Workers Meeting of AICRP on Arid Zone Fruits, 10-12 May 2006 S.D. Agricultural University, S.K. Nagar, Gujarat.

Awasthi, O.P., Singh, I.S. and Sharma, B.D. 2006. Effect of mulch on soil hydro thermal regimes, growth and fruit yield of brinjal under arid conditions. *Indian Journal of Horticulture,* 63(2): 192-194.

Awasthi, O.P., Saroj, P.L., Singh, I. S. and More, T. A. 2007. Fruit Based Diversified Cropping System for Arid Regions, CIAH Tech. Bull. No. 25, CIAH, Bikaner, 18p.

Arora, Y.K. and Mohan, S.C. 1988. Water harvesting and moisture conservation for fruit crops in Doon valley. In. National Seminar on Dryland Horticulture, 20-22 July, 1988 CRIDA, Hyderabad.

Chundawat, B.S. 1990. Arid Fruit Culture. Oxford & IBH Publication Co. Pvt. Ltd., New Delhi, India

Dhandar, D.G., Saroj, P.L., Awasthi, O.P. and Sharma, B.D. 2004. Crop diversification for sustainable production in irrigated hot arid eco-system of Rajasthan. *Journal of Arid Land Studies,* 148: 37-40.

Gupta, J.P. 1995. Water losses and their control in rainfed agriculture. In: (ed. Singh, R.P.,) Sustainable Development of Dryland Agriculture in India. Jodhpur, India: Scientific Publishers, pp. 169-176.

Jones, R.J. and Mansfield, T.A. 1971. Antitranspirant activity of the methyl and phenyl esters of abscisic acid. *Nature,* 231: 331-332.

Muthana, K.D., Yadav U.S., Mertia, R.S. and Arora, G.D. 1984. Shelterbelt plantations in arid regions. *Indian Farming,* 33: 19-21.

More, T.A. and Singh, R.S. 2008 Conserving biodiversity in different areas, The Hindu Survey of Indian Agriculture, Chennai, pp 50- 54..

More, T.A., Samadia, D.K., Awasthi, O.P. and Hiwale, S.S. 2008. Varieties and Hybrids of CIAH Tech. Bull. No. 30, Bikaner, 11p.

Pareek, O.P. 1978. Quicker way for raising ber orchards. *Indian Horticulture,* 23: 5-6.

Pareek, O.P. and Sharma, S. 1991. Fruit trees for arid and semi-arid lands. *Indian Farming,* 41: 25-30.

Raturi, G.B. and Hiwale, S.S. 1988. Horticulture based cropping systems for drylands. In: National Seminar on Dryland Horticulture, 20-22 July, 1988, CRIDA, Hyderabad.

Saroj, P.L., Dubey, K.C. and Tewari, R.K. 1994. Utilization of degraded lands for fruit production, *Indian Jour. Soil Conservation,* 22: 162-176.

Sharma, K.D., Pareek, O.P. and Singh, H.P. 1986. Micro- catchment water harvesting for raising Jujube orchards in arid climate. *Trans. ASAEI,* 29: 112-118.

Shukla, A.K., Singh, D., Meena, S.R., Singh, I.S., Bhargava, R. and Dhandar, D.G. 2006. Enhancement of water use efficiency in aonla through double ring system of irrigation under hot arid agro-ecosystem. In: Abstract of National Seminar on Input Use Efficiency at IIHR, Banglore, August 9-11, 2006, p.99.

Singh, A.K., Singh S., Apparao, V.V., Meshram, D.T., Bagle, B.G. and More, T. A. 2009. High density planting systems in Aonla, CIAH Tech.Bull. No. 34., Bikaner.

Vishal Nath, Saroj, P.L., Singh, R. S., Bhargava, R and Pareek, O. P. 2000. *In situ* establishment of ber orchards under hot arid eco-system in Rajasthan. *Indian Journal of Horticulture,* 57(1): 21-26.

2

Potential of Horticultural Crops in Integrated Farming System for Higher Income in Arid Zone

P.R. Meghwal

ICAR-Central Arid Zone Research Institute, Jodhpur, Rajasthan-342003

The hot arid regions of India are spread over 32 million hectares in the states of Rajasthan (61%), Gujarat (20%), Punjab (5%), Haryana (4%), Karnataka (3%) and Andhra Pradesh (7%). The arid regions of western Rajasthan experience an annual rain fall between 100-500 mm with a coefficient of variations varying from 40-70 per cent. Low and erratic rainfall coupled with extreme temperature results in complete or partial crop failure, considerably affects the agricultural economy of the region. High wind velocity causes wind erosion and shifting of sand dunes covers the fertile agricultural land and canals (Rao and Singh,1998). The rise in temperature begins from March onward and attains its peak in May as high as 46°C. The transpiration is low during winters but during summer in months (May & June), it reaches up to 16 mm per day. The annual estimated potential evapotranspiration values ranges from 1600 mm in eastern part and 1800 mm in western part of the region.

In the arid regions of Rajasthan, Haryana, Punjab and eastern part of arid Gujarat the soils have developed from alluvial and aeolian parent materials. In context to the type of soils light brown sandy soils with dune dominance is 30.6 per cent and associated others 34 per cent area, light brown soils 1.7 per cent, Grey brown soils 13.6 per cent hard pan in 5.9 per cent and seirozems in 1.7 per cent area (Dhir, 1977). Kolarker *et al.* (1989) reported that in 12 district of Rajasthan sandy soils dominated in 14.3 million ha. These soils are poor in fertility very low organic carbon (0.03-0.1%), light in texture called desert soils (north western) in the order of aridisols. It has undulated topography with the dominance of sand dunes inter-dunal plains. About 80 per cent area falls under rain fed cropping and moth bean, cluster bean, pearl millet, and sesame are the main crops grown during kharif season after the onset of monsoon. Under the vagaries of monsoon and recurrent drought the rain fed cropping becomes a gamble, which creates severe food and fodder scarcity in the region.

Ground water resources in the region is not only limited but water quality is also very poor which cannot be used for longer time to irrigate the crops. The depth of water table is very high ranging from 300 to 550 feet. The surface water potential of western Rajasthan is also very low.

Though the problems of arid areas are immense, it has certain strength as well. These include vast land resource, plenty of sun light, wind energy, cheap labor, less problems of pest and diseases, increase in infrastructure and irrigation facilities due to increase in tube well and canal command areas. These strengths can be better converted into opportunities by integration of horticultural crops in the farming system. Since per unit returns from horticultural crops are always higher, their inclusion in the system will improve the net return of farmer besides other advantages due to perennial and multipurpose nature of such crops. Moreover, horticulture based farming system is demand driven system approach by integrating multi-enterprise based on resource situation on a land management unit for sustainability and finally improving the living standard of the people. This approach also cater to the multiple needs of the farmer/enterprise including food, fruit, fodder, fuel wood, fibre etc. by utilizing both natural resources i.e. soil, water vegetation, solar energy and farm inputs like seed, fertilizer, chemicals, labour etc, economize productivity, nutritional security, employment generation with increase in export potential. The additional advantages of horticultural based diversification are labour forces can be utilized year round because of integration of diverse nature of crops/enterprises, ecologically and economically synergistic influence against adverse climatic conditions, establishment of secondary enterprise like bee keeping, dairying, poultry and processing unit establishment besides other advantages of such diversification.

Integrating perennial horticultural crops in farming system in arid zones have added advantages due to their special adaptive mechanism like deep tap root sysyem, leaf shedding in summer, summer dormancy, waxy leaves, sunken stomata in certain species, tolerance to drought and salinity, and synchronisation of flowering and fruiting during the maximum moisture availability period. Huge area is available under degraded land and wasteland such as sand dunes, ravines, salt affected soils etc. which are not much suitable for arable crops alone but by integrating salt tolerant horticultural crops over long period, such soils can be made amenable to crops as well. Monocropping is generally risky and non-remunerative, and hence integration of certain indigenous arid fuits in the farming system can lead to enhancement of overall productivity of the system. Moreover, most of the perennial horticultural crops are multipurpose meeting not only the requirement of fresh fruits but also of vegetable, fuel, fodder and fencing materials. A variety of vegetable crops have also been identified for growing in resource constrained environment of arid zones.

Lists of different horticultural crops that can be grown either as rain fed or with supplementary irrigation in combination with arable crops are given below (Table 1). Many of these like khejri, kumat, jhar ber, ker, peelu etc. are naturally grown and have been utilized by the farmers as components of agro forestry system. However, these are now being considered as multipurpose horticultural crops as they provide, fruits/vegetable, fodder, fuel wood, timber etc. Efforts are now under way to plant such species under a systematic plan with proper spacing and canopy management so as to derive maximum benefits of their association.

Agri-Horticulture

After two decades of experiments on research farms and farmer's fields, a new agri-horticulture system of jujube (*Ziziphus mauritiana*) inter-cropped with arid legumes e.g. clusterbean/mothbean/greengram has been developed for the areas receiving rainfall of more than 250 mm. This agri-horticulture system has been found to give better and earlier production, year-round work and resilience to erratic rainfall. In this system initially the jujube saplings are raised in the farmer's nursery by stored rainwater (farm pond) and then saplings are transplanted at a distance of 5 x 10 meters irrigated twice in first two weeks; thereafter the trees receive only harvested rainwater. Arid legume crops are raised successfully in the interspaces of jujube. It is a short duration (50-75 days) crop, sown at the onset of monsoon and harvested in September/October, i.e. before the full growth and blooming of jujube. This offers least competition and both is complimentary as the jujube provides a favorable environment for arid legumes while legumes fix nitrogen, and leaf fall adds organic matter to the soil. Besides fruits and wood from pruning (for fuel or fencing), leaves as fodder are the additional products from jujube.

Table 1: Suitable horticultural crops for dry land farming

Fruits	Vegetables
Ber (*Ziziphus mauritiana*),	Water melon/Mateera (*Citrullus lanatus)*
Ker (*Capparis decidua)*	Musk melons(*Cucumis sativus)*
Lasoda (*Cordia myxa)*	Bottle gourd*(Lagenaria ciceraria)*
Kumat(*Acacia senegal)*	Round melon(*Citrllus lanatus var.fistulosus*)
Tamarind(*Tamarindus indica)*	Long melon(*Cucumis melo var.utillismus)*
Bordi(*Ziziphus rotundifolia)*	Snap melon(*Cucumis melo var. momordica)*
Phalsa (*Grewia subinaequalis)*	Bittergourd (*Momordica charantia)*
Pilu (*Salvadora oleoides)*	Kachri (*Cucumis callosus)*

Contd.

Khejri(*Prosopis cineraria*)

Aonla (*Emblica officinalis*)

Jamun (*Syzygium cumini*)

Bael (*Aegle marmelos*)

Jhar ber (*Ziziphus nummularia*)

Karonda (*Carissa carandas*)

Natal plum(*Cariss grandiflora*)

Prickly pear(*Opuntia ficus-indica*)

Date palm (*Phoenix dactylifera*)

Drumstick(*Moringa oleoiedes*)

Fig(*Ficus carica*)

Papaya(*Carica papaya*)

Guava(*Psidium guajava*)

Grape(*Vitis vinifera*)

Custard apple(*Annona squamosa*)

Sour lime(*Citrus aurantifolia*)

Kinnow(*Citrus spp*)

Sweet orange(*Citrus sinensis*)

Cucumber(*cucumis sativus*)

Ridge gourd(*Luffa acutangula*)

Sponge gourd(*Luffa cylindrica*)

Cow pea(*Vigna unguiculata*)

Clusterbean(*Cyamopsis tetragonoloba*)

Amaranthus(*Amarnathus spp.*)

Brinjal(*Solenum melongena*

Chillies(*Capsicum annum*)

Tomato(*Lycopersicon esculentum*)

Carrot(*Daucus carota*)

Onion(*Allium cepa*)

Garlic(*Allium sativa*)

Radish(*Raphanus sativus*)

Okra(*Abelmoscus esculentus*)

Cabbage(*Brassica oleracia var.capitata*)

Cauliflower(*Brassica oleracia var botrytis*)

Potato(*Solenum tuberosum*)

Pea (*Pisum sativum* var *hortans*)

Overall productivity of the system in normal and drought years shows less variation in yield due to variations in rainfall. This system can provide economic returns of Rs 28000- 37000/yr as compared to the sole cropping of arid legumes (Rs. 8000-12000/-) with low input supply, besides non-tangible environmental improvements which give resilience to drought.

Experiment conducted on integrated farming system involving ber and other agrofrestry species revealed maximum net returns in ber based production system(Rs.61,520/-) and minimum in arable farming system(Table 2). The economics of mung bean was higher in *Z. mauritiana* based agri-horti system as compared to sole cropping system (Table 3).In the same study the soil moisture utilization from 0-15 cm soil layer was 1.92 and 1.05% in mung bean and 4.32 and 2.58% in cluster bean during seed development phase at 1 and 2 meter distance from tree trunk.

Table 2: Economic evaluation of various farming system component (Anon, 2010-11)

Farming system component	Net returns (Rs. ha^{-1})	Contribution of crop/grasses	B:C ratio
Arable farming	22460	100	2.2
Agro forestry with *P. cineraria*	30180	80	2.61
Agri-horti with *Z .mauritiana*	61520	38	2.67
Farm forestry with *H.binata*	3015	77	3.04
Sivipasture with *C. ciliaris+C. mopane*	40862	92	3.66
Sivipasture with *C.ciliaris+Z.rotundifolia*	37235	90	3.62

Table 3: Performance of dry land crops under various agro forestry systems (Anon, 2010-11)

Crop	Grain yield (kg ha⁻¹)	Fodder (kg ha⁻¹)	Net returns (Rs. ha⁻¹)	B:C ratio
Sole pearl millet(HHB-67)	1055	5005	22460	2.24
Sole mung bean(SML668)	540	2160	23161	2.71
Sole cow pea(V-585)	650	3565	24120	2.78
Sole moth beanCZM-3)	550	3000	24070	2.79
Pearl millet+P.cineraria	920	5065	21880	2.23
Pearl millet+A. tortilis	1425	5215	26390	2.38
Mung bean+Z.mauritiana	566	2150	23580	2.74

In agri horti system involving *Ziziphus* and *V.radiata* during sub normal year when rainfall was 51% less than long term average of 360 mm, the yield of mung bean was reduced by 51%. The inventory of the system showed that this agri horti system can provide round the supply for 5 goat/sheep per hectare and fuel wood for a family of 4 members besides efficient nutrient cycling and increase in economic stability (Faroda, 1988). Gupta *et al.,* (2000) reported that 3 year old plantation of *Z. mauritiana* (400 plant/ha.) in association with green gram performed well with seasonal rainfall of 210 mm and fruit yield from intercropped increased net profit up to Rs.288.6 ha⁻¹. This shows that agri horti-system minizes risk in arid regions and thus helps in imparting economic stability. According to Singh *et al.,* (2003) intercropping of legume with ber orchard produced higher grain yield of intercrops by 5-20% over the sole cropping and intercropping was found promising particularly during juvenile period of fruit plantation.

Aonla based horti-pasture system with *Dicanthium annulatum* revealed that after 10 years of association, there was significant improvement in organic carbon (92%), available N(20.8%), P(9.0%) and K(58%) with very high B:C ratio 3.7 as compared to only 1.85 in pure grass (Table 4). As has been mentioned earlier, horticulture based farming system has much more employment generation potential. The employment generation in aonla based horti-pasture system was found to be 4.74 man days per month as against 2.07 man per month in pure pasture (Kumar *et al.*, 2009). Rainfed farming system results in large scale production of fruits and vegetables and the surplus produce are generally sun dried for use in off season. Good quality dehydrated products could be prepared by pre-treatment of kachra fruits , guar pods , tinda fruits, khejri pods and cow pea pods with sulphur, potassium metabisulphite or by blanching(Meghwal, 2008).

Table 4: Changes in organic carbon and major nutrients (30 cm) over 10 years period under aonla based horti pasture system (Kumar *et al.*, 2009)

Composition	Initial(1996)	2005		% improvement after 10 years
		Aonla+*Dicanthium*	Sole *Dicanthium*	
Organic-C (%)	0.23-0.27	0.41-0.55	0.38-0.42	92
Available N (kg ha⁻¹)	140-148	159.9-188.2	150-152.6	23.42
Available P (kg ha⁻¹)	3.8-4.1	4.3-4.9	3.9-4.0	9.0
Available K (kg ha⁻¹)	110-124.3	174.9-196.4	164.2-168.1	58

References

Dhir, R.P. 1977. Western Rajasthan soils, their characteristics and properties. In: *Desertification and its control*, ICAR, New Delhi. Pp. 102-115.

Faroda, A.S. 1998. Arid Zone Research: An Overview. *In: Fifty Years of Arid Zone Research in India* (Eds. A.S Faroda and M. Singh), CAZRI Jodhpur. pp. 1-16.

Gupta, J.P., Joshi, D.C. and Singh, G.B. 2000. Management of arid ecosystem. In: *Natural resource management for agricultural production in India*. (Eds. J.S.P. Yadava and G.B. Singh) pp.551-668. Indian Society of Soil Science, New Delhi.

Kolarkar, A.S., Jain, S.V., Dhir, R.P. and Singh, N. 1989. Distribution, morphology and land use-Rajasthan. *In Reviews of research on sandy soils in India*.CAZRI, Jodhpur, pp.1-20.

Kumar, S., Kumar S. and Choubey, B.K. 2009. Aonla based hortipasture system for soil and nutrient build up and profitability. *Annals of Arid Zone*, 48(2): 153-157.

Meghwal, P.R. 2008. Dehydration of fruits and vegetable from traditional farming systems of arid Zone. In: *Diversification of Arid Farming System* (Eds. Pratap Narain, M.P.Singh, Amal Kar, S.Kathju ,and Praveen Kumar), Arid Zone Research Association of India and Scientific Publishers (India), Jodhpur, pp.405-407.

Rao, A.S. and Singh, R.S. 1998. Climatic features and crop production. In: *Fifty Years of Arid Zone Research in India* (Eds. A.S Faroda and M. Singh), CAZRI Jodhpur. pp. 18

Singh, R.S., Gupta, J.P., Rao, A.S. and Sharma, A.K. 2003. Microclimatic quantification and drought impacts on productivity of green gram under different cropping systems of arid zones. In: *Human Impact on Desert Environment* (Eds. P. Narain, S. Kathju, A. Kar, M.P. Singh and Praveen Kumar). Scientific Publisher, Jodhpur. pp.74-80

3

Vegetable Production: Prospects and Challenges in Rajasthan

A.K. Shukla, B.L. Jangid, D.K. Gupta, Keerthika A.
M.B. Noor mohamed and R.S. Mehta

ICAR-Central Arid Zone Research Institute, Regional Research Station
Pali, Rajasthan-306401

Rajasthan has geographical area of 3.42 lakh sq. km has attained the status of being the largest state of India. The state represents 10.4% land surface area with 5.5% population of India, 66% is dependent on agricultural for their livelihood. The state is divided into 33, districts, which are further subdivided into 338 Tehsils and 295 Panchayat Samitis. The total cultivable area is around 220.00 lakh ha. The average rainfall of the State is 575 mm, out of which about 532 mm precipitation occurs in the rainy season i.e. June to September. The average rainfall of eastern Rajasthan is about 704 mm and that of western Rajasthan is about 310 mm which reflects a vast variation. Rivers in Rajasthan are non-perennial except for Chambal and Mahi rivers. The State is presently divided into 33 administrative districts and has 10 agro-climatic zones. Agriculture in Rajasthan is primarily rain fed covering country's 13.27 per cent of available land. Groundwater is getting depleted as well as polluted. In general, every third year is a drought year. Despite these, the State has made significant achievements since independence and has attained self-sufficiency in food-grains. Rajasthan is located in the north-western part of the subcontinent. It is bounded on the west and north-west by Pakistan, on the north and north-east by the states of Haryana, Punjab, and Uttar Pradesh, on the east; in the south-east by the states of Uttar Pradesh and Madhya Pradesh, and on the south-west by the state of Gujarat. The Tropic of Cancer passes through its southern tip in the Banswara district. In the west, Rajasthan is relatively dry and infertile: this area includes some of the Thar Desert, also known as the Great Indian Desert. In the south-western part of the state, land is wetter, hilly, and more fertile.

Potential of Horticulture in the State

The diverse agro-ecological conditions prevailing in State is amenable for growing fruits, vegetables, spices, flowers, root and tuber crops, medicinal and aromatic crops. Out of the net cultivated area of about 165 lakh hectares in Rajasthan, horticultural crops are grown in an area of about 18.34 lakh hectares with annual production of about 45.80 lakh tonnes Area, production and productivity of horticultural crops in Rajasthan is given in table (1).

Table 1: Area, production and productivity of horticultural crops in Rajasthan (2018)

Horticultural Crops	Area (000ha)	Production (000 t)	Productivity (t/ha)
Fruits	59.78	895.90	13.5
Vegetables	168.5	1767.5	10.2
Flowers (loose)	4.0	7.5	1.54
Aromatics and medicinal plants	480.0	450.0	0.89
Spices	1122.1	1459.1	1.99
Total	1834.38	4580.0	26.56

Source: NHB Database, 2018

Vegetables play a vital role in balance diet as they supply all main components of human diet. Presently India is the second largest producer of vegetables next to China with a total production of 1.76 million tonnes from an area of 1.68 Lakh ha during 2018-19 (NHB Data base, 2018). Hence it is necessary to enhance the production and productivity of vegetables to meet the demand of growing population to ensure nutritional security. The major natural constraints in arid vegetable production are very low and variable rainfall (100-450mm), very high potential evapo-transpiration rate, intense solar radiations (320-619/ cm^2/day), high wind velocity (20km/hr), high infiltration (9cm/hr), extremes of temperature (0 to 48^{0}C), high soil salinity, unavailability of ground water, etc. which limit the scope of high productivity. However, these conditions favor successful cultivation of several vegetable crops such as watermelon/*Mateera* (*Citrullus lanatus*), *Kachri* (*Cucumis callosus*), snap melon/*Phoot* (*Cucumis melo* var. *momordica*), long melon/*Kakri* (*Cucumis melo* var. *utilissimus*), round melon/*Tinda* (*Praecitrullus fistulosus*), cluster bean (*Cyamopsis tetragonoloba*), bottle gourd (*Lagenaria siceraria*), ridge gourd (*Luffa acutangula*), etc. Other vegetable crops like solanaceous (tomato, brinjal, chilli), bulb crops (onion, garlic), root crops (carrot, radish), cole crops (cabbage, cauliflower, knol khol), leafy (palak, *Chenopodium*, fenugreek, *Amaranthus*), legumes (cowpea, Indian bean) and okra also have good potential under limited irrigation water facility by adopting suitable production and protection technologies.

Particularly traditional cucurbitaceous vegetable crops have great potential in arid pockets of country because they are well adapted to existing adverse environmental conditions and possess unprecedented ability to withstand against stresses like moisture stress, high soil pH and load of pests and diseases. Snap melon is good source of carbohydrate (3.0%), protein (0.3%), fat (0.1%) and vitamin A (265 IU) and minerals (0.4%). It is much liked by the people suffering from sugar related disorders. *Kachri* contains carbohydrate (7.45%), protein (0.28%), fat (1.28%), fibre (1.21%), vitamin C (29.8 mg) and iron (0.18 mg). It is one of the components of the delicious vegetable popularly known as *Panchkuta* in the desert districts of north-western India. The mature fruits of snap melon and *kachri* are usually cooked with various vegetable preparations, *Chutney*, pickles and are also used for garnishing the vegetables or as salad. Spine gourd is good source of protein (3.1g/100g edible part) and iron (4.6 mg/ 100g edible part). It is beneficial to those suffering from diabetes, piles, ulcer and problems related to digestion. *Mateera* fruits contain 90-95% water, 3.5-7.4% carbohydrate, 0.18-0.25% protein, 0.10-0.20% fat, 0.25-0.35% fibre, 0.2-0.3% minerals and good source of lycopene. Rind contains citrulline, an amino acid which is involved with atheletic ability and immunity. Besides the nutritional and medicinal value, *Mateera* has a significant role in quenching thirst of dessert masses and keep them cool during hot summers. Its seed kernel contains 35-40% fat and consumed by arid population.

These vegetables are the main source of crop diversification in desert areas and commonly grown as inter crop in pearl millet and ber orchards. These worthy vegetables should address the challenge of sustaining their diversity under the changing scenario of climate to improve their performance according to consumers' preference.

Agro-climatic zones of Rajasthan

i) Arid Western Plain

In this zone, Bikaner, Jaisalmer and Barmer districts, are included and are the most arid part of the state where the annual rainfall varies from 10 to 40 cm, summer temperatures are high upto 49° C. Vegetables suitable for this region are chilli, onion and cucurbits.

ii) Irrigated North-Western Plains

Ganganagar district, part of this region, which is arid, is the northern extension of the Indian Thar Desert covered with wind-blown sand. High summer and low winter temperatures, is its usual climatic characteristic. The average annual rainfall is about, 40 cm. Network of Ganga Canal and Bhakhra Canal and

Indira Gandhi Canal, has made the entire area green and productive. Major vegetable crops suitable for growing in this region are Tomato, Chilli, Brinjal, cucurbits and cluster bean

iii) Semi-arid Eastern Plain

This region comprises four districts namely, Jaipur, Dausa, Tonk and Ajmer. Banas, with its several tributaries, forms a rich fertile plain. On the western side, the region is flanked by the low Aravalli hills which extend from the south-west to the north- east.

The annual rainfall of the region varies from 50 to 60 cms. Summer and winter temperatures are not as extreme as in the arid west but the summer temperature may reach around 45° C and in the winter, minimum may be 8°C. In the total gross cultivated area of this zone, bajra, sorghum and pulses are grown in the kharif season, and wheat, barley, gram, mustard in the rabi season. Productivity of all crops in this zone is better than that of the agro-climatic zones that are to the west of the Aravalli range. The vegetables suitable for this region are tomato, brinjal, chilli, peas, melons, cucumber, cauliflower, radish.

iv) Flood Prone Eastern Plains

This region comprises the districts of Alwar, Bharatpur and Dhaulpur and the northern part of Sawai Madhopur (Mahuwa, Todabhim, Hindon, Nadauti, Bamanwas, Gangapur, Karauli, Sapotra and Bonli tchsils). Except for few low hills which exist in Alwar and Sawai Madhopur districts, the entire region is a flood plain of the Banganga and the river Ghambhiri. The region has rich alluvial soils the fertility of which is replenished every year by the flood water of the rivers. Vegetables that are grown in this zone are okra, kharif onion, tomato, brinjal, cucurbits.

v) Sub-humid Southern Plains & The Aravalli Hills

Bhilwara district, all tehsils of Udaipur district, except Dharyiawad, Salumber and Sarada, all tehsils of Chittaurgarh district, except Chotti Sadri, Pratapgath, Arnod and Bari Sadri and Abu Road and Pindwara tebsils of Sirohi district form this agro-climatic zone. The region has a moderately warm climate in summers and with mild winters. The annual rainfall varies from 50 to 95 cms. Temperature in summer goes up to 38.6 ° C in summer. Vegetables suitable for this zone are tomato, brinjal, chilli, spongegourd, carrot, peas, okra and Cole crops.

vi) Humid Southern Plains

The districts of Dungarpur and Banswara, parts of Udaipur (Dhariyawad, Salumber and Sarada tehsils) and Chittorgarh (Chotti sadri, Bari Sadri,

Pratapgarh and Arnod tehsils) are included in this region. The area comprises of low Aravalli hills.The area has humid climate with an average rainfall of more than 70 cm per year. Vegetables suitable for cultivation in this region are okra, brinjal, chilli, peas and cucurbits.

vii) Humid South-Eastern Plains

The districts of Kota, Baran, Bundi and Jhalawar and two tehsils of Sawai Madhopur namely Khandar and Sawai Madhopur. Canal irrigation system with a series of dams and barrages on the Chambal, has made this area rich in agricultural production. Gandhi Sagar, Rana Pratap Sagar and Jawhar Sagar dams together with Kota Barrage have generated enough resources of electricity and canal water for irrigation. The region has warm summers but mild winters. Summer temperatures sometimes touch 45 °C. The relative humidity is generally high in this zone. The annual rainfall varies from 60 to 85 cm. Vegetables suitable for this zone are okra, brinjal, chilli, and cucurbits.

Varieties / F1 hybrids recommended by AICRP for growing in Rajasthan

1. **Tomato (OP varieties):** Co-3 Determinate (TNAU), Punjab Kesari Determinate (PAU), Pant T3(Pantnagar), DVRT-2(IIVR), BT116-3-2 Determinate. (OUAT), DT 10 Indeterminate(IARI), SEL-3 Determinate (HAU).

2. **Tomato (F1 hybrids):** TH-01462 Det. (Syngenta), BSS 20(Bejo Sheetal), Arka Rakshak and Arka Samrat(IIHR).

3. **Brinjal (OP varieties):** Punjab Sadabahar (PAU) Long, PLR-1(TNAU) small round, DBSR-31(IARI) Long, H-8(HAU) round.

4. **Brinjal (F1 hybrids):** VNR 51(VNR Seeds) small round, VRBHR-1 (IIVR) round, PBH-6 (Pandey beej) long, ARBH 541 (Ankur seeds) Long, Pusa hybrid-9 (IARI) round.

5. **Chilli (OP varieties):** BC 225(OUAT), BC 14-2 (OUAT).

6. **Chilli (hybrid):** CCH-2(IIVR), ARCH- 228(Ankur seeds).

7. **Capsicum (hybrid):** KTCPH-3 (IARI-Katrain)

8. **Peas:** Arkel - Ealy (IARI), VL 7 early (VPKAS, Almora), VRP-2(IIVR). Midseason varieties: PC 531(PAU), CHP – 2 (HARP), Ageta – 6(PAU), Arka Ajit (IIHR, resistant to PM, Rust), PRS 4 (CSAUAT), Oregon Sugar (PAU).

9. **Onion:** Arka Kalyan (IIHR) AFLR(NHRDF), PBR -5 (PAU), All for rabi season,

10. **Cauliflower:** DC 76(IARI) Mid group

11. **Bitter gourd hybrid:** Pusa hybrid -2 (IARI)

12. **Bottle gourd (Varieties):** PBOG-61 (Pantnagar), NDBG-132(NDUAT), NDPK -24(NDUAT)

13. **Cucumber (hybrid):** PCUCH – 1 (Pantnagar)

14. **Muskmelon :** Pusa Madhuras(IARI),

15. **Watermelon :** Durgapura Meetha(Durgapura), MHW-6 Hybrid (MAHYCO).

16. **Sponge gourd :** Sel 99 (IARI)

17. **Okra:** DVR -2 (IIVR), JNDOH (Junagadh), Arka Anamika(IIHR)

    Resistant to YVMV: P-7 (PAU), PB 57 (MAU Parbhani), IIVR 11(IIVR), HRB 107-4 (HAU), VRO-5 (IIVR), HRB 9-2, HRB 55 – Varsha Uphar(HAU).

18. **Pumpkin:** NDPK -24 (NDUAT), Arka Chandan(IIHR).

Vegetable varieties

There are several varieties of vegetable crops developed from SAUs and ICAR institutes are becoming popular among the vegetable growers. A list of the important vegetable varieties suitable for cultivation in the state is given in the table (2).

Table 2: A list of the important vegetable varieties suitable for cultivation in the state

Crop	Variety released	Characteristics
Mateera	AHW-19	Medium-early maturing variety, produces 3.0-3.5 fruits per vine, flesh dark pink, solid (firm) with good eating quality and taste having 8.0 to 8.4% TSS. High yielder (460-500 q/ha) and tolerates high temperature.
Mateera	AHW-65	Very early maturing variety, produces 3-4 mature fruits per vine and gives yield of 375-400 q/ha. The flesh is delicious, pink, solid (firm) having 8.0-8.5 per cent TSS.
Mateera	Thar Manak	Developed through selection from the local land races found in arid region. It is very early. For fruit quality improvement in drought hardy *mateera* (watermelon), intensive breeding work was taken at CIAH, Bikaner, and resulting to this a new high yielding and better quality variety named as Thar Manak has been released in 2007 for the cultivation in arid region. It is very early for first marketable harvesting (75-80 DAS).

Contd.

Bottle gourd	Thar Samridhi	The yield potential is 50-80 tones/ha under arid conditions. It is exhibited high yield potential (3.82 – 5.82 kg/plant) under hot arid environment. Fruits weighing 450 – 700 g are ready for first harvesting after 50 to 55 days from sowing. The fruit yield potential is 240-300 q/ha.
Kachri	AHK-119	Fruits are small, egg shaped weighing 50-60 g. Fruits are ready for picking in 68-70 days after sowing, 22 fruits per vine, and yields of 95-100 q/ha.
Kachri	AHK-200	The fruits are 100-120 g in weight, become ready for harvest in 65-67 days after sowing, bears about 20 fruits per vine, yield of 115-120 q/ha
Snap melon	AHS- 10	Fruits can be harvested 68 days after sowing, fruits are oblong and medium in size (900 g), flesh whitish pink, sweet in taste having 4.5-5.0 per cent TSS. Bears 4.0-4.5 fruits vine each giving a yield of 225-230 q/ha under arid conditions.
Snap melon	AHS- 82	Fruit harvest starts 67-70 days after sowing, each vine bears 4.5-5.0 fruits giving an yield of 245-250 q/ha. The flesh is light pink, sweet having 4.3-4.9 per cent TSS.
Kakdi	AHC- 2	It is a very early maturing variety bearing uniform, long fruits. Fruits have light green skin without furrows. Fruit weighs 275-300 g, suitable for slicing. The crop yields 175-202 q/ha under arid situations.
Kakdi	AHC- 13	It is a very early and highly productive variety with profuse hermaphrodite flowers. On an average 2.15 kg tender fruits can be harvested per vine giving an yield of 85-125 q/ha. The variety also has high heat tolerance
Cluster bean	Thar Bhadavi	High yielding vareity. Pod yield poentital is 75 g/plant and 9.0 cluster/plant. Plant height is 65-70 cm. First harvest is 55-60 days after sowing.
Cluster bean	Goma Manjiri	Plants are erect, single stemmed, non-branching, bearing from base in cluster upto 35 with 8 to 10 pods per cluster, yielding 88 to 103 q/ ha, photo insensitive, , crop period 75 to 80 days, tolerant to drought.
Khejri	Thar Shobha	In khejri (*Prosopis cineraria*), the first high yielding and better quality variety Thar Shobha has been recommended for uniform tender pod harvesting for vegetable use. A five year grafted khejri plant yields a harvest of about 4.25 kg tender pods (*sangri*) and 6027 kg fodder (loong) per year.

Source: CIAH, Bikaner

District wise horticulture plan especially for vegetables/ tuber crops

In horticultural crops, vegetables are playing major role not for economic upliftment of rural families but also for nutritional security. In the state, vegetables are grown in 1.68 lakh hactare area with the annual production of 16.99 lakh tonnes. The productivity of vegetables in the state is 10.2 t/ha. Area, production and productivity of vegetables grown in Rajasthan state during last three years is given in table (3).

Table 3: Area, production and productivity of vegetable crops in Rajasthan

Years	Area (000ha)	Production (000MT)	Productivity (t/ha)
2013-14	149.64	1107.61	7.4
2014-15	153.91	1433.22	9.3
2015-16	194.64	2020.95	10.9
2016-17	170.13	1812.84	10.6
2017-18	166.23	1699.58	10.2

Source: Directorate of Horticulture, Rajasthan

In Rajasthan, tomato, okra, onion, peas and cucurbits are mainly grown. Cole crops occupy important place during winter season. Cauliflower and cabbage are mainly grown in Jaipur, Ajmer and Alwar areas. Recently, some progressive growers have started to produce broccoli, though, presently in small areas, however, it is expected that it will spread in larger areas in coming years owing to its medicinal and nutritional importance. A list of vegetables suitable for growing in Rajasthan is given in table (4)

Table 4: A list of vegetables suitable for growing in different districts of Rajasthan

Vegetables	Districts
Onion	Jodhpur, Ajmer, Sikar, Jaipur, Alwar
Tomato	Jaipur, Sirohi, Jalore
Cole crops	Ajmer, Jaipur, Alwar
Pea	Jaipur, Bundi
Okra	Jaipur, Alwar, Ajmer
Tinda	Jaipur, Alwar, Tonk

In Rajasthan, more than 40 types of vegetables are grown, and among them 30 are indigenous in nature. Crop wise adoptions of different vegetables are given in table (5).

Protected cultivation

Vegetable production is significantly influenced by the seasonality and weather conditions. The extent of their production cause considerable fluctuations in the prices and quality of vegetables. Striking a balance between all-season availability of vegetables with minimum environmental impact, and still to remain competitive, is a major challenge for the implementation of modern technology of crop production. The main purpose of protected cultivation is to create a favourable environment for the sustained growth of plant so as to realize its maximum potential even in adverse climatic conditions. Greenhouses, rain shelters, plastic tunnels, mulches, insect-proof net houses, shade nets etc. are used as protective structures and means depending on the requirements and cost-effectiveness.

Table 5: Crop wise area production and productivity of vegetables in Rajasthan (Area in ha and Production in MT)

Crop	2015-16			2016-17			2017-18		
	Area	Production	Productivity	Area	Production	Productivity	Area	Production	Productivity
Onion	86306	1435112	16628	62499	1149291	18389	64760	996733	15391
Tomato	20507	83286	4061	20366	90224	4430	18115.6	88732	4898
Potato	14322	229829	16047	14552	234552	16118	13819	278519	20155
Brinjal	6078	25789	4243	5881	31715	5393	5423.2	29602.8	5459
Okra	3282	10340	3151	3702	15379	4154	4153	21393	5151
Tinda	6645	14650	2205	3865	11190	2895	3838.5	14273.5	3719
Cucumber(Khira)	2334	7018	3007	1184	5239	4423	576	2646	4594
Long Melon	1130	3021	2673	2240	10193	4550	2214	13989	6318
Pumkin	1273	16449	12921	819	8539	10426	629	5623	8940
Bottle Gaurd	4435	15131	3412	4673	27168	5814	4539	23825.3	5249
Ridge Gaurd	1495	3973	2658	1161	4525	3898	1219	5893	4834
Bitter Gaurd	822	2331	2836	723	2306	3192	524	1599	3052
Cole crops	13857	56083	4047	12250	77992	6367	11455	63393	5534
Carrot	1530	8904	5820	3745	19771	5279	3390	27039	7976
Radish	851	3403	3999	1094	5071	4634	514	3103	6037
Peas	14219	31280	2200	13831	36375	2630	12923	33358	2581
Cluster Bean	694	976	1406	979	2664	2721	1232	3644.5	2958
Sweet Potato	1003	9602	9573	520	6516	12531	554	8659	15630
Spinach	2909	15974	5491	4274	20808	4869	2515.43	15814	6287
Colocassia	378	1346	3561	413	2836	6867	317	2522	7956
Water Melon	2929.94	27564	9408	2518	23607	9374	2013	19504	9689
Musk Melon	981.78	6661.5	6785	1186	5679	4789	1001.1	6095	6088
Others	6663	12234	1836	7657	21208	2770	10509	33624	3200
Total	194645	2020957	10383	170132	1812848	10656	166233.83	1699584.1	10224

Source: Directorate of Horticulture, Rajasthan

Besides modifying the plant's environment, these protective structures provide protection against wind, rain and insects. Protected cultivation has very high entrepreneurial value and profit maximization leading to local employment, social empowerment and respectability of the growers. Environmentally safe methodologies involving IPM tactics reduce the hazards lacing the high value products. Fertigation has been found to be one of the most important production technologies for hi-tech horticulture and protected cultivation. It helps in achieving higher productivity and enhancing the quality of horticultural produce. Precision application of water and nutrient is possible through drip fertigation to attain very high crop water and nutrient use efficiency mainly in protected cultivation.

Rajasthan state government has made provision of additional subsidy for green house construction. Now subsidy available for green house in the state as follows:

- 75 % for SF/MF farmers instead of 50% under NHM

- 50 % for other farmers instead of 33% under NHM

Targeted area under different protected structure and allocation of funds are given in table (6).

Table 6: Targeted area under different protected structure and allocation of funds

Component	Phy. Targets	Fin. (Rs. lac)
Greenhouse	30000 Sq. Mtr.	81.00
Mulching	50 ha.	3.50
Shade Net	50000 Sq. Mtr.	3.50
Plastic tunnel	50000 Sq. Mtr.	2.50
Total		90.50

Table 7: District wise proposed area under greenhouse of important vegetables

District	Greenhouse units (Each 500 sq.mt)	Crops
Jaipur	15	Capsicum, Tomato, Cucumber, Rose, Chili, flowers
Alwar	15	Capsicum, Tomato, Flowers
Ajmer	10	Gerbera, Rose, Capsicum, Tomato
Kota	10	Tomato, Capsicum, flowers
Sri-Ganganagar	10	Capsicum, Tomato, flowers
Total	60	

i. **Acrylic:** This material has long service life, good light transmittance (80%), moderate impact resistance, but prone to scratches. It has a high coefficient of expansion and contraction. Being inflammable and costly, it is not a preferred material.

ii. **Corrugated / Multi Wall Polycarbonate Sheet:** It is available in single or double wall sheets of different thickness. A new polycarbonate sheet has good light transmittance of about 78%, but reduces with age. It has excellent impact resistance and low inflammability. High cost limits its use on large scale.

iii. **Fiberglass Reinforced Plastic Panels (FRP):** These plastics consist of polyester resins, glass fibers stabilizers etc. It has a initial light transmittance of about 80% and has high impact resistance with a service life ranging from 6-12 years. Good quality FRP materials for greenhouse coverings are not quite assured.

iv. **Polyethylene Film:** A clear, new polyethylene sheet has about 88% light transmittance. Its higher strength and low cost have made it most popular replacement to glass. An ultra-violet (UV) stabilized plastic sheet can have a service life of 3 years. These sheets are generally available in 7 and 9 meter widths with 200 micron (0.2 mm) thickness.

v. **Thermal and Shedding Net**

Types of Greenhouses

The greenhouses design and cost range from a simple plastic walk-in tunnel costing about Rs.100/m^2. meter to a climate-controlled, saw-tooth greenhouse with automatic heating, ventilation and cooling, costing more than Rs. 3000/m^2. The selection of the greenhouse design should be determined by the grower's expectations, need, experience, and above all its cost-effectiveness in relation to the available market for the produce. Obviously, cost of greenhouse is very important and may outweigh all other considerations. Greenhouses are classified in different shapes, which also determine their cost, climate control and use in terms of crop production. Commonly used structural designs are briefly described below.

(i) **Gable:** This is the most basic structure similar to a hut-like construction and was perhaps the first version of a greenhouse with glass as the covering material. The roof-frame can be inclined at any angle to present an almost perpendicular face to the sun to minimize losses due to external reflection. The structure also allows large openings in the side-walls and at the ridge for high rates of natural ventilation. Modern gable-shaped greenhouses are multi-span units with bay widths of 6-12 meters.

(ii)   **Gambrel:** These structures are similar to the gable but have high strength to withstand high wind loads during storms. This design is more suitable where wood or bamboo are to be used for the greenhouse construction.

(iii)   **Skillion:** In this kind of structure, the roof consists of a single sloping surface. This is because the greenhouse is built as the southward extension of a building with a solid wall on the northern side. Such greenhouses have the advantage of low structural requirements.

(iv)   **Curved-roof – Raised High Arch / Raised Arch:** The semi-circular tunnel greenhouse structures appeared with the introduction of polyethylene film as the covering material. These structures, besides being most simple and easy to construct, have the advantage of high strength with a relatively light frame due to inherent strength of the curved arch. But these structures have the disadvantage of poor ventilation efficiency since the curved roof is not amenable to the incorporation of ridge ventilators. In an attempt to improve the ventilation efficiency of curved roof greenhouses, raised arch type of structures have been adopted. This design has vertical side-walls, which permit high head room and improved ventilation due to the wind velocity.

(v)   **Saw-tooth:** In these structures, the roof consists of a series of vertical surfaces separated by a series of sloping surfaces, all of which are pitched at the same angle and facing in the same direction. The vertical surfaces consist entirely of ventilating area. These types of greenhouses are most efficient from ventilation point of view. Such greenhouses are also suitable for multi-span structures. Orientation of saw-tooth greenhouses can also be used as a means of maximizing natural ventilation. By facing the open vertical ventilation areas away from the wind, airflow over the greenhouse roof creates a negative pressure, which facilitates in sucking out warm greenhouse air. However, this air dynamics relies on the premise that there are large ventilation areas in the greenhouse walls on windward side.

(vi)   **Plastic Low Tunnels / Row Covers:** These structures are laid in open fields to cover rows of plants with transparent plastic film stretched over steel hoops of about 50 cm height spaced suitably along the rows. Polyethylene film of 30-40 micron thickness, without UV stabilization, is used which is perforated in situ as the season gets hotter. Row covers used in vegetable production have different purposes in temperate and tropical regions. In cold conditions, they are used to conserve warmth, stimulate germination and early growth, protect plants from frost injury, and improve the quality of the crops. Other beneficial effects, such as maintaining soil structure, and protecting crops from the attacks of birds and pests, can also be expected. The main advantage of these covers in northern India is to grow vegetables, especially cucurbitaceous crops, ahead of normal season in winters. Experiments on muskmelon have proved it a highly profitable proposition. The muskmelon seedlings could be transplanted under such covers in the last week of January. The crop growth was sustained during the cold period. Temperature profile inside the cover indicated a difference of about 7°C averaged over 24-hour cycle. This rise in temperature provided necessary warmth to sustain the growth of plants. In hot season, however, materials used as row covers need to have adequate permeability to air and moisture, to prevent the accumulation of excessive heat inside the covers. The covering materials used in summer are woven polyester wind-break nets, cheese-cloth, and insect-proof screens. These types of covers are generally laid over the planted rows without the support of steel hoops, and are also called as floating covers. To provide adequate space, the seedlings are planted in the furrows and the covers are laid over the ground. But such planting should only be adopted in light textured soils having high infiltration rates.

Table 8: District wise area and allocation of funds for mulching and shade net

Mulching			Shade net		
District	Area (ha.)	Fin. Req. (Rs. lac)	District	Shade net units (Each 500 sq.mt.)	Fin. Req. (Rs lac)
Sriganganagar	10	0.70	Jaipur	10	0.35
Jaipur	10	0.70	Kota	10	0.35
Jodhpur	5	0.35	Ganganagar	10	0.35
Pali	5	0.35	Ajmer	10	0.35
Jalore	5	0.35	Alwar	10	0.35
Nagaur	5	0.35	Chittor	10	0.35
Ajmer	5	0.35	Jhalawar	10	0.35
Barmer	5	0.35	Jodhpur	10	0.35
Total	50	3.50	Pali	10	0.35
			Nagaur	10	0.35
			Total	100	3.50

Strategy and Development of Disease Free Quality Planting Material and Availability of Quality Seeds

Inadequate availability of quality planting material is one of the important deterring factors in development of sound vegetable cultivation. It is of special significance especially in annual perennial vegetable crops which has a long gestation period and effects are known only in later stages. A number of Government nurseries also exist in different states. Planting material is also being produced by the ICAR institutes and SAUs, Private Seed Company also play important role to meet the requirement of the growers and at present the number of small and medium scale seed producers is over 6300. Presently only 30-40% demands of seed material is being met by the existing infrastructure. Generally, farmers do not have access to good quality certified disease free seed material of true to type varieties as a result of which production, productivity and quality of the of produce suffers heavily. At present, most of the dependence is on the unregulated Private Seed Company in most of the states which lacks modern infrastructure such as greenhouse, mist chamber, efficient nursery tools and gadgets, implements and machinery.

There are several constraints in the existing system of seed multiplication. There are several Private Seed Company operating in the country playing important role in multiplication of planting material of vegetable crops and many of them follow traditional methods and lack adequate infrastructure and sell seed material of unknown pedigree. Of many other constraints, un-availability of non-maintenance of healthy stocks of elite varieties are worth mentioning. Further, concerted efforts be done for quality seed production of vegetable crops which are self-pollinated in nature e.g. pea, cowpea, methi and guar.

- Tuber crops like turmeric, sweet potato, yam etc should be promoted.
- Vegetable seed production can be taken as intercrops in fruit orchard
- Federating farmers for quality seed production of improved vegetable varieties
- Capacity development programmes be takenfor farmers to encourage for quality seed production.

4

Genetic Diversity of Vegetables in Arid Region

Pradeep Kumar, P.S. Khapte and P.R. Meghwal

ICAR-Central Arid Zone Research Institute, Jodhpur, Rajasthan-342003

Vegetables are of short duration, have higher productivity and give more return per unit area, besides giving nutritious food to human being and can be fitted in different farming systems. They also play a major role in balancing human diet, thereby helping in alleviating hunger, poverty and malnutrition. Because of the public awareness for the significance of vegetable consumption, the demand of vegetables in the global market has increased immensely. Thus, vegetable farming is becoming profitable venture globally, opening avenues of self employment and revenue generation (Anon., 2009). At global level, India holds second position, next only to China in vegetables production with a share of about 14%. The present annual vegetables production of country is about 175 million tons from an area about 10.3 million ha with a productivity of 17.1 t/ha. India grows largest number of vegetables in the world, as having a spectrum of wide range of agro-climatic situations, ranging from dry temperate to humid tropics between the altitudes from sea level to snow line.

The Indian hot arid region is one of the most important arid ecosystems of the world in terms of biodiversity it possesses. This arid region spreads over 31.7 mha, occupies major part of north western India including Rajasthan, Gujarat, Punjab and Haryana, which together contribute 73% of arid region. High temperature, wind velocity, low humidity, low and erratic rainfall and high potential evapotranspiration are important characteristics of this region which narrowed down the choice for vegetables, besides permit their low productivity. Despite the various constraints, arid region holds various strengths such as available vast land resources, surplus family labours, plenty of solar radiation, increasing canal command area, developing infrastructure facilities, opportunity to establish ago-based and allied cottage industries. Several well established as well as indigenous vegetables used for fresh and multiple uses in processing industries have long been commercially grown and adapted in the region. They are of short duration and hardy crops and being grown during kharif as rainfed mixing with other commercial field crops. These hardy vegetables mostly belong to cucurbitaceous (kachri, mateera, snap melon, round melon, etc.) and leguminous group (cluster bean), besides some other vegetable are also being grown as rainfed in region (Samadia, 2004; Meena, *et al.*, 2009).

There are some indigenous perennial hardy and well established species of arid region which have been used either as fresh or in processed form in a variety of ways in vegetables. They are khejri (immature pods-*sangri*), ker (fruit), lasora/gonda (immature fruit), drumstick (flowers, leaves and pods), curry leaf (leaves), guarpatha (succulent leaves), cactus pear (immature leaves and fruits), etc. In last few decades, area under irrigation increased with the increase in canal command area, wells, tubewell and water storage devices. Increase in irrigated areas coupled with modern tools and technologies has made the growers aware of greater utilization of available vast land resources and focused their attention towards high value vegetable crops which give more yield per unit land and time, besides realizing their potential in crop diversifications. Increasing population, people's purchasing power, variety in tastes, and growing awareness towards nutritional security resulted in quantum jump in demands of vegetables in arid region accelerated their cultivation. In irrigated areas, intensive cultivation of vegetables have been followed where large number of vegetables are grown which include both indigenous and introduced crops, they often regarded as irrigated crops. Irrigated vegetable crops include almost all the crops grown in other parts of country except only few requiring typical climatic condition.

Important horticultural flora and their parts used as nutritious vegetables in arid region

Parts used	Horticultural flora/ vegetable cops of economic use
Fruits	*Fresh*: Tomato, Brinjal, Chilli, Melons, Gourds, Ker, Gonda
	Dried: Chilli, Kachri, Snap melon
Pods	*Fresh*: Okra, Peas and Beans, Drumstics, Khejri (*sangri*)
	Dried: Cluster bean, Khejri
Leaves	Methi, Palak, Mustard green, Green onion, Curry leaf, Guarpatha
Roots & Tubers	Carrot, Radish, Sweet potato, Knol-Khol (*knob/tuber)*
Bulbs	Onion & Garlic (*cloves*) – largely fresh, also dried
Head	Cauliflower (*curd*), Cabbage
Seed	Kumat (dried)

Further, it was observed that people made useful selection in several vegetable crops, which were eventually domesticated and cultivated. Several weedy species were never or only temporarily domesticated, remaining as weed but often hybridizing by chances with the cultivated ones and thus, enhancing the diversity in cultivated plants (Ram, *et. al*, 2007)

Genetic diversity in vegetable crops

Genetic diversity found in vegetable crop species offers a great opportunity for successful breeding of improved cultivars with added value and desirable resistance to diseases and pests. Assemblage, evaluation, documentation, conservation of plant genetic resources are important not only to support the present days vegetables improvement programmes but they will also be needed to meet the aspirations of future generations who may require new sources of genes when faced with unforeseen challenges such as hostile changes in global climate, severe abiotic stresses and virulent forms of pests and pathogen (Sharma and Pandey, 2009). The diversity in vegetable crops occurs in the form of land races, traditional cultivars and primitive types. The Indian sub continent, one of the Vavilonian centres of origin of crop plants endowed with wide range of diversity in several vegetable crops and the distribution of variability in the major agro-climatic zones of India is given in table1.

India occupies a unique position among the major gene-rich countries and recognized one of the world's top twelve mega diversity nation. Indian sub continent is well known as important centre of origin for brinjal, smooth gourd, ridge gourd, ivy gourd, pointed gourd, cucumber, taro and elephant phoot yam whereas, centre of diversity for okra, melons, chayote, chilli, *Cucurbita* species. Appreciable diversity exists in several introduced crops such are tomato, french bean, cowpea, leafy brassiceae, coles, amaranths, yam (*Dioscoria* spp.), cucurbits (bottle gourd, bitter gourd, ivy gourd, ash gourd, snake gourd), and in bulbous crops like onion and garlic (Malik, *et al.*, 2001).

Distribution of vegetable crops variability in different agro-climatic zones of India

S.No.	Ago-ecological region	Geographical range	Variability in major vegetable crops
1.	Humid western Himalayan region	Jammu & Kashmir, Himachal Pradesh and Uttarakhand	Cucurbits, radish, carrot, turnip, cowpea, chilies, brinjal, okra, spinach, fenuegreek, amaranths, *Solanum khasianum, S. hirsutum, Schium edule, Basella rubra.*
2.	Humid Bengal/ Assam basin	West Bengal and Assam	Cucurbits, radish, cowpea, chilies, brinjal, spinach, beet, amaranths, okra, *Abelmoschus manihot* sp. *Manihot, Solanum indicum, S. khasianum, S. surrattense, Cucumis sativus* var. *vikkimensis, Edgeria darjelingensis,*

Contd.

			Meloththria assamica, Momordica cochichinensis, Schium edule, Tauladiantha coirdifolia, Basella rubra.
3.	Humid eastern Himalayan region and bay island	Arunachal Pradesh, Nagaland, Manipur, Mizoram, Tripura, Meghalaya and Andaman & Nicobar island	Cucurbits, radish, peas, cowpea, chillies, brinjal, okra, spinach beet, amaranth, *Abelmoschus manihot* sp. *tetraphyllus, S. khasianum, S. torvum, S. sysimbrifolium, S. ferox, S verbascifolium, Cucumis histrix, Luffa echinata, Sechium edule*
4.	Sub-humid Sutlej Ganga alluvial plains	Punjab, Uttar Pradesh and Bihar	Cucurbits, radish, carrot, peas, brinjal, okra, spinach beet, fenuegreek, onion, garlic, *Abelmoschus manihot* sp. *tetraphyllus* var. *pungens, A. tuberculatus, S. indicum, S. khasianum, S. surattense, S. hispidum, S. torvum, Cucumis hardwickii, C. trigonus.*
5.	Humid eastern and south eastern uplands	East Madhya Pradesh, Orissa and Andhra Pradesh	Cucurbits, radish, cowpea, chillies, brinjal, okra, spinach, amaranth, garlic, carrot, *Abelmoschus manihot* sp. *manihot, S. surattense, S. torvum.*
6.	Arid western plains	Haryana, Rajasthan and Gujarat	Cucurbits, cauliflower, radish, carrot, peas, cowpea, chillies, brinjal, okra, spinach beet, fenuegreek, onion, garlic, *Abelmoschus tuberculatum, A. ficuleus, A. manihot* ssp. *tetraphyllus, S. surattense, S. nigrum. Citrullus colocynthis.*
7.	Semi arid lava plateau and central highlands	Maharashtra and west Madhya Pradesh	Cucurbits, cauliflower, radish, carrot, cowpea, chillies, brinjal, okra, spinach beet, fenuegreek, amaranth, onion, *S. surattense, S. nigrum. S. torvum, S. khasianum, Cucumis seosus, Luffa acutangula* var. *acutangula.*
8.	Humid to semi arid western Ghats and Karnataka plateau	Karnataka, Tamil Nadu, Kerala and Lakshdeep island	Cucurbits, chillies, brinjal, okra, Amaranth, *Abelmoschus crinitus, A. angulosus, A. ficuleus, A. moschatus, A. manihot* var. *tetraphyllus, S. triobatum, S. indicum, S. insianum, S. pubescens, S. surattense, S.torvum, Luffa acutangula* var. *acutangula, Melothria angulata, Basella rubra.*

Source - Sharma, S.K., 2010

The Arid Western Region falls in the northwestern region of the Indian subcontinent possesses wide range of diversity of a number of vegetable crop species. Amongst which few are endemic to the regions and several others have adapted there over the years in cultivation. The great deal in variability occurs owing to most of their cross pollinated behavior and means of their regeneration through seeds. The commonly occurring species include *Cucumis, Citrullus* and *Momordica* species including their wild relatives and weedy forms, cluster bean, peas, cowpea, chilli, brinjal and wild species of *Solanum (nigrum, khasianum, torbum* and *surattense)*, fenuegreek, spinach, onion, garlic, carrot, radish, okra and relative species of *Abelmoschus (tuberculatus, ficulneus* and *manihot* ssp. *tetraphyllus)*. In addition, several perennial indigenous species of horticultural value of which economic plant parts (pods, fruit, leaves, etc) are either fresh or in processed form used in vegetable or pickle, such are khejri, ker, lasora, gurpatha, cactus pear, etc. having great variability in their natural occurring populations. Ram, *et al.* (2007) reported that in Rajasthan, the whole plant of *Gisekia pharnaceoide* (family Mulluginaceae) is widely consumed during food shortage but in the South and West (Deccan Region) the leaves are used as greens, as are leaves of *Glinus trianthemoides*.

Cucurbits diversity

Cucurbits form an important and large group of vegetables and Indian sub continent is considered to be the centre of origin and diversity of a number of wild and cultivated cucurbits. There is tremendous genetic diversity within the family, and the range of adaptation for cucurbit species includes tropical and subtropical region, arid deserts, and temperate regions. The region wise distribution of cucurbits in India is given in table given below. The genetic diversity in cucurbits extends to both vegetative and reproductive characteristics and considerable range in the monoploid (x) chromosome numbers viz., x=7 (*Cucumis sativus*); x=11(*Citrullus* spp., *Momordica* spp., *Lagenaria* spp., *Sechium* spp. and *Trichsanthes* spp.); x=12 (*Benincasa hispida, Coccinia cordifolia, Cucumis* spp. other than *C. sativus,* and *Praecitrullus fistulosus*); x=13 (*Luffa* spp.) and x=20 (*Cucurbita* spp.) (Gupta, 2009).

Distribution of cucurbits species in India

Cucurbitaceous Species	Distribution
Cucumis hardwickii, C. trigonus, Luffa graveolens, Trichsanthes multiloba, T. himalensis	Western Himalayas
Cucumis trigonus, Luffa graveolens, Neoluffa sikkimensis	Eastern Himalayas
Cucumis hystrix, C. trigous, Luffa graveolens, Momordica cochichinensis, M. macrophylla, M. subangulata, Trichsanthes anguina, T. dioica, T. dicaeleosperma, T. khasiana, T. ovata, T. truncate	North-eastern region
Luffa echinata, Momordica cochichinensis, M. cymbalaria, M. dioica	Gangatic plains
Citrullus colocynthes, Cucumis prophetarum, M. balsamina	Indus plains
Cucumis setosus, C. trigonus, Luffa graveolens, M. cochichinensis, M. subangulata, Trichosanthes anamalaeiensis,	Western peninsular tract

Contd.

T. bracteata, T. cuspidate, T. nervifolia, T. perotteliana *Cucumis hystrix, C. setosus, Luffa acutangula* var. *amara,* *L. graveolens, L. umbellate, Momordica cochichinensis,* *M. cymbalaria, M. denticulate, M. dioica, Trichsanthes bracteata,* *T. cordata, T. himalensis, T. multiloba*	Eastern peninsular tract

Source: Ram and Srivastava (1999)

Cucurbit group is the most important and diverse group which represents the hardiest vegetable species grown in arid condition and some of which are endemic to the region viz., kachari (*Cucumis melo* var. *agrestis*), snap melon (*Cucumis melo* var. *momordica*), spine gourd (*Momordica dioica*) and some are the wild relatives viz., bitter melon (*M. balsamina*), tumba (*Citrullus colosynthis*) and hill colocynth (*Cucumis hardwickii*) which grow naturally during rainy season and generate good source of income for the locals. These vegetables possess very good nutritive and medicinal value with resistance to biotic and abiotic stresses. Some others that have been adapted to the region due to natural crossing and selection that might have happened during course of their cultivation, they are musk melon, mateera/watermelon, round melon, etc. Considerable genotypic and phenotypic variations at both the species and genera level present in their natural as well as in form of farmers varieties across the region. In this region good amount of variability of certain cucurbits have also been found in their natural/wild form of which economic value yet to explore, such are *Cucumis prophetarum, C. trigonum, C. setosus,*

The region is rich in diversity for drought hardiness and sweetness, better shelf life for most crops, but particularly for *Citrullus lanatus* and *Cucumis melo* var. *momordica*. In *Cucumis sativus*, specific variability has been recorded for small (tender) fruit size, drought tolerance, and yield. Using this genetic variability a number of varietal products have been developed through selection from local landraces; for example, in watermelon, Durgapura Kesar and Durgapura Madhu; in musk melon, Durgapura Madhu, Akra Rajhans, Pusa Madhu; in *Momordica charantia*, Pusa Do Mausmi (Sirohi *et al.*, 2005)

Cucumis species

The genus *Cucumis* comprises about 30 species distributed over two distinct geographic areas (South-east of Himalayas group and African group), however, major species of economic importance are only *C. sativus* (cucumber), *C. melo* (musk melon), *C. melo* var. *utilissimus* (long melon), *C. melo* var. *momordica* (snap melon/phoot) and *C. melo* var. *agrestis* or *C. callosus* (kachri). Indian sub continent is the centre of origin for *C. sativus* and centre of diversity for *C. melo* (Zeven and de Wet, 1982). However, north western arid part of Rajasthan is rich in the land races of *C. melo* var. *momordica* and *C. melo*. As a result of natural crossing among different species of *Cucumis*, several new forms have stabilized which are quite different from the traditional forms. The highest genetic diversity has been observed in *kachri* and snap melon. However, large phenotypic variations are also found in the land races, open pollinated varieties and farmers varieties of musk melon, cucumber and long melon in the regions. In addition, considerable amount of diversity exists in natural forms of several wild species of the *Cucumis* which includes *C. hardwickii, C. prophetarum, C. trigonum, C. setosus*, which might carry sources of resistance for various pests and diseases and hardiness ability that could be explored and utilized for improvement of commercial crops.

Threats to biodiversity are increasing in India due to population pressure on cultivatable land. Sweet melon evolution under domestication has resulted in better productivity and fruit quality. However, this process narrowed the genetic basis. Natural variation among the sweet melon relatives *i.e.*, snap melon (var. *momordica*), for increased acidity, high tolerance to biotic and abiotic stresses and kachri (var. *agrestis*), for large number of fruits per vine, drought tolerance and disease resistance, provides an opportunity to enrich the gene pool of sweet melon with novel alleles that eventually could improve productivity, quality and adaptation and lessen the risk of genetic vulnerability. (Dhillon, *et al.*, 2007)

Muskmelon (*Cucumis melo*) is the most important species of *Cucumis* largely grown in dry and semi dry areas. Its ripe fruits are eaten as desert. Melon diversity in the region represents characteristic features of higher sweetness and hardiness with wide variability in its plants and fruit characters. The variability was observed in vine length, branches, fruits size, -shape, -weight, -no./plant, rind colour, cavity size and pulp content, no. of seeds/ fruit, etc. A wild relative of musk melon 'Jangli Indrajan' collected from Sirohi, Rajasthan which was having very sweet taste.

Cucumber (*Cucumis sativus*) is indigenous vegetable commercially grown all over the country. The immature fruits are used as salad and for pickling. Tender leaves are also reported to be used as vegetable. In north western India, considerable diversity found in its land races particularly in its fruits morphological characteristics. Chandra (1995) reported that locally adapted landraces in cucumber (Sawanariya, Balamkakdi) had more variation in Sirohi (Abu) and Shekhavati areas of Rajasthan for fruit shape, fruit skin colour, fruit skin mottling, colour of stripes, fruit skin glossiness, fruit length/width, weight and stem-endlblossom-end fruit shape, flesh texture and flesh colour. The wild species *C. hardwickii* R., a small bitter cucumber with sparse and stiff spines is probable progenitor of cucumber has been found in foothills of the Himalayas where it is thought to be originated, has also been reported to be found in Sirohi area in Rajasthan.

Kakdi (*Cucumis melo* spp.), the natural combinations of *Cucumis* species resemble cucumber or long melon and are commonly used for salad and sometimes tender fruits can be used for garnishing the vegetables. The unripe mature fruits are cooked as vegetable (Annon, 2007). Good amount of variability observed in the collected germplasms from different parts of arid and semi arid regions. The variability exists in morphological characteristics of plant and fruits.

Kachri (*Cucumis callosus* or *C. melo* var. *agrestis*) is an important drought hardy cucurbit vegetable of arid ecosystem. It has long been extensively grown under rainfed conditions during *kharif* season with various field crops like pearl millet, cluster bean, moth bean, etc. Its mature unripe fruits are usually cooked with various vegetable preparations, *chutnies* and pickles, whereas ripe fruits are eaten raw as salad or dessert. *Kachri* is one of the components of the delicious vegetable popularly known as *Panchkuta* in the desert districts of north-western India. Fruits are known to contain vitamin C. Residents of the region cut the fruits in small pieces, dry in shade and use them during rest of the year. Their dehydrated powdered products either used directly or for making other products. A large number of collections have been made by different research institutes/centers from different parts the regions and wide range of variability found in the collected germplasms with regards growth, maturity, flowering, fruiting and fruit characters besides resistance to biotic and abiotic stresses. It is known to carry resistance to CGMMV (Sheshadri & More, 2002). An exploration programme was undertaken by NBPGR in 2006 for the germplasms collection of different *Cucumis* species from different parts of Rajasthan including Dungarpur, Pali, Sirohi and Udaipur districts covering Aravali range. Out of total collections (41) were made, 21 of *C. callosus*. Wide range of variability was observed in plant habit; shape, colour, texture of leaves; size, shape, colour and weight of fruits; fruit skin texture and ornamentation; fruit length and width; thickness of mesocarp and pulp; pulp colour; size shape an colour of seeds and 100 seed weight. Two promising accessions (NKD/OPD-3345 and NKD/OPD-3346) collected from tribal village Ghata Bhardaria of Dungarpur (Rajasthan) had dark blackish green fruits which become orange at maturity (Annon, (2006-07). An evaluation study of 558 collected samples revealed a wide genetic variation in characters such as days to open first female flower (28.2-48.7 day after sowing, DAS), days to first harvest (54.5-80.0 DAS), fruits/plant (4.0-102.4), vine length (0.90-2.50 m), no. of branches / plant3.0-14.5), fruit weight (10.2-220.5 g), fruit length (2.30-12.01 cm), fruit diameter (2.21-6.50 cm) and flesh thickness (0.11-1.64 cm). Variability in fruit shape (oval, round, oblong, long, spindle or pear shaped), size (very small, small, medium or large), skin colour (light green to green, light yellow or orange to whitish, yellow, striped or mottles) and flesh colour (whitish to whitish yellow or light orange to yellow orange) were also observed. Some of the promising genotypes of desirable traits have been identified at CIAH, Bikaner are AHK 5, AHK 26, AHK 109, AHK 119, AHK 155, AHK 200 and AHK 202 (Samadia, 2004). The potential areas with wide variability are western parts of Rajasthan (Barmer, Bikaner and Nagaur), besides, north-eastern and Aravalli hills of Rajasthan also possess good variability.

Snap melon (*Cucumis melo* var. *momordica* Duthie & Fullar) is commonly known as 'phoot' which means to split. This is a hardy cucurbit vegetable extensively grown under rainfed condition in the arid region. Its immature fruits are cooked or pickled, whereas ripened fruits are eaten raw as salad or as dessert. Ripe fruits invariably crack. It has wide distribution in northern parts of India particularly arid and semi arid tracts. Considerable diversity of snap melon land races have been observed/collected with wide variations in morphology and fruit characteristics, during different explorations and collections in this region. Based on an exploratory survey, covering various parts of Rajasthan, Chandra (1995) stated that Jodhpur, Nagaur and Bikaner areas were veritable 'goldmine' of snap melons, where they were widely distributed (including in the sand-dunes habitats) and the collected germplasms were highly variable for fruit characters. Snap melon has been reported to carry resistance to powdery mildew (*Sphaerotheca fuliginea*), downy mildew (*Pseudoperonspora cubensis*), cucumber green mottle mosaic virus (CGMMV) (Sheshadri & More, 2002); Zucchini yellow mosaic and Papaya ring spot viruses, insect (*Aphis gossipayi*) and nematode (*Meloidogyne incognita*) (Dhillon, *et al.*, 2007). RAPD primers based grouping analysis of 36 diverse snap melon landraces collected from different parts revealed that Indian snap melon was rich in genetic variation and region and sub-region approach should be followed across India for acquisition of additional melon land races. The comparative analysis of the genetic variability among Indian snap melons and an array of previously characterized reference accessions of melon from Spain, Israel, Korea, Japan, Maldives, Iraq, Pakistan and India using SSRs showed that Indian snapmelon germplasms contained a high degree of unique genetic variability which was needed to be preserved to broaden the genetic base of melon germplasms available with the scientific community (Dhillon, *et al.*, 2007).

Citrullus species

Genus *Citrullus* has two important edible species of economic importance i.e., *C. lanatus*, water melon and *C. lanatus* var. *fistulosus*, round melon/ tinda widely found to occur in arid region. The wild species *C. colosynthis*, known as tumba, is endemic to the arid region naturally grown on desert sands.

Mateera [*Citrullus lanatus* (Thunb.) Mansf.], a water melon of desert, is one of the most important cucurbitaceous vegetables in the hot arid ecosystem where it appears to have acquired the drought hardy characteristic. Here, its ripe fruits are predominantly eaten as a dessert while immature small green fruit (*loyia*) are used as vegetable. Its seeds are dried and made into flour which then mixes with *bajra* to prepare *roti*. The roasted seeds with salt are also eaten by the local inhabitant. Enormous diversity in its land races are found in the region. The widely distributed populations in the region have significant variations in their morphological characteristics. An evaluation study made very early at CAZRI, Jodhpur in which around 40 genotypes including wide range of collected variations of local *mateera*, heterozygous materials and some released varieties showed wide variations in vine length, no. of fruits and fruit weight per vine, flesh colour, TSS, rind colour and thickness, etc. (Annon, 1977). The typical form with drought hardy nature and high sweetness are found to occur in the hot arid regions. A promising land race 'Chinimatiri' has been reported to be collected from Deoli, Pali (Raj.) having oval shaped fruits with thin skin and very sweet and was tolerant to salinity.

Round melon (*Citrullus vulgaris* var. *fistulous* or *Praecitrullus fistulosus*) is locally known as 'tinda'. It is another important cucurbit vegetable grown largely in the region. Its tender fruits are used as vegetables, sometimes also pickled. Enormous diversity exists in its available populations distributed throught the warmer parts of India including arid and semi arid regions. Variability occurs in its vine growth habit and fruit characteristics. A promising accession of tinda named '*Moritinda*' having large sized fruits weighing 500 g each and 350 seeds/fruit collected from Kharchia region of Sirohi (Rajasthan) (Annon, 2005).

In arid region, numerous populations of *Citrullus* exist, hence genetic diversity could be utilized in improvement of the commercially cultivated plants for their better utilization. There are a large number of strains of *Citrullus colocynthis*, locally known as 'tumba', a perennial natural growing creeper found throughout the desert, spreading on loose sand. It has a great survival value under extreme xeric conditions and could be stated as a characteristic 'drought enduring' plant of this area (Sen, 1973). However, in the past several years, ever since its seeds began to be collected for oil, it has become a threatened plant in the desert. A natural hybrid '*Ta-tumba*' crossed between tumba (*C. colocynthis*) and commercial water melom (*C. lunatus*) has been found in the desert (Singh, 1978; Bhandari and Shirangi, 1986). Owing to cross compatibility between these two, there is the possibility of transferring disease-resistant and hardiness genes to the water melon. The genetic diversity of the desert adapted plants is likely to be of immense value in increasing the yields and for coping with changing environmental conditions.

Luffa species

The genus *Luffa* has two important cultivated species, sponge or smooth gourd (*Luffa cylendrica* Roem. syn. *aegyptiaca*) and ridge or ribbed gourd (*L. acutangula* Roxb. L), locally known as *ghiya tori* and *kali tori*, respectively. Fruits of former type contain higher protein and carotene than later type. The *Luffa* has essentially old origin in subtropical Asian region particularly India (Kalloo, 1993). Based on cytogenetic investigations, Roy *et al.* (1970) suggested that the wild species, *L. graveolens* as the prime species which has given rise to these two cultivated species. *L. hermaphrodita,* also originated from *L. graveolens,* is another potential species which is mentioned as rare species distributed in parts of north-central India and *L. acutangula* var. *amara*, a wild relatives of *L. acutangula*, grown in peninsular India (Ram, *et al.*, 2007) have also been found and collected from the parts of the region.

Coccinia spp.

Ivy gourd or *kundru* (*Coccinia grandis*) is a semi perennial cucurbit vegetable of Indian origin and largely grown under tropical and subtropical regions. Its natural growing creepers have also been found to spread all through the shrubs, small trees and boundary walls in some of the areas of Aravali ranges particularly of Sirohi district. The NBPGR, RS, Jodhpur has collected some of the germplasms of *kundru*. In recent, collection and evaluation studies of *kundru* initiated at CAZRI to identify the suitable genotypes for commercial cultivation in arid condition. The germplasms include introduced as well as collected materials from various parts. Considerable variations were observed with regards plant spread (3.1- 4.8m), leaf shape (lobes), fruit skin colour and shape, fruit length (4.33 to 5.66 cm), fruit girth (2.33 to 2.53 cm) and average fruit weight (16.67 to 21.43 g).

Solanacious vegetables

Tomato, brinjal and chilli are important solanaceous vegetables widely grown in the region with limited or assured irrigation. The region encompasses enormous variability in chillies germplasms. Several wild relatives or weedy forms of genus *Solanum,* to which brinjal belongs have also been found to occur in the regions.

Chilli (*Capsicum annum* L.) is one of the most important vegetable crops of the irrigated arid areas. Although originated in Southern America, India houses great deal of variability in its cultivated forms and recognized as secondary centre of origin of chillies. Arid climate greatly favours its high quality fruits especially red ripe fruits which are used as spices. Long fruited green chillies are also in good demands and largely used as salad, vegetables and also pickled in western arid region. Being a often cross pollinated crop, chance, though lesser in arising of variability is obvious that might have led a good deal of variability which exists in its open pollinated populations in form of its land races and other cultivated forms. Samadia (2004) reported that an exploration was undertaken in 1998 in chilli growing areas of western Rajasthan and 132 collections were made. These were evaluated during *kharif* season of 1999 at CIAH, Bikaner. On evaluation of collected germplasm, in general, four types of land races in cultivation in chilli growing areas of arid regions and these were *Mathania* or desi, Haripur Raipur, Mehsana and Mandoria type. A wide range of variability was observed for all the agromorphological characters under study. The range of variations were recorded for days to first flowering (45.2 – 60.5 DAT), days to 50 % flowering (52.5 – 68.4 DAT), days to first harvest (green 85.7-95.5, red ripe 110.4-122.2 DAT), number of fruits (21.4-98.4), plant height (46.5-78.4 cm), number of branches per plant (3.2-9.8), fruit weight(green 3.42-20.29, red ripe 4.22-22.68 and dry 0.50-2.33 g), fruit length (9.2-17.5cm), fruit diameter (1.12-2.42 cm), number of seed/fruit (25.2-105.4) and seed weight/fruit (0.195-0.943 g). Variability study of with 30 chilli genotypes done by Chaudhary and Samadia (2004) under hot arid environment and reported high phenotypic and genotypic variations for red ripe fruit yield per plant (44.20 and 42.91), green fruit yield per plant, weight of seeds per fruit, fruit weight and number of fruits per plant. Variability studies of 30 chilli genotypes done by Chaudhary and Samadia (2004) under hot arid environment. They reported high phenotypic and genotypic variations for red ripe fruit yield per plant (44.20 and 42.91), green fruit yield per plant, weight of seeds per fruit, fruit weight and number of fruits per plant.

Brinjal (*Solanum melongina*) is an important vegetable crop of Indian origin and widely grown throughout the country. In arid region, the variability in plants (tall, dwarf, spiny, spineless stem and leaf, etc.) and fruit (long, small, round and oval shaped; purple, green, whitish green colour) characteristics can be seen in its widely distributed populations. Enormous diversity exists within the genus *Solanum* and a numbers of non cultivated or wild species with wide variations, some of which possess medicinal significance are reported to be widely distributed in the regions such are *S. surattens, S. albicule, S. incanum, S. indicum, S. trilobatum, S. gigantium,* from different parts of Rajasthan and Gujarat. Amongst these, *S. gigantium* and *S. albicule* have been mentioned as rare species.

Leguminous vegetables

A number of leguminous vegetables are grown in the region under rainfed or irrigated condition such are cowpea (*Vigna unguiculata*), pea (*Pisum sativum*),

Indian bean (*Dolichos lablabs*), mucuna bean (*Mucuna pruriens*), cluster bean (*Cyamosis tetragonoloba*). However, only few species have ability to withstand considerable drought and have primarily being grown as rainfed crop during kharif under arid condition.

Cluster bean (*Cyamopsis tetragonoloba* L.) commonly known as *guar*, is most drought hardy leguminous vegetable widely grown as rainfed crop during *kharif*, however under irrigated condition it can be taken both in summer as well as rainy season. Considerable amount of variability observed in the collected germplasms from the region. The variability exists in its plant height, branching habit, photo sensitivity, length and width of pods, number of pods/plant, fruit yield, drought hardiness, etc.

A land race of cluster bean called as *'vakodia guar'* (IC 538002) collected from Jam Knaria of Kachchh region of Gujarat is considered as superior vegetable type by the local farmers of region (Annon, 05-06)

Cowpea (*Vigna unguiculata*) is another important leguminous vegetable crop of arid ecosystem owing to its hardy nature. It is widely grown in the region in both summer and kharif season. Considerable amount of variability has been noticed in its morphological characters of both plant growth and fruiting behavior.

Abelmoschus species

The genus *Abelmoschus* includes a most common warm season vegetable crop i.e., okra (*Abelmoschus esculentus*), which withstand appreciable drought while grown in *kharif*, but is mostly grown as irrigated crop both in summer as well as kharif season in the region. *Abelmoschus tuberculatum, A. ficuleus, A. manihot* ssp. *tetraphyllus* are important wild relatives of okra which are widely distributed and naturally occur throught the arid regions and showing great variations in their phenotypic characters including plants and fruits.

Leafy vegetables

The important leafy vegetables grown in arid regions are *palak*/beet leaf (*Beta vulgaris* var. *bengalensis*), *methi*/fenuegreek (*Trigonella foenum-grecum* L.), coriander (*Coriandrum sativum* L.), amaranth (*Amaranthus* sp.) and mustard green (*Brsica juncia*). There are some other crops of which leaves are also eaten such as drumstick, curry leaf, etc. The region produces high quality fenugreek and coriander, which are grown for spices (seeds) as well as the greens (leaf). *Palak* and *methi* have wide range of variability with respect to leaf size, shape, thickness, glossiness and varied degree of adaptability to wide range of temperature, moisture and soil conditions.

Moringa species

Genus *Moringa* belongs to family Moringaceae of which three species occur in India. Of these, only two species have been found in the region. *M. oleifera* Lam. (*sahjana*) is one of the most important tree vegetable species of Indian household of which edible immature tender pods, leaves and flowers are used in vegetables, besides having medicinal properties. *M. concanensis* Nimmo ex Dalz. & Gibs. (*sarguro*) is a another species which has been listed as rare and found to naturally occur in some rocky and gravelly terrain of western Rajasthan. Its unripe pods are known to be eaten by natives, besides its various parts having a number of medicinal properties (Kumar, *et al.*, 2010). Good amount of variability found in its available populations in the region which includes plant height, spreads, pod size, shape, colour, bitter/sweet-ness, seed no./ pod, seed weight, etc. Seeds from their natural habitats have been collected by various institutes include NBPGR, RS, Jodhpur, CAZRI, Jodhpur, CIAH, Bikaner, etc. for their *ex-situ* conservation and further utilization for improvement programme.

Aloe vera or Indian aloe (*Aloe barbadensis* Mill.) is a succulent perennial herb locally known as *gurapatha* (in Rajasthan and Gujarat) and *ghritkumari* (Hindi/Sanskrit). The thick, fleshy leaves can hold onto a great deal of water, allowing it to withstand short periods of drought. Owing to its draoght hardiness and medicinal value, it is being grown in arid regions for its multiple use. Edible succulent fleshy leaves of sweet type, free from bitterness, are used in variety of ways including vegetables preparation. Rich diversity exists in its

Khejri [*Prosopis cineraria* (L.) Druce.], a common indigenous multipurpose tree species is known as 'lifeline' of the desert. Owing to having deep and extensive root system well established plant can withstand drought for many years (Mann and Saxena, 1980). Tender pods (*sangri*) are used fresh or after drying in vegetables, and major ingredient in *Panchkuta*, a special dish of *Marwar* region. Mature pods with sweet pulp are eaten as fruit. It played an important role in the amelioration of the conditions in arid and semi-arid lands and in providing livelihood support in the Thar Desert. *Khejri* is cross pollinated and its natural regeneration occurs through seed and, therefore, wide range of variability exists in its natural populations with respect to various variability indicators. In general, there are two distinct ecotypes, one having drooping branches with smaller leaves while the other having erect branches with bigger leaves. In some plants, upright branching pattern was also found. Similarly some plants are more thorny and some were observed as thorn less.

Wide diversity for vegetative growth, yield, and quality attributes such as cluster bearing, larger pod size, sweet taste and precocity, etc. in khejri occur in the region.

Ker [*Capparis decidua* (Forsk.) Edgew] is a succulent, xerophytic shrub of Thar desert found extensively on almost all the land forms of arid region. It is widely distributed as natural wild in the arid and semi arid regions of north-west India mainly in the Indian desert, which covers the parts of western Rajasthan. Besides, this species is widely distributed in the drier parts of Gujarat, Haryana and Punjab (Malik, *et al.*, 2010). This species is highly tolerant and adapted well to extreme temperatures and drought conditions. It is a very common thorny shrub or tree of the desert. Unripe fruits are edible and used as fresh vegetable and also pickled in various ways. Ker makes the major component in the delicious dish *Panchkuta* of Marwar region. Singh (2008) reported immature flower buds and flowers of ker also used in vegetables and pickling purpose. Fully ripen fruits are sweet and eaten raw by local people. Immature fruits are also dried for subsequent use as vegetables in off season (Sen, 2004). Apart from these various plant parts have immense medicinal properties to cure several disease and disorders.

In the natural occurring populations in its potential diversity area i.e., western Rajasthan, adjoining area of Gujarat, Haryana and Punjab, enormous variability have been reported from collected germplasms in plants growth habit, spiny nature, branching pattern, foliage colour, flowering and fruiting behavior, fruit shape, size and colour, pulp content etc. Prolific fruiting type and genotypes with less or no spines have been identified during exploration in the parts of Haryana and Rajasthan by NBPGR. Accessions IC345829, IC345837, IC345840, IC345842 and IC345845 have been found to be with less or no spines, while accessions numbers IC345819 and IC561789 have been identified for prolific bearing and bold fruits (Malik, *et al.*, 2010).

Conservation of ker germplasm is presently being undertaken using *ex situ* conservation approach at ICAR institutes and state agricultural universities located in Rajasthan and Haryana. Germplasm in the field genebank is being maintained at CAZRI, Jodhpur (20 accessions), CIAH, Bikaner (65 accessions), CCSHAU, Regional Research Station, Bawal and NBPGR Regional Station, Jodhpur (22 accessions), besides these a total of 88 diverse accessions have been successfully cryostored at NBPGR, New Delhi (Malik, *et al.*, 2010) as the species mentioned with short seed viability and limited establishment of new seedlings in the nature (Deora and Shekhawat, 1995).

Lasora (*Cordia myxa* L.) locally known gonda, is an underutilized crop. It is thought to be native of Northwestern India (Stewart and Brandis, 1992) and is distributed throughout country mainly in warmer regions where it is found as natural wild, occasionally cultivated. It is grown in the homestead gardens, backyards and farmers fields as isolated tree or few in numbers. However, recently some progressive farmers have started small commercial orchards in

Rajasthan and Haryana using local selections. Its unripe fruits are largely used for vegetables purpose or pickled. Other *Cordia* species are *C. rothii* Roem. syn. *C. gharaf* (Forst.f.) Ehrenb. and Asch., *C. crenata* Delile Fl.

A large number of diverse collections of germplasms of different *Cordia* species have been made by NBPGR in collaborations with other ICAR institutes, SAUs and state agencies from different parts of Rajasthan, Haryana, Gujarat, Madhya Pradesh, Himachal Pradesh and Uttar Pradesh. Collected germplasms represented the sizable diversity in fruit weight, shape, size, surface feature, pulp content, seed size, weight and shape. Germplasm of lasora is being conserved in the field gene bank at CCSHAU, Regional Research Station, Bawal (30), CIAH, Bikaner (65), NBPGR Regional Station, Jodhpur (73) and NDUAT, Faizabad (Malik *et al.*, 2010).

Genetic erosion/ extinction and conservation of vegetable crops

Valuable genetic resources of vegetable crops are vanishing very fastly. Several factors are responsible for loss of genetic diversity like shrinking of natural land resources, population pressure, urbanization, deforestation, monoculture, changing cropping pattern including use of hybrids/ improved cultivars, exotic vegetables. Looking to the importance of presence of wide range of diversity in growing populations of vegetables in their natural as well as cultivated forms, tremendous efforts have been made by various research organizations in the past for collection, characterization and documentation for further utilization of these variability for future improvement programme. A large number of exploratory survey have been conducted time to time by various institutes including NBPGR, Regional Station, Jodhpur; CAZRI, Jodhpur; CIAH, Bikaner; various SAU's and many others separately or in collaborations with others and large numbers of germplasms in form of landraces, primitive cultivars, wild relatives and weedy types have been collected from different tracts of arid and semi arid region. The collected germplasms are being conserved in various forms (field gene banks, medium term seed storage, national gene bank, cryogene bank or tissue culture) at NBPGR, New Delhi and at various other institute/centres in the field gene banks. After their evaluation, characterization and identification for unique traits, as promising types, they have been registered with NBPGR for their future utilization in research endeavour. NBPGR, the nodal national organization responsible for plant genetic resources management in India, has been undertaking explorations and collections in consultation and collaboration with scientists of relevant horticultural institutes of ICAR and SAUs for a holistic approach for PGR collection and utilization, besides, acting as pivotal role in the exchange of germplasms both at national and international level. As on 31.12.2009 a total of 24,112 accessions of vegetables are conserved as base collection in National Gene bank at NBPGR, New Delhi (at -18 C).

In last ten year (2001-2010), as per the NBPGR annual reports, its regional station at Jodhpur located in arid region has collected a large number of germplasms of vegetables or the crop plants of which economic parts are used as vegetable, and their wild relatives during explorations from different parts of Rajasthan and Gujarat, albeit adjoining areas also included. The collected species include *Cucumis melo* (49), *C. sativus* (14), *C. melo* var. *agrestis/ callosus* (31), *C. melo* var. *momordica* (28), *C. melo* var. *utilissimus* (18), wild species of *Cucumis* i.e., *hardwickii, prophetarum* and *trigonum* (46), *Citrullus lanatus* (75), *C. vulgaris* var. *fistulosus* (15), *C. colosynthis* (22), *Momordica dioica* (1), *M. charantia* (2), wild variety/species of *Momordica i.e., charantia* var. *muricata* (6), *M. balsamina* (3), *Aloe barbadensis* (21), *Allium cepa* (3), *A. sativum* (4),*Capsicum annum* (56), *Trichosanthes dioica* (3), *T. anguina* (3), *T. cucumerina* (12), *Cucurbita moschata* (4), *C. argyrosperma* (1), *Luffa cylendrica* (5), *L. acutangula* (9), *L. acutangula* var. *amara* (2), *L. hermophrodita* (1), *Benincasa hispida* (1), *Lagenaris siceraria* (11), *Coccinia grandis* (8), *Solanum melongena* (2), *S. surrattense* (1), *Lycopersicon esculentus* (3), *Raphanus sativus* (1), *Spinacea oleracia* (2), *Chenopodium album* (1), *Abelmoschus esculentus* (14), *A. manihot* (1), *A. moschata* (2), *A. ficulens* (2), *A. tuberculatus* (2), *Dolochos labab* (1), *Canavalia gladiata* (3), *Vicia acountifolia* (4), *Vigna unguiculata* (6), *Colocasia esculenta* (1), *Dioscoria alata* (1), *D. deltoidia* (1), *Cordia myxa* (3), *C. crenata* (3), *C. rothii* (4), *Capparis deciduas* (2), *Murraya koengi* (5). In addition, several other institutes/research centres have made good collections of different vegetable crop species.

Khan (1997), suggested that many endemic species require immediate attention to save them from the dangers of extinction. The densities of khejri (*P. cineraria*), ker (*C. decidua*), and *S. surattense* are reducing very fast. These will very soon become endangered species of the Thar Desert.

References

Annonymous (2005). Sustainable management of plant biodiversity- a success story. Document of NATP mission mode project. (Eds. S.K. Purohit, D.C. Bhandari, Anjula Panedy and D.S. Dhillon).

Anon (2005-06). NBPGR *Annual Progress Report.* pp. 91-95

Anonymous (2006-07). NBPGR *Annual Progress Report.* pp. 101-105

Anonymous (2007). CIAH- *Perspective Plan Vision 2025.* (Eds. More,T. A., Bhargava,R., Sharma,B. D., Singh,R. S., Awasthi,O.P. and Nallathambi,P.), Central Institute for Arid Horticulture Bikaner, Rajasthan.

Anonymous (2009). Vegetable Variety Development and Evaluation (Eds. Kalia, P., Joshi, S., Behra, T.K., Pandey, S.), Division of Vegetable science, IARI, New Delhi, India; vi+ 256.

Bhandari, M.M. and Shirangi, O.P. (1986) Conservation of threatened plants of western Rajasthan. In Environ- mental degradation in western Rajasthan. Eds. S.M. Mehnot and M.M. Bhandari. pp. 129±32. Jodhpur: Scientific Publishers.

Chandra, Umesh (1995). Diversity of Cucumber and Melons from Rajasthan. *Indian J. Plant Gen. Resour.* Vol 8; No 1

Chaudhary, B.S. and Samadia, D.K. (2004). Variability and Character Association in Chilli Landraces and Genotypes under Arid Environment. *Indian Journal of Horticulture*. 61(2):

Deora NS and Shekhawat NS (1995) Micropropagation of *Capparis deciduas* (Forsk.) Edgewa tree of arid horticulture. *Plant Cell Reports* 15: 278-281.

Dhillon, N. P. S., Ranjana, R., Singh, K., Eduardo, I., Monforte, A. J., Pitrat, M., Dhillon, N. K. and Singh P. P.(2007). Diversity among landraces of Indian snapmelon (*Cucumis melo* var. *momordica*). *Genet. Resour. Crop. Evol.* 54:1267–1283

Gupta, H.S. (2009). Genetic diversity and its exploitation in vegetable improvement. In *Vegetable Variety Development and Evaluation* (Eds. Kalia, P., Joshi, S., Behra, T.K., Pandey, S.), Division of Vegetable science, IARI, New Delhi, India. pp 6-10.

Kalloo, G. (1993). Loofah-*Luffa* spp. In Genetic improvement of vegetable crops (Eds. G. Kallo and B.O. Bergh). Pergamon Press. pp 265-266.

Khan, T.I. (1997). Conservation of biodiversity in western India. *The Environmentalist* 17, 283-287.

Kumar, S., Purohit, C.S. and Kulloli, R.N. (2010). *Ex-situ* conservation of rare plant- *Moringa concanensis* Nimmo ex Diaz.& Gibs. *J. Econ. Taxon. Bot.* Vol. 34(3). 693-698.

Malik, S.S, Srivastava, U., Tomar, J.B., Bhandari, D.C., Pandey, Anjula, Hore, A. and Dikshit, N. (2001) Plant exploration and germplasm collection. In National Bureau of Plant Genetic Resources- a compendium of achievement. (Eds. Dhillon, B.S., Varaprasad, K.S., Srinivasan, K., Singh, M., Archak, S., Srivastava, U. and Sharma, G.D) pp 31-68.

Malik, SK, Chaudhury R, Dhariwal OP and Bhandari DC. 2010. Genetic Resources of Tropical Underutilized Fruits in India. NBPGR, New Delhi, p.168.

Mann, H.S. and Saxena, S.K. (1980). Khejri (*Prosopis cineraria*) in the Indian desert. *CAZRI, Monograph* No. 11, IV:25.

Meena, S.R., More, T. A., Singh, D. and Singh, I. S. (2009). Arid Vegetable Production Potential and Income Generation. *Indian Res. J Ext. Edu.* 9(2). pp 72-75

Ram, D. and Srivastawa, Umesh (1999). Some lesser known minor cucurbitaceous vegetables: their distribution, diversity and uses. *Indian J. Pl. Genet. Resour.* 12: 307-316

Ram, D., Rai, M. and Singh, M. (2007). Temperate and subtropical vegetables. In *Biodiversity in Horticultural Crops*. (eds. Peter, K.V. and Abraham, Z.); Daya Publishing House, New Delhi. pp-71-108.

Roy, R.P., Mishra, A.R., Thakur, R. and Sing, A.K., (1970). *J. Cytol.Genet.* 5:16-20.

Samadia, D.K. (2004). Improvement of arid vegetables. In *Advances in Arid Horticulture, vol. 1*. (Eds. Saroj, P.L., Vashishtha, B.B. and Dhandhar, D.G.). pp 185-201.

Sen, D.N. (1973). Ecology of Indian desert III- Survival adaptations of vegetation in dry environment. *Vegetutio*. 27: 201-265.

Sen, N.L. (2004). *Kair* : Production technology of underutilized fruit crop. Yash Publishing House, Bikaner.

Seshadri, V.S. and More, T.A (2002). Indian Land Races in *Cucumis melo*. *Acta Horticulturae*. No 588.

Sharma, S.K. (2010). Management of vegetable genetic resources for designing of trait specific plants in vegetable crops. In *Designing Nutraceutical and food colorant vegetable crop plants: Conventional and Molecular approaches*.(Eds. Kalia, P. and Behra, T.K.). Division of Vegetable Science, IARI, New Delhi. pp 1-10.

Sharma, S.K. and Pandey, S. (2009). Plant genetic resources in vegetable crops: Present status and future prospective. In *Vegetable Variety Development and Evaluation* (Eds. Kalia, P., Joshi, S., Behra, T.K., Pandey, S.), Division of Vegetable science, IARI, New Delhi, India. pp 11-15.

Singh A. K. 1978. Cytogenetics of semi-arid plants. III. Anatural interspecific hybrid of *Cucurbitaceae (Citrullus colocynthis* x *Citrullus vulgaris* Schrad). Cytologia 43: 569-574.

Singh, R.S. (2008). Genetic resources of underutilized fruits and opportunities for their exploitation. In *Hi-tech production of arid horticulture* (Eds. More, T.A., Singh, D., Awasthi, O.P., Samadhia, D.K. and Singh, I.S). pp 77-88.

Sirohi, P.S., Kumar, G., Munshi, A.D. and Behera, T.K. (2005). Cucurbits. In: Plant Genetic Resources: Horticultural Crops (Eds. Dhillon, B.S., Tyagi, R.K., Saxena, S. and Randhawa GJ). Narosa Publishing House, New Delhi, India. pp. 34–58.

Stewart JL and Brandis D (1992) *The forest flora of North-West and Central India.* Reprinted by Bisen Singh and Mahendra Pal Singh, New Connaught Place, Dehradun, p. 602.

Zeven, A.C. and J.M.J. de Wet (1982). *Dictionary of Cultivated Plants and Regions of Diversity.* Wageningen. p 259.

5

Prospects of Vegetables Grafting in Arid and Semi Arid Regions

Pradeep Kumar, P.S. Khapte, Akath Singh and P.R. Meghwal

ICAR-Central Arid Zone Research Institute, Jodhpur, Rajasthan-342 003

Vegetables are short duration crops and provide more returns per unit area and time besides, they play a major role in balancing human diet, thus, helping in alleviating hunger and malnutrition (Annon., 2009). In the previous few years, India has attained quantum jump in vegetable production and present estimated vegetable production is around 162.9 mt in an area of about 9.4 mha (NHB, 2014), and holds second position at global level. But this level of vegetable production is insufficient to meet the recommended daily intake (~300 g per capita per day) for a huge population of the country. This is mainly due to lower average productivity (17.5 t/ha) of the nation. There are several crop diminishing factors such as soil borne- and foliar- diseases, insect-pests and nematodes as biotic factors, and soil or water salinity, alkalinity, metal toxicity and heat and moisture stress as abiotic factors, which greatly affect vegetable crops yield in almost all growing areas, especially in fragile arid and semi arid regions.

Why Vegetable Grafting?

The availability of cultivars with genetic potential (resistant cultivars/hybrids) to resist pests and diseases is primary prerequisite for successful commercial crop production, particularly for constraint conditions of (semi) arid regions. Though, the development of resistant lines in vegetable crops for biotic or abiotic stresses by breeding or genetic engineering is possible, it takes considerable time before commercial varieties are available for production (Nilsen *et al.*, 2014). Moreover, the cross-incompatibility beyond the species or genera may limit the scope of resistance gene transfer through breeding. However, the resistance in shoot of existing high-yielding commercial cultivars is possible by 'grafting' them onto the robust/ hardy rootstock genotypes, belonging to same or different species, genera and even possible onto the wild species of same or different genus. For instance, cucumber (*Cucumis sativus*) could be successfully grafted onto genotypes of pumpkin (*Cucurbita* spp.); muskmelon onto squash (*Cucurbita* spp.), sponge gourd (*Luffa* spp.), and wax gourd (*Benincasa* spp) and watermelon (*Citrullus lanatus*) onto wild watermelon (sp. *citroids*, sp. *colosynthes*), bottle gourd (*Lagenaria* spp) or squash (*Cucurbita* spp.; King *et al.*, 2010). Grafting is a sustainable approach and now regarded as a faster alternative tool to the relatively slow breeding methodologies in increasing plant tolerance to different biotic or abiotic stresses of important fruiting vegetables of solanaceae (tomato, eggplant and capsicum) and cucurbitaceae (cucumber, watermelon and muskmelon) families (Flores *et al.*, 2010).

History of Vegetable Grafting

Although, grafting has been employed for centuries in woody-perennials, it is relatively a new innovation in herbaceous vegetables. The first record of vegetable grafting comes from Japan, where a watermelon farmer reportedly developed watermelon [*Citrullus lanatus* (Thunb.) Matsum.&Nakai] to tackle disease and increase yield by grafting with squash rootstock (*Cucurbita moschata* Duch.) in the 1920s (Tateishi, 1927). For members of the *Solanaceae*, the first record was of eggplant (*Solanum melongena* L.) grafted on scarlet eggplant (*Solanum integrifolium* Poir.) in the 1950s (Oda, 1999). Grafting tomato (*Solanum lycopersicum* L.) was introduced commercially in the 1960s (Lee and Oda, 2003). In fact, vegetable production using grafted seedlings at commercial scale was started in Japan and Korea three decades ago (Lee *et al.*, 2010). It was re-introduced in Europe in the early 1990s as a means to control root diseases of soil or hydroponic systems (Lee *et al.*, 2010). The adoption of vegetable grafting in the Western world (American, European and Middle East countries) has increased significantly since the ban on the fumigant methyl bromide in 2005 by the Montreal Protocol (Cohen *et al.*, 2012). Because of this, grafting has been primarily used in fruiting vegetables to improve plant tolerance against soil borne pathogens (Crinò *et al.*, 2007) but its application dramatically increased over the years in mitigating the negative effects of other biotic and abiotic stresses with expanding the reasons of grafting. Though, grafting was initially practiced to overcome the problems of soil borne diseases and nematodes in East Asia and Europe, especially under intensive vegetables production system (Lee and Oda, 2003), lately its potential was realized and exploited against several abiotic stresses. Recent published reports claimed that grafting could efficiently alleviate tolerance to abiotic stresses such as salinity, alkalinity, nutrient deficiency, low temperature stress, toxicity of heavy metals and drought stress in different fruiting vegetables (Venema *et al.*, 2008; Colla *et al.*, 2010; 2013; Savvas *et al.*, 2010, 2013; Sa´nchez-Rodrý´guez *et al.*, 2013; Kumar *et al.*, 2015a; 2015b).

Prospects in Arid and Semi-arid Regions

In the recent years, grafting has become an effective tool and the cultivated area of grafted vegetables in many parts of the world, especially in Western and South-East Asian countries, has increased tremendously. However, its application in India is in nascent stage, where several stress factors cause huge crop loss every year, thus affecting economy and nutritional security. In particular, in arid region, where water is a scarce resource and the mostly available water is of poor quality, this technique can be a boon for vegetable production under such conditions. Besides, the under-utilized resources such as marginal quality land or water (saline or sewage water) could also possibly be used under cultivation of fruiting vegetables, which will enhance economy as well as nutritional security of the region. In addition, grafting can also minimize the crop loss occurred due to major pests and diseases and nematodes prevailing under intensive vegetable cultivation either under field or protected conditions. In fact, the influence of the rootstock on the mineral content in aerial parts of the plant was attributed to physical characteristics of the root system, such as lateral and vertical development, which resulted in enhanced uptake of water and minerals, this being one of the main motives for the widespread use of grafted rootstocks (Heo, 1991).

Despite several adversities, (semi) arid regions of India are bestowed with a great deal of genetic potential in locally available genetic materials in terms of land races and ecotypes of economic vegetable species e.g., watermelon (*Citrullus lanatus*), round melon (*Citrullus fistulosus*), cucumber (*Cucumis sativus*), muskmelon (*C. melo*) and related under-utilized species such as local water melon/mateera (*Citrullus lanatus*), snap melon/kachra (*C. melo* var. *momordica*) and kachri (*C. melo* var. *agrestis* or *callosus*) and their wild relatives such as *C. hardwikii, C. prophetarum,* tumba/bitter melon (*Citrullus colosynthes*), 'tatumba'-natural cross of tumba and mateera. The availability of vast and diverse genetic resources including those available in this region could serve a potential genetic pool to improve resistance to certain biotic or abiotic stresses. The available vast genetic potential in terms of landraces or related under-exploited or wild species of vegetable crops in arid regions could successfully be exploited by using as rootstocks in grafting to enhance resistance to biotic or abiotic stresses. Among important groups, cucurbitaceae is the most important and diverse group which represents the hardiest vegetable species grown in arid condition and some of which are endemic to the region viz., kachari (*Cucumis melo* var. *agrestis*), snap melon (*Cucumis melo* var. *momordica*), spine gourd (*Momordica dioica*) and some are the wild relatives viz., bitter melon (*M. balsamina*), tumba (*Citrullus colosynthis)* and hill colocynth (*Cucumis hardwickii*) which grow naturally during rainy season. In this region good amount of variability have also been found in their natural/wild form of *Cucumis prophetarum, C. trigonum, C. setosus*. Many of these species are reported to carry resistance to biotic and abiotic stresses including drought (Dhillon *et al.*, 2007; Singh and Sarkar, 2014), and many others need to be explored. Some of these species have been used to incorporate disease resistance into commercial vegetable species by breeding in India and abroad (Dhillon *et al.*, 2007). Recently, some of these were used in grafting as rootstocks to increase the resistance in cucumber and bitter gourd shoot genotypes against nematodes (TamilSelvi *et al.*, 2013; Punithaveni *et al.*, 2015).

This shows that grafting technology in fruiting vegetables has huge potential in promotion of cultivation in non-traditional and fragile agro-eco system such as arid and semi-arid regions of India, where water stress, salinity, heat stress and metal toxicity are common constraints for crop production. The efforts are to be made to harness the potential indigenous cucurbit genetic materials for increasing resistance to such stresses in commercial vegetables such as watermelon, muskmelon, cucumber through the way of this rapid and sustainable tool of grafting.

Procedure of Vegetable Grafting

There are a number of grafting methods have been tried in different vegetables. The most common methods are splice, tongue, hole-insertion, tube and cleft. Splice or side grafting in solanaceous and hole-insertion in cucurbitaceous vegetables are commonly used because of their higher success. For the production of grafted plants, rootstock seeds are sown prior to the sowing of scion seeds. After attaining appropriate stem thickness of rootstock and scion seedlings grafting is done. Immediately after grafting, grafted plants are placed to the healing chamber (25-30 °C and 85-95% RH) providing 24-48 hours dark and subsequently to the hardening chamber, keeping relatively lesser humidity and shade, for the acclimatization of grafted plants. In this way grafted seedlings are ready to transplant within a month span. Though, the cultivation cost will increase due to grafted seedlings, which can be compensated by extra income gained through increased production under constraints conditions.

References

Anonymous. 2009. Vegetable Variety Development and Evaluation (Eds. Kalia, P., Joshi, S., Behra, T.K., Pandey, S.), Division of Vegetable science, IARI, New Delhi, India; vi+ 256.

Cohen, R., Omari, N., Porat, A., Edelstein, M. 2012. Management of Macrophomina wilt in melons using grafting or fungicide soil application: Pathological, horticultural and economical aspects. *Crop Protection*, 35: 58-63.

Colla, G., Rouphael, Y., Cardarelli, M., Salerno, A., Rea, E. 2010. The effectiveness of grafting to improve alkalinity tolerance in watermelon. *Environ. Exp. Bot.*, 68: 283–291.

Colla, G., Rouphael, Y., Jawad, R., Kumar, P., Rea, E. and Cardarelli, M. 2013. The effectiveness of grafting to improve NaCl and $CaCl_2$ tolerance in cucumber. *Sci.Hortic.*, 164: 380–391.

Crinò, P., Lo Bianco, C., Rouphael, Y., Colla, G., Saccardo, F., Paratore, A. 2007. Evaluation of rootstock resistance to fusarium wilt and gummy stem blight and effect on yield and quality of a grafted 'Inodorus' melon. *Hort Science*, 42: 521–525.

Dhillon, N. P. S., Ranjana, R., Singh, K., Eduardo, I., Monforte, A. J., Pitrat, M., Dhillon, N. K. and Singh, P.P. 2007. Diversity among landraces of Indian snapmelon (Cucumis melo var. momordica). *Genet Resour Crop Evol.*, 54: 1267–1283.

Flores, F.B., Sanchez-Bel, P., Estan, M.T., Martinez-Rodriguez, M.M., Moyano, E., Morales, B., Campos, J.F., Garcia-Abellán, J.O., Egea, M.I., Fernández-Garcia, N., Romojaro, F. and Bolarín, M.C. 2010. The effectiveness of grafting to improve tomato fruit quality. *Sci. Hortic.*, 125: 211–217.

Goyal, M. and Sharma, S.K. 2009. Traditional wisdom and value addition prospects of arid foods of desert region of North-west India. Indian J. Traditional Knowl., 8(4): 581-585.

Heo, Y.C. 1991. Effects of rootstocks on exudation and mineral elements contents in different parts of Oriental melon and cucumber. MS thesis, Kyung Hee University, Seoul, South Korea, P. 53.

King, R. S, Davis, A.R., Zhang, X. and Crosby, K. 2010. Genetics, breeding and selection of rootstocks for Solanaceae and Cucurbitaceae.Scientia Horticulturae, 127: 106–111.

Kumar, P., Rouphael, Y., Cardarelli, M. and Colla. 2015b. Effect of nickel and grafting combination on yield, fruit quality, antioxidative enzyme activities, lipid peroxidation and mineral composition of tomato. *J Plant Nutri Soil Sci.* online June, 2015.

Kumar, P., Lucini, L., Rouphael, Y., Cardarelli, M., Kalunke, R.M. and Colla, G. 2015a. Insight into the role of grafting and arbuscular mycorrhiza on cadmium stress tolerance in tomato. *Front. Plant Sci.*June, 2015. doi:10.3389/fpls.2015.00477.

Lee, J. M., Kubota, C., Tsao, S.J., Bie, Z., Echevarria, P.H., Morra, L. and Oda, M. 2010. Current status of vegetable grafting: Diffusion, grafting techniques, automation. *Sci. Horti.*, 127: 93-105.

Lee, J.M. and Oda, M. 2003. Grafting of herbaceous vegetable and ornamental crops. *Hortic. Rev.*, 28: 61–124.

McCreight, J.D., Staub, J.E., Lopez-Sese, A.I. and Chung, S.M. 2004. Isozyme variation in Indian and Chinese melon (*Cucumis melo* L.) germplasm collections. *J Amer Soc Hort Sci.*, 129: 811–818.

NHB, 2014. Horticulture database 2014. National Horticulture Board. Gurgaon, India.

Nilsen, E.T., Freeman, J., Grene, R. and Tokuhisa, J. 2014. A Rootstock Provides Water Conservation for a Grafted Commercial Tomato (Solanum lycopersicum L.) Line in Response to Mild-Drought Conditions: A Focus on Vegetative Growth and Photosynthetic Parameters. PLoS ONE 9(12): e115380. doi:10.1371/journal.pone. 0115380.

Oda, M., 1999. Grafting of vegetables to improve greenhouse production. Food and Fertilizer Technology Center, Extension Bulletin. 480: 1-11.

Punithaveni, V. P., Jansirani and Sivakumar, M. 2015. Screening of cucurbitaceous rootstocks and cucumber scions for root knot nematode resistance (Meloidogyne incognita Kofoid and White). Electronic J. Plant Breeding, 6(2): 486-492.

Sa´nchez-Rodrý´guez, E., Romero, L. and Ruiz, J.M. 2013. Role of Grafting in Resistance to Water Stress in Tomato Plants: Ammonia Production and Assimilation. *J. Plant Growth Regul.*,32: 831–842

Sain, R.S. and Joshi, P. 2002. Cross compatability between wild and cultivated species of Citrullus. *Indian J. Genet.*, 62(2): 171-172.

Savvas, D., Colla, G., Rouphael Y. and Schwarz. D. 2010. Amelioration of heavy metaland nutrient stress in fruit vegetables by grafting. *Sci.Hortic.*, 127: 156–161.

Savvas, D., Ntatsia, G., Barouchas, P. 2013. Impact of grafting and rootstock genotype on cation uptake by cucumber (*Cucumis sativus* L.) exposed to Cd or Ni stress. *Sci. Hortic.*, 149: 86–96.

Sen, D.N. 1973. Ecology of Indian desert III- Survival adaptations of vegetation in dry environment. *Vegetutio*, 27: 201-265.

Singh, D.K and Sarkar, M. 2014. Biodiversity of Cucurbitaceous Vegetable Crops in India. In. Cucurbitaceae 2014 Proceeding. Bay Harbor, Michigan, USA. 64-68.

TamilSelvi, N.A., Pugalendhi, L. and Sivakumar, M. 2013. Screening of cucurbitaceous rootstocks against root knot nematode Meloidogyne incognita Kofoid and White. The Asian J Hortic., 8 (2): 720-725

Tateishi, K., 1927. Grafting watermelon on squash. Japan. *J. Hort.*, 39: 5-8.

6

Production Technology of Less Known Fruit Crops of Arid Region

P.R. Meghwal

ICAR-Central Arid Zone Research Institute, Jodhpur, Rajasthan-342003

India is a natural reservoir of the several fruit species. The several less known fruit species, which have the potential for commercial exploitation, are yet to be utilized to their potential. An attention to the less known fruit would play a significant role in increasing the income, providing employment opportunities, uplifting of the poor, small and marginal farmers and the development of value added products. These fruits are well known for their medicinal and nutritive value. There is a great need to develop and popularize not only for their domestic market but also for their export. Some of the less known fruits are described here with respect to their cultivation technology.

1. Karonda (*Carissa carandas L.*)

It belongs to family Apocynaceae. It is indigenous to India and is a hardy shrub having prominent spines on branches. Fruits are very nutritious and very useful to cure anaemia. Fruits are sour astringent in taste contains 87 to 90% pulp, 13.0 to 14.5% TSS, 4 to 6% acidity. The fruit pulp contains Vitamin C (201 to 555mg/100g), carbohydrate (67.1 g), minerals (2.8g), iron (39.1mg) and protein (2.3g) per 100g on dry weight basis (Gopalan *et al.*, 1985). Mature fruits are used for making syrup, jelly, pickle, chutney, preserve and candy. Fruits may also be dried for preparing the condiments and cooked as vegetable.

Climate and Soil

Karonda is cultivated in tropical and sub-tropical climate but not suited to higher hills of the Himalayas. Owing to its xerophytic nature, plants are well adapted to the arid and semi-arid climates. Since, it is a hardy plant; it can thrive well under drought conditions. Karonda can be grown on wide range of soil. It can also tolerate saline and sodic soils to some extent.

Varieties

The varieties are classified according to fruit colour viz. green, pink and white. Some important varieties developed recently are Pant Manohar, Pant Suverna, Pant Sudarshan, CZK-2011, CZK-2022, CZK-2031 etc.

Propagation

Karonda generally propagated by seeds. Seeds may be treated with GA_3 200 ppm for 24 hrs before sowing. Seed should be extracted from fully ripened fruits during August-September sown immediately for better germination. Under good management, seedlings become ready after 9-12 months of sowing for transplanting.

Vegetative Propagation

Stem cutting

Propagation by stem cutting is also a viable option of vegetative propagation. Though cuttings are difficult to root type, they may be planted under polyhouse for early sprouting and better rooting. Where mist facilities are available, soft and semi-hard wood cutting are more successful. Significant rooting was found with the aid of 100 ppm IBA + 4 % sugar and 100 ppm IBA + 5 % sugar. Treating the cuttings in 500 ppm of IBA, before planting results in significant increase in rooting.

Air layering

Air layering is quite successful in karonda with the use of growth regulators. It is done in the beginning of monsoon. Rooted layers are obtained after 3.5 months of layering. IBA application is more effective than NAA in promoting rooting in air layers.

Planting and Canopy Management

Karonda is suitable for independent orchard as well as for hedge row plantation for live fencing. Orchard can be laid out in square or rectangular system. Spacing may differ from 3 to 5 meters either way depending upon soil and climatic conditions. In hedge row planting, distance varies from 0.90 to 2.0 m between the plants. The pits of 60 cubic cm sizes should be dug one month before planting and filled with a mixture of soil and well rotten FYM in the ratio of 3:1. The plants should be placed in the centres of the pits. After planting, watering should be done immediately and repeated at 3-4 days intervals till the plants are established. June-July is the best time for planting but it can also be done in February-March, where the irrigation facility exists. At the time of planting, each plant is supported with the help of a wooden stakes so that it remains in erect position. Unwanted laterals are removed from time to time. Bearing trees usually do not require pruning. However, to give desired shape, additional twigs are pruned. Diseased or broken and dried twigs are removed from time to time. Old hedge may be headed back to induce new growth.

Manure and Fertilizers

Manuring is beneficial in karonda for satisfactory growth. Well rotten FYM at the rate of 10- 15 kg per plant should be applied regularly once in a year. Fertilizers should be applied in three splits doses according to age of the plant starting from monsoon as given in table 1.

Table 1: Dose of fertilizer and manure in Karonda

Age of plant (year)	FYM (kg)	Urea (g)	Single super phosphate (g)	Murate of potash (g)
1	-	100	150	75
2	10	200	300	150
3 and above	20	300	450	225

It is advisable to supplement the fertilizer application with organic manures and micronutrient sprays. Karonda responds well to Zn, copper and boron as foliar sprays. The fertilizers should be applied closer to the tree trunk and it should be mixed at 15-30 cm depth for better absorption of mineral nutrients.

Irrigation

Karonda is a hardy plant and once established can thrive well without irrigation for a long period. Irrigation after planting and manuring is essential. In winter, irrigation should be provided for protecting plants against frost. However, if there is no rain during the fruit development, the plants should be supplemented with irrigation which helps in increasing the fruit size.

Major Pests, Diseases and their Control Measures

It is damaged by anthracnose, leaf eating caterpillars and termites. The details regarding insect pest and disease, nature of damage and their control measures are given table 2.

Harvesting and Yield

Karonda starts bearing in third year of its planting. The plants flower during February-March and fruit set takes place during March end and fruit development completes by 100-110 days after fruit set. At this stage, fruits

Tabl 2: Majer pests and disease of Karonda

Insect-pests	Nature of damage	Control
Leaf eating caterpillars (*Digama hearseyama*)	Caterpillars eat out new leaves, defoliate the plants.	Caterpillars can be controlled by spraying of nuvacron (0.05 %).
Bihar hairy caterpillar (*Spilosoma oblique*)	The larvae of this insect feed extrovertly and scrap the young leaves but at later stage they starts to disperse and feed the leaves totally.	They may be hand-picked and destroyed. Spraying of 0.04 % methylparathion.
Diseases		
Anthracnose	The disease is caused by *Colletotrichum inamdarii*. *Anthracnose* is a very common disease. The fungus produces brown spots which results in the premature death of leaves and thin twigs.	The disease can be controlled by spraying the plants with Blitox-50 or Phytolan (0.2 %). Wounds on the branches may be pasted with Bordeaux paste.
Dieback	Dieback is caused by *Phytophthora* sp. and *Rhizoctonia solani*. Disease produces symptoms of dieback in trees through rotting of rootlets and defoliation. The fungus is soil inhabitant.	The disease can be controlled by summer ploughing and drenching with Benomyl or Carbendazin (0.25%).
Immature fruit drop.	Immature fruit drop is one of the major constraints. There is no definite reason for this problem but it is thought to be due to the over bearing and deficiency of nutrients.	Foliar sprays of micronutrients especially boron, zinc sulphate and copper sulphate twice a year help to minimize the loss.

develop their natural colour. Unripe fruits are harvested in July-August. Karonda requires 2-3 pickings. Fruit yield varies from 15-20 kg per plant depending on variety, location and management.

Post-harvest and Storage

Fruits should be kept in shade after harvesting. They are sorted and graded and packed in baskets. Harvested fruits can be stored for a week whereas ripe fruits are perishable and can be stored only for 2 days. Raw/ mature fruits can be processed for making pickle, jelly, candy and preserve while ripe fruits can be processed for making squash, RTS. They can also be dried.

2. Khejri (*Prosopis cineraria* Druce)

It is an evergreen, slow growing multipurpose tree found extensively in arid and semi-arid parts of Rajasthan, Gujarat, Punjab, Haryana and southern states of India. Systematic orcharding is lacking. It is multipurpose tree which not only provide shade besides nourishing the soil and environment but also provide vegetable, fruits, fodder, firewood, timber, medicine, gum and fencing material. Green leaves are very nutritive and contains crude fibre (13- 22%), crude protein (11.9 to 18.0%), nitrogen free extract (43.5%), ash (6- 8%), calcium (2.1%), Phosphorus (0.4%) and 2.9% ether extract. The unripe pods (Sangri) are rich in pulp are used as a vegetable or in pickle. The pods contain 18% crude protein, 26% crude fibre, 56% carbohydrate, 2% fat, 0.4% phosphorus, 0.2% iron and 0.4% calcium (Pareek *et al.*, 1998). The ripe pods are known as khokha and are edible and used to prepare flour.

Climate and Soil

It has deep and extensive root system which can withstand drought. Plant can sustain temperature as high as 49°C. It can control soil erosion and reclaim sand drift areas. Sandy to sandy loam soil favours good growth and development. Plants are very much tolerant to highly saline condition and can grow at salinity level 10.0 to 25.0 EC dsm^{-1} (Rao *et al.*, 1993-94 and 1994-95).

Propagation

Khejri plant is available in natural status. The seeds are collected from fully ripe pods during June- July and are sown in polythene bag in nursery beds. The plants can be raised successfully by air layering (Solanki *et al.*, 1986). It can be propagated through cutting with the use of plant growth regulators. Immense variability found in natural population offers scope for selection of desirable types and sustained them by clonal propagation. Thorn less canopy, higher green leaf and pod yield are desirable attributes. For vegetable purpose the pod should be long, thin, less fibrous and sweet in taste. However, particular desirable types may be lost forever due to uprooting of trees by wind storms, flood, other natural calamity, by natural death or by the attack of pest and diseases. Techniques have been developed to perpetuate desirable pod or any other character by budding (Meghwal and Harsh, 2008).

Plantation

Seedlings of one year age are used for planting in the pits of 60 x 60 x60 cm at a distance of 8m x 8m. The pits should be filled with FYM and top soil in equal proportion. The plants are watered after planting.

Manure and Fertilizers

About 20 kg FYM per plant per year can be recommended in general, however there is no recommendation based on research findings.

Irrigation

Depending upon soil and climate the plant can be irrigated once in 15 days during winter and at 7-10 days during summer during initial three years of planting. Once the plants are established well they may be encouraged to survive under natural rainfed condition

Pruning

The branches of hand thickness are cut annually and are allowed to dry in sun. The dried leaves are collected to use as fodder. Srivastava (1998) suggested annual pruning for higher leaf fodder yield.

Harvesting and Yield

The plants raised vegetatively start bearing pods after 3 years of planting whereas bearing in seedling plants start after 8-10 year. The flowering starts during February- March and fruit set takes place in April- May. For vegetable purpose, tender green pods are harvested by collecting the pod in bunches from the tree. There is no reliable data on the yield. However, 10-15 kg green pods can be harvested from a well grown tree up to 50-60 years. The harvested pods are dried either in sun or in dehydrator after blanching which are known as Sangri. The recovery of sangri from tender green pods varies from 25- 30%. The dehydrated pods/sangri is sold in the local market at the rate of Rs 150-200 per kilogram depending upon the quality.

3. Lasora or Gonda (*Cordia myxa* L.)

It belongs to family Boraginaceae. It is smaller, medium size tree with crooked stem grows throughout in India in arid and semi-arid regions. Its systematic cultivation is largely lacking and finds place as wind break in orchards, planting along street, roads etc. and are found growing without care and management. The fruits are used medicinally as diuretic, demulcent and expectorant. Leaves are used for making pattal, bark is used for making paper pulp wood is durable in water and is used for building boats, furniture, agricultural implements and as fuel. It is rich source of carbohydrate; pulp content varies from 94 to 97%. Pulp contains TSS from 6-7%, acidity from 0.08 - 0.1 % and Vitamin C from 32 to 48 mg/100g (Singh, 2001). Green mature fruits are acidic and are eaten as vegetable or for making pickle. Ripe fruits are also eaten freshly and may be used for the 'liquor'. Chandra and Pareek (1992) reported 14 average fruiting clusters per branch, 5.0g average fruit weight with 69.6 per cent pulp content at Bhojka in Jaisalmer district fruit bearing gonda trees.

Climate and Soil

The lasora plants are tolerant to drought, frost and moderate shade. It can tolerate temperature as high as 49°C. Rainfall to the extent of 250-300 mm is sufficient to meet out the requirement. Lasora is not very strict in soil requirement and can be grown successfully in all type of soil. However, it grows well on sandy to clay loam soil and is tolerant to salinity.

Varieties/germplasm

As such there are no improved varieties. Recently a large fruited type 'Paras gonda' has been identified in Gujarat. More than 110 genotypes of lasora have been collected by NBPGR, RRS, Jodhpur in collaboration with HAU, RRS Bawal from Rajasthan and adjoining areas (Pareek *et al.*, 1998). At CAZRI, Jodhpur, survey and collection of gonda germplasm was initiated in the year 2002. An exploration tour was conducted in 2002 covering Pali, Udaipur, Chittorgarh, Bhilwara and Ajmer districts. The plant materials in the form of seedlings, bud wood as well as ripe fruits were collected from different location/ places of natural occurrence. During the survey it was observed that it was mostly planted in on filed boundaries. In some places it was also planted as block plantation. In Rajasthan, it is more popular fruit vegetable in Jodhpur, Barmer, Jaisalmer, Ajmer, Pali, Bikaner, Ganga Nagar and Hanuman Garh districts. In Jodhpur district, it is mostly cultivated in and arround Jodhpur. It is also cultivated in tube well irrigated areas of Balesar, Phalodi Pokran and Jaislmer. Sporadic plantation of gonda can also be found in Nagaur, Pali, Jalore, Sikar, Churu as well. Gonda fruits of Pushkar region of Ajmer is also famous, though, it is grown in other parts of Ajmer as well. It was found growing wild in Udaipur, Chittor, Rajsamand and Bhilwara districts and commercial plantations/cultivated plantations were rare in these areas. In all 18 germplasm accessions were collected and evaluated for its performance.After long term evaluation four high yielding genotypes viz. CAZRI-G2011, CAZRI-G2012, CAZRI-G CAZRI-2021 and CAZRI-G2025 have been identified for release

Propagation

It is propagated by seeds and by vegetative means. The freshly harvested ripe seeds are extracted and sown in the beds during April- May. Among vegetative propagation, T budding has been found successful and even 96.7% success was obtained during the month of August. Twelve-month-old stock and 2 months old scion was found to be the most congenial for grafting. *Cordia gharaf* was found superior rootstock over *Cordia dichotoma* (Pundir, 1987). Recent studies have revealed that small fruited types have higher germination (50-60%) and it is also compatible with commercial big fruited *gonda* for propagation by budding and it can be used as rootstock (Meghwal *et al.*, 2014) Seed treatment with GA3(250ppm for 2 hours) improved germination to 50% as compared to only 10 % in control. However, the highest germination was recorded when this treatment was preceded by mechanical scarification (Meghwal, 2007). Evaluation of different genotypes on three types of rootstocks revealed small fruited gonda as the best rootstock for commercial big fruited gonda irrespective of genotypes (Meghwal *et al.*, 2014). Maximum bud take was observed when it was done on 15th August (95%) on gonda rootstock and, though it could be done till 15th September with identical success.

Plantation

Under arid conditions, the planting is best out during August- September. For under taking plantation, the pits of 60 x 60 x 60 cm are dug out during May-June (to ensure natural sterilization of soil through intense solar radiation. It is filled using FYM and upper layer soil in the ratio of 1:1. Considering the vegetative growth a space of 5-7 from row to row and plant to plant seems to be appropriate depending upon soil and climatic condition of a particular region.

Training and Pruning

Trees should be trained and pruned in order to have strong framework. Pruning should be done from January to mid of February. Dead and diseased woods should be removed as and when observed.

Irrigation

Newly planted plant requires regular irrigation. In general, irrigation one at 15 days during winter and at 7-10 days interval during summer is sufficient for newly planted plants. Irrigation should be provided during active growth period. Heavy irrigation should be avoided during flowering to prevent excessive fruit drop. Agrotechnique has been developed to take early crop by defoliation and irrigation scheduling. Regular irrigation is given during establishment phase (3-4 years) after which no irrigation is given during monsoon and winter season except occasional life saving irrigation. This facilitates leaf fall during Decmeber-January. The natural leaf fall is supplemented with manual defoliation to complete the process. Chemical defoliation may also be done with foliar spray of 1000 ppm of 2-chloroethyle phosphonic acid during first week of January. The harvested leaves of each tree is spread in the basin of same tree and covered with thin layer of soil. This helps to conserve moisture besides bringing down the soil temperature. Manuring and irrigation is started from 15th February onwards with the rise in temperature. This ensures new growth and flowering which occur almost simultaneously. Regular irrigation is given after this till middle of May or when fruit harvest is completed. The fruits attain marketable maturity after 40-50 day of fruit setting. Mature green fruits are harvested before ripening. On ripening fruits turn yellowish brown. Ripened fruits are harvested for extraction of seeds for raisning nursery

Manure and Fertilizers

The amount of fertilizer largely depends upon the age and size of tree. The nutrient requirement of the plant has not been standardized. However, application of well rotten FYM before flowering and N, P, K should be applied after pruning and fruit set in split doses.

Disease and Insect

Information on disease and insects is lacking. However, tree leaves are seen infected with fungal disease. Fruits have problem of hole in stones. by insects. For precautionary measure tree should be sprayed with copper fungicide once before initiation of new growth and 1-2 times during active growth period and fruit development.

Harvesting and Yield

The plants come into bearing during fourth year and flowering is reported to change from place to place during March- April in Rajasthan. The duration' of flowering varied from 41 to 50 days. The fruits are ready for harvest by the middle of May. These ate harvested when fruit colour is still green and pulp is properly formed. The fruits are plucked along with fruit stalk so that fruits remain fresh for a long period. These are packed in bamboo basket. These can not be stored for long period as these turn yellow. After adopting proper cultural practices, 30- 50 kg fruit per plant can be harvested.

4. Ker (*Capparis decidua* (Forsk.) Edgew)

It belongs to family Capparaceae. It is found growing in arid, semi-arid regions under subtropical and tropical climate. It is a desert plant which grows widely without much care in the Thar Desert of western Rajasthan. It is a much branched, straggling, glabrous shrub having green zigzag thorny branches. Ker is a boon for desert. It's wood is very strong and durable and is used to make the pivotal of stone mill. Branches are used for fencing and mulching. It is effective to stabilize sand dunes and checking soil erosion. Fruits of ker are rich in nutritive value. Unripe green fruits are rich source of protein (18.6%). Fruits after processing are eaten as vegetable and also used for making pickle. Mature ripe fruits are eaten raw by local people. Ker possesses high medicinal value. It is very good for relieving stomach and cardiac troubles. Its seeds are rich in oil (20.3%) of which 68.6% is unsaturated fatty acid, which is useful for skin diseases.

Climate and Soil

Ker can withstand extremely high temperature. Thus, it is found to grow in arid regions having extreme hot dry conditions. This high temperature in these regions is associated with low and erratic rainfall conditions which vary from an average of 100 to 400mm per year. Ker is very hardy tree and it thrives well even in wastelands, which are sandy and poor in organic matter and nutrients and even in saline alkali soils. Ker is the main shrub in Jodhpur, Nagaur, Shekhawati zone of western Rajasthan where soils are heavy in saline depressions with 50 to 100 dsm^{-1}. Besides, it is extensively growing in wastelands in Bikaner and Jaisalmer districts. It has been established that there is marked improvement in fertility status and reduction in alkalinity in dune sand under ker compared to barren dune sand.

Cultivars

There is no named cultivar of ker. However, efforts were made by NBPGR, CAZRI, RAU and CIAH scientists to identify promising types of ker. These were collected and multiplied. Central Institute for Arid Horticulture, Bikaner has maintained 32 accessions of ker.

Propagation

Multiplication through root suckers is considered as natural propagation of kair. However, not much success has been achieved when such suckers were separated from mother plants and attempted to establish at another site. Seed germination is quite good but post germination mortality hinders large scale multiplication. The germination of seed started after 10 days of sowing and continued upto 27 days with mean germination of 72.4% while field survival after one year of planting ranged from 36.1-64.3% (Mahla, *et al.*, 2012) in different accessions. Vegetative propagation is essential for producing true to type plants. Planting of hard wood cuttings during July-August showed significantly higher rooting per cent irrespective of rooting hormone treatment (Meghwal and Vashishtha, 1998). Propagation through hard wood cutting using quick dip with Indole Butyric Acid(1000ppm) in the month of July resulted in poor success in rooting(Vashistha,1987). Tyagi and Kothari (1997) also reported in vitro clonal propagation of kair using nodal explants from mature trees as well as seedling derived cotyledonary node cotyledon and hypocotyl explants. M.S. medium enriched with 5 mg L^{-1} BAP showed maximum shoot proliferation from nodal explants of seedling and mature plants. Maximum rooting occurred on half strength MS medium supplemented with1 mg L^{-1} IBA.

Plantation

Presently ker is not cultivated through systematic and scientific procedures. The plantation is almost natural in association with other natural bushes and small trees.

Harvesting, Storage and Yield

In ker, flowering occurs twice a year during February- March and July- August. Flowering is more during February- March and also fruit quality is superior during February- March as compared to later. In Ker, green unripe berries are harvested by picking. Mature green berries can be dried and recovery is almost 20%. Dried berries are utilized in making pickle and also for medicinal uses. According to Pareek (1978), 20- 80q/ha fresh fruits can be harvested. On an average Rs. 10,000/hectare income can be obtained from ker plantation by proper care.

5. Pilu *(Salvadora oleoides* Decne)

It is found in deserts of Rajasthan and Gujarat. It is an evergreen, small to medium size twisted bushy plant found in mixed xeromorphic woodlands. It makes an excellent fodder for Camels. Pilu fruits are edible and seed yields non edible oil (40-50%) and is used for soap and candle making. It is potential substitute for coconut oil used in Rheumatism. The wood is used for making agricultural implements.

Climate and Soil

Arid climate of tropics and semi tropics is suitable for it. Pilu is suitable for planting in sandy soils and even in barren rocky wastelands. It can tolerate drought and salinity.

Propagation

Pilu is propagated by seed and root suckers. In situ raising is preferred as in given circumstances the plantations are rainfed and little after care is possible. Seeds can be sown in situ during rainy season.

Plantation

One year old plants can be used for planting in the beginning of monsoon at 6m x 6m distance in the pits of 60 x 60 x 60cm. The pits are filled with well rotten FYM and top soil in 1:1 ratio.

Harvesting and Yield

A plant takes about 5 years come into bearing. Flowering starts March- April and fruits are thus available during May-June. The fruit turn yellow on ripening and are picked by hand. On an average yield is about 5-10 kg/tree.

References

Mahala, H.R., Singh, V.S., Singh, D. and Singh, J.P. 2012. Capparis decidua (Forsk.) Edgew.: an underutilized multipurpose shrub of hot arid region: Distribution, diversity and utilization. Genetic Resources and *Crop Evolution*, 60: 385-394.

Meghwal, P.R., Singh, A. and Pradeep-Kumar. 2014. Evaluation of selected gonda genotypes (*Cordia myxa* L.) on different rootstocks. *Indian J. Hort.*, 71(3): 415-418.

Meghwal, P.R.and Vashishtha, B.B. 1998. Effect of time of planting and auxin on rooting of cuttings of Kair (*Capparis decidua* L. Forsk). *Annals of Arid Zone*, 37 (4): 401-404.

Meghwal, P.R. 2007. Propagation studies in lehsua (Cordia myxa). *Indian J. Agric. Sci.*,77(11): 765-767.

Meghwal, P.R. and Harsh, L.N. 2008. Budding in khejri: an important technique for conservation and propagation of elite germplasm. *Green Farming*, 1(4): 49.

Meghwal, P.R., Singh, A. and Pradeep-Kumar. 2014. Evaluation of selected gonda genotypes (*Cordia myxa* L.) on different rootstocks. *Indian J. Hort.*, 71(3): 415-418.

Pareek, O.P. 1978. Horticultural development in arid regions. *Indian Hort.*, (2): 25-30.

Pareek, O.P. 1998. Arid zone fruit research in India. *Indian J. Agric. Sci.*, 68 (8): 508-514

Pundir, J. P. S. 1987. Studies on growth, flowering and fruiting behavior of lehsua. Ph.D. Thesis. Sukhadia University, Udaipur.

Tyagi, P and Kothari, S.L. 1997. Micropropagation of Capparis decidua through *in vitro* shoot proliferation on nodal explant of mature trees and seedling explants. *J. Biochem. Biotech.*, 66: 19-23.

Vashishtha, B.B. 1987.Vegetative propagation of Capparis decidua. *Annals of Arid Zone*, 26(1&2): 123-124.

7

High Density Planting in Fruit Crops

*A.K.Shukla, B.L. Jangid, D.K. Gupta, Keerthika A.
and M.B. Noor mohamed*

*ICAR-Central Arid Zone Research Institute, Regional Research Station
Pali, Rajasthan – 306401*

With the culmination of green revolution, the yield in agriculture is almost static with the use of even great deal of inputs and resources with appearing decadence in nature. The agriculture, if continuous on with same trend, will no longer productive and sustainable. Heterogenity being vested with best survival and sustenance, attempt is underway towards crop diversification deploying remunerative crops and techniques so as to keep pace with rising demand of mankind. Added to this, dwindling land : man ratio (0.13 ha during 1990s), harvesting more return from a unit area of land, disguised unemployment, improper availability of fruits and vegetables (40 g and 120 g respectively as against 120 g and 280 g for fruits and vegetables, respectively), generating employment (143 man-days/ha in cereals, 860 man-days/ha in fruit production and 1000-2500 man-days/ha in cultural intensive fruit crops), are major concerns for high density planting (HDP). It is a holistic production system which helps in promoting and preserving dynamic agro-ecosystem. HDP is also an aid to biodiversity, biological cycles and organic farming. As an alternative to raise production and productivity especially to compete global trade, HDP is worth adoption.

Types of Planting Density

Low density planting

It is a non-intensive system of cultivating fruit crops. It is an age old planting system in which trees are planted at wide spacing accommodating about 100-250 trees on a hectare. Pruning is resorted at a minimal level and orchards are nurtured so as to favour maximum development of tree. Dwarfing rootstocks don't find place in planting. Trees acquire commercial production potential after 10-15 years of planting. The output from the orchard during early 10-15 years is less. The system being less input and care intensive holds popularity among growers.

Medium density planting

Trees are planted at highly minimized distance covering 250-500 plants/ha. Proper pruning is undertaken to manage the tree in desirable shape. The system being more care-intensive the labour requirement is more and so also the yield obtained is more. This system is taking lead in output conscientious growers to produce amenable fruits crops like pomegranate, citrus, guava, papaya, banana etc. with assured production management system.

High density planting

This system is very condensing in planting accommodating 500-100000 plants on a hectare basis. Different types of high density planting are as under (Mohammed and Wilson, 1984):

- Medium high density   :   500-1500 plants/ha.

- Optimum high density   :   1500-10000 plants/ha.

- Ultra-high density   :   100000 plants/ha.

This system relies heavily on rigorous training and pruning. Maintenance pruning is very heavy in this system. Use of dwarfing rootstocks and chemicals find place in the high-density planting. Passo et al., (1979) reported inoculation of some viruses and viroids in controlling the size of plant. The expenses per unit are high and so is the yield.

Meadow Orcharding

The term meadow literally means grassland. Analogous look of grassland covering every bit of land forms the basic for naming. This is ultra-high density planting system in which 20000-100000 plants/ha is accommodated in order to maintain tree form, severe top pruning is practised similar to mowing of grassland. The plant intended to produce yield after two years age, heavy use of grow regulator is made in meadow orchard.

In case of apple during first year attempt is made to induce shoots to reproductive stage and during second year, plants flower and produce fruits. When fruiting is over, they are cut back severely in stump form and again new shoots appear and repeat the biennial cycle of fruiting. Mechanized harvesting is followed using combine which separate the fruit from the shoot. However, the system being capital intensive, it may not suit the small growers. Suiting to the requirement, Erez (1982) devised two systems of meadow orcharding;

mechanized system for a large farm and intensive system for the small farm. In the mechanized system only one stem is allowed to bear fruit. When the fruiting is over, this is cut from ground level and over new shoot, hormonal spray is made to accomplish fruiting in two years cycle. In intensive system the tree is developed to two shoots. Out of these two shoots, one is allowed to produce fruits and another is pruned to a stump. The pruned shoot regenerates to produce new growth and flower bud during the growing season. Thus every shoots produce fruits every second year. The tree managed by intensive system yields more in comparison to mechanized system due to increased flower bud, fruit set and yield per tree. Fruit plants which are very easy to propagate by cutting are especially suitable for HDP. The species which are propagated by deploying rootstock, removal of sprouting appearing out of it, makes meadow orcharding difficult.

Plant Competition in HDP

Increasing number of plants per unit area produces more yields in proportion to the increased number of plants until the competition between plants sets in. When the plant population per unit area increases and the competition sets in for recourses, there is reduction in yield in proportion to the degree of competition between plants. With increase in population beyond the optimal, the plants respond to population pressure in form of decreased number, reduced size, reduced colour of fruits etc. Even though with reduced fruiting surface there is progressive decline in per plant yield, the yield per unit area is higher in HDP owing to higher plant population.

Component of HDP

HDP is resorted by adopting (a) dwarf scion varieties (b) dwarf rootstock (c) training And pruning (d) suitable crop management practices (e) use of bio regulators (Singh, 2008).

Dwarf scion varieties

Apple   :   Red Spur, Star Crimson Spur, Gold Spur, Well Spur, Oregon Spur, Silver, Spur, Red Chief, Hardi Spur and Hardi Spur.

Peach   :   Red Heaven, Candor

Cherry   :   Compact Lambert, Compact Stella

Pecan   :   Western, Desirable, Cheyenne,

Walnuts   :   Hartely, Chico, Vina, Howard

Sour cherry :   English, Morello

Papaya   :   Pusa Nanha

Banana   :   Dwarf Cavendish

Mango   :   Amrapali

Sappta   :   PKM-1, PKM-2

Dwarf Rootstock Varieties

HDP is possible using dwarf growing rootstocks. These rootstocks for different fruits are as under:

Apple   :   M9, M26, M4, M7, M9, MM106, M 27 (ultra-dwarfing.)

Pear   :   Quince

Cherry   :   Mahaleb

Mango   :   Vellaikolumban (for alphonso),Olour (for Himsagar & Langra)

Guava   :   Pusa Srijan, *Psidium friedrichsthalianum*, Aneuploid No.82

Citrus   :   Trifoliate orange, Sour orange, Citranges

Avocado   :   Mexican rootstocks.

Loquat   :   Quince

Ber   :   *Zizyphus nummularia*

Physiology of Dwarfing

The dwarfing effect of rootstock is impacted by the translocation of auxin. The translocation of auxin from shoots and young leaves to roots is governed by the amount of IAA oxidase, peroxidase and phenols and other compounds present in the phloem and cambial cells. The levels of these compounds are genetically controlled and may vary from one species to another. Such variation is responsible for varying growth performance. The rootstocks with thicker bark and high per cent of starch indicate low levels of auxins. Dwarfing rootstock or interstock may have dwarfing effect by controlling flow of auxin through bark of rootstock or interstock (Lockard and Schneider, 1981). The barks of different genetic compositions allow different amounts of auxin to pass through. It affects root growth. This ultimately affects the cytokinin synthesis. The level of cytokinin synthesic affects shoot growth leading to small or large tree.

Training and Pruning

Training and pruning are effective tools in HDP by virtue of their impact on shape and size control of the tree. HDP amenable training and pruning methods are described as under:

Bush system

In this system, the height of the plant is kept to 2.0 metres. During first year, the plant is cut at a height of 70 cm. No shoot is allowed to grow up to a height of 25 to 30 cm. Above this height, 3 to 4 branches are allowed to grow over which number of branches emerge out. The plants acquire the shape of bush. The centre of the plant is kept open. This system is suitable for apple.

Pyramid system

In this system the plants are trained in a fashion so that lower branches may remain longer and higher branches gradually smaller. The alternative tiers of horizontal branches radiating from main stem scattered all around, gives the plant an appearance of pyramid. The branches are allowed to grow on main stem at 20 cm height from ground level. The plants are pruned from the tip of main stem and branches to maintain pyramid shape.

Espalier system

The word espalier is French in origin meaning a fence, a fruit wall or paling. It refers to the support used for training trees especially apples and pears. The tree trained through this system consists of three to six tiers of horizontal branches trained to grow one foot apart from one another at right angles to the main stem. Thus, the branches grow parallel to the ground. In this system using poles, three to six rows of wires are stretched one above the other. The first row of wire is stretched at a height of 60 to 70 cm, the second row at 130 to 140 cm height and the third row is stretched at a height of 200 cm from ground level. Over these wires, the branches of the trees are trained in both the directions parallel to the ground. In this system, the line to line distance of the plant is kept less as the plants are grown only in two directions along with wire.

Cordon system

Cordon refers to closely spurred single stemmed tree tied to a support e.g. wires or bamboo canes, either in vertical, oblique or horizontal position. This system usually finds favour in apples and pears. The trained plants bear early crop as compared to dwarf pyramid and bush system. The plants are planted at a distance of 1 to 1.5 m. The stem of the plant is tied with wire. The wires of 12 to 13 gauges are fixed to the ground using cement and concrete at 4.5 to 6.0 metres interval. The plants are maintained single stemmed by practising severe pruning of emerged branches during winter and summer. Depending upon the number of main stem trained along with wire, the system is known as single cordon and triple cordon

Tatura trellis

This is system of training fruit trees along with trellis of wire to harvest early and high yield without use of dwarfing rootstock. The alignment of tree lies in the centre of trellis and its branches are trained along the wires of the trellis. The system was developed by David Chalmers, Ban Van den Ende and Leo van Heek during the year 1973 at Irrigation Research Institute, Tatura, Victoria, Australia. The trellis of wire is erected using iron pole of 10.5 ft height over which $12^1/_2$ guage high tensile steel wire is stretched. In between two trellises, gap of 7 ft is maintained. The orientation of trellises is kept to north-south directions. Common orchard cultural practices are followed. Canopy of tree is maintained under desired frame using mechanical hedger. Harvesting is done mechanically. However, the system is highly labour intensive, as it requires hand training and hand thinning also, to get more yields of large, high quality fruit. The system favours high yield owing to optimum light interception and close planting. This is suitable for training and maintaining orchard of apple, pear, peach, plum, apricot, nectarine, sweet cherry, kiwifruit, grape etc. which sustain the rigour of pruning.

Hedge row system

As the name indicates, the system accommodates plants similar to as in hedge planting. The distance between trees is set 50 cm^2 and that between rows 3-4 metres. The close planting of tree develop in form of compact wall looking like hedge. The orientation of row should be North-South so as to have ensured availability of light in both sides of the hedge walls of the tree. The tree can be arranged in single, double or multiple rows system. However, arrangement in single row is favorable in view of maximum availability of solar radiation to the trees. Reduction in spacing within row between plants reduces yield and also the size of fruits. Multi-row system though favours light interception and so also the yield but the fruit colour is less desirable. Added to this, the canopy being too compact, it becomes difficult to follow intercultural operations and the chances of invasion of pests/ diseases increases.

Use of bio- regulators

Different growth retardants are used to control shape and size of trees besides favouring reproductive growth at the expense of vegetative growth. They find uses in manipulating different physiological activities as under:

Prolonging dormancy

Tender growth of plant is susceptible to winter injury if appears during severe winter. Growth retardants like Abscisic acid (ABA), SADH, (Succinic acid-2,

2-dimethylhydrazine, also known as B-9, B-995 or Alar), CCC (Chlormequat chloride, also known as cyocel, chlorocholine or 2-chloroethyl ammonium chloride) when applied to foliage, delays bud break. They are of use in citrus and grape.

Reducing vegetative growth

In order to facilitate better management practice in orchard, reduction in vegetative growth is envisaged. AMO 1618 (ammonium, 5-hydroxycarvacryl) trimethly chloride piperidine carboxylate, CCC, Paclobutrazol, B-9 have been found effective in reducing the growth of pear, peach, lemon, litchi, mango, apple, plum and apricot.

Flowering

Many bio- regulators play key role in induction of flowering. Ethylene 200-400 ppm induces flowering in pineapple. SADH promotes flowering in apple, pear, peach, blue berry and lemon. Paclobutrazol is effective in promoting flowering in lower nodes of papaya.

Reducing fruit drop

SADH when sprayed at full bloom reduces fruit drop. The chemical also minimizes development of water core and cold strong scales in fruits. SADH at 1000-2000 ppm, ccc at 2000-3000 ppm, paclobutrazol (Cultar or Bonzi or PP333) 1000-2000 ppm, MH at 1000-1500 ppm have produced desired results in HDP. The bio-regulators alter assimilate partitioning more in favour of reproductive shoot. Less shoot growth favours reducing cost incurred in pruning. In fruit trees in which neither dwarfing rootstock nor compact scion cultivars are available, bio-regulators are especially useful.

Adoption of suitable crop management practices

HDP being productive, it always calls adoption of proper crop management practice Mulching, fertigation, organic farming, INM, IPM etc. like hi-tech strategy prove boonful in harvesting proper yield and potential of a particular fruit crop.

Comparison between traditional and HDP orcharding

HDP orcharding being labour and cost intensive though yield obtained is more (table 1 and 2), it has not yet trickled to majority of farm families. The detail of various features of HDP and Traditional system of planting is produced below:

S.No.	Parameters	Traditional	HDP
1.	Plant population	Less (150-200/ha.)	More (5w-100000/ha.)
2.	Production	Low (15-25 t/ha)	High (30-50 t/ha)
3.	Management	Large tree size difficult to control to manage	Small tree size convenient
4.	Labour	Requirement more	Requirement less
5.	Harvesting	Difficult manual	Easy by machine
6.	Quality	Large canopy, poor sunlight penetration, poor quality	Small canopy better sunlight
7.	Establishment	Cost less	Cost high
8.	Machinery	Doesn't require expensive machine	Requires expensive machine
9.	Bioregulator	Use is not essential	Use is essential

Source: Ray, P. K. (1999)

Table 1: Fruit productivity in traditional (TO) and high-density orchards (HDO)

Fruit crop	Plant distance (m)	Trees (/ha)	Tree age (yr)	Cultivar	yield (t/ha)*
Almond	HDO: 5.4 x 5.4 -7.5 x 7.5	177-342	5	Tuono	5.0-10.0
	TO : 5.0 x 2.5	800			Higher
Apple	5.0 x 5.0-8.0 x 8.0	227-400		Jonathan	30-50
	3.0 x 0.75	2222			68.75
Apricot	6.0 x 6.0	2771481	10-13	Tyrinthos	13-22
	4.5 x 1.5				64.40
Avocado	7.0 x 7.0	204		Hass	10-15
	5.0 x 2.5	800			34.4
Banana	2.0 x 2.0	2500		Robusta	35-36
	1.2 x 1.2	6944			174.39
Ber	6.0 x 6.0 - 8.0 x 8.0	227-227	10-20	Jinsixlaozao	25 - 40
	2.0 x 2.0	2500			250 - 400
Cherry	8.0 x 5.0	250	3	Van	6.3-9.0
	4.0 x 2.0	1250			16.0
Grape	5.0 x 3.0 - 6.0 x 3.0	555-66	9-14	Grenache Noir	15 - 20
	1.8 x 0.8	77000			Higher

Fruit crop	Plant distance (m)	Trees (/ha)	Tree age (yr)	Cultivar	yield (t/ha)*
Grape fruit	6.0 x 8.0 - 8.0 x 8.0	156-227	7	Redblush	10-15
	3.3 x 1.5	2020			58.25
Guava	6.0 x 6.0	277	5	Sardar	15.70
	2.0 x 2.0	2500			58.30
Kiwi fruit	5.0 x 5.0 - 6.0 x 6.0	277-400	1	Hayword	2537
	4.5 x 1.5	1481			
Lemon	6.0 x 6.0 - 8.0 x 8.0	156-227	4		10-15
	3.47 x 3.47	832			68.2
Litchi	10.0 x 10.0	100	10	Mauritius	9-10

Contd.

	5.0 x 5.0	400			12
Mandarin	6.0 x 6.0-8.0 x 8.0	156-227	5		10-15
	3.0 x 2.2 - 2.5 x 2.0	1500-2000			55
Mango	7.5x7.5-12.5 x 12.5	64-177	1424	Dashehari	4.25-9.0
	3.0 x 2.5	1333		Dashehari	34.74
Orange	6.0 x 6.0 - 8.0 x 8.0	156-227	7	Valencia	10-15
	3.3 x 1.5	2020			23-75
Papaya	2.0 x 2.0	2500			40-50
	1.2 x 1.2	6944			75
Pomegranate	5.0-6.0 x 5.0-6.0	277-400			7-12
	5.0 x 2.0	1000			30
Sapota	10.0 x 10.0	100	6-12	Kalipatti	15-20
	5.0 x 5.0	400			13.61
Strawberry	45 x 90 (cm)	24691			17.5
	25 x 60 (cm)	66666			30
Tanger	6.0 - 8.0 x 6.0-80	156-227	7		10-15
	11 x 5 feet	800			52
Walnut	10.0 x 10.0	100	5	Tulare	7.8
	6.1 x 3.0	519			

*The upper value in each column connotes to yield under traditional planting while the lower under HDP. *Source* : Adapted from Singh (2006)

References

Erez, A. 1982. *Hort. Science*, 17: 138-142

Lockard, R.G. and Schneider, G.W. 1981. Stock and scion growth relationship and the dwarfing mechanism of apple. *Horticulture Reviews* Jules Janick (Ed.) Avian publishing Company, *Westport*, 3: 315-75.

Mohammed, S and Wilson, S.A. 1984. *Tropical Agric.*, 61: 137-42.

Passos, O.S.. Almas, C.des and Boswell, S.B. 1979. *Citrograph*, July, pp. 211-218.

Ray, P.K. 1999. Orchard establishment. In *Tropical Horticulture,* Vol-1, Naya Prakash,Kolkatta, P.22.

Singh, Jitendra. 2008. *Basic Horticulture*, Published from Kalyani publishers, Ludhiyana, pp.14-16.

Singh, C. P. 2006. Establishment and management of high-density orchards. In: Integrated *crop management in subtropical fruits* (Eds. J. P. Tiwari and R. Srivastava), published from Agrotech Publishing Academy, Udaipur, pp. 53-61.

8

Rejuvenation of Old and Senile Orchards

Akath Singh and P.R. Meghwal

ICAR-Central Arid Zone Research Institute, Jodhpur, Rajasthan-342003

India has witnessed tremendous increase in the production of horticultural crops especially fruits since independence. Due to specialized scientific interventions and application of latest research tools, the country is now among the top fruit producing countries of the world ranking second next only to China. However, the productivity has still remained low as against area under the fruit cultivation. Several neglecting issues related to the production technology have remained unattended so far. The old and senile orchards are now reverting towards a declining trend of production because of plant age factor, non-compatible varieties and poor canopy management (Baba *et al.,* 2011).

In the recent past declining productivity of old and dense orchards existing in abundance has become a matter of serious concern for the orchardists, traders as well as scientists. In India 30-35 per cent area under fruit crops is occupied by old, dense and diseased orchards. For overcoming the problem of unproductive and uneconomic orchards existing in abundance, large scale uprooting and replacement with new plantations (rehabilitation) will be a long term and expensive strategy. Therefore research efforts were initiated to standardize a technology for restoring the production potential of existing plantations by a technique called Rejuvenation. The term 'Rejuvenation' means renewal or making new or young again. As applied to the orchard tree it would mean restoring the productive capacity of the fruit trees. The meaning of 'Rejuvenation' according to Chamber's dictionary is 'to recover youth character or to grow again'. Obviously, this would apply to those plants which have attained a stage where they are no more profitable from the grower's point of view.

Need for rejuvenation

The old fruit orchards need to be rejuvenated as they show decline in yield and quality of produce which may be attributed to any one of the following factors:

- Reduction in the photosynthetic surface area
- Non availability of productive shoots
- Increased incidence of insect paste and diseases
- Less penetration of sunlight due to overcrowding of branches as a result of which the fruits on the interior areas of the tree do not develop proper colour
- Growth of wild shrubs and grasses in between the rows
- Unsystematic planting, aging and poor orchard management practices
- Inferior varieties and planting material
- Damage due to adverse weather conditions, rodents and other enemies

Rejuvenation methodology
- Identification of old orchards
- Top and Frame Working by Power Pruning Saw
- Procurement of bud wood and scion from genuine source
- Frame working with latest available varieties
- Operations under expert guidance viz. training and pruning, orchard soil management, manuring and fertilization
- Proper control measure of insects-pests and diseases

Status of success of rejuvenation in fruit trees

Based on research investigation, it was observed that tree architecture engineering, canopy density and photosynthetic efficiency play important role in governing the fruiting potential (Lal, 2008). There are several reports on rejuvenation by pruning, canopy management, dehorning and top working in different fruit crops like mango, guava, aonla, litchi, peach, apple and ber. Rejuvenation technology with heading back of branches in mango during December at a height of 4 meters from the ground level and effective after care management could give new lease of life to the unproductive orchard for another 25-30 years and make it productive and economic after four years of rejuvenation (Lal, 2008). Similarly in Guava yield enhancement in the range of 70-90% over the unpruned trees can be recorded after first year of rejuvenation. Lal and Misra (2007) reported that cumulative yield of mango was recorded highest with heading back up to secondary branches followed by centre opening. Fruit size and other quality parameters was found superior in centre opening and lowest values in un-pruned trees. Success percentage of different mango varieties during top working revealed that significantly higher percentage of success was recorded in cultivar Dashehari followed by Mallika and Ambika and minimum with Chausa and Langra (Misra and Lal, 2007). Misra *et al.*, 2004 reported 50.0-77.7% success due to top working in aonla with commercial cultivars on seedling trees. Pathak *et al.*, (1996) also found 40-100% success rate by top working on seedling aonla trees. They also reported that varietal variation in success percentage might be due to presence of different levels of endogenous phytochemicals and physiological status of different cultivars. There is a loss of two fruiting season from rejuvenated trees and fruiting started from third year as meager fruiting but a lot of improvement in fruiting of rejuvenated trees in fourth years. The quality of fruits obtained from top worked trees was significantly better than fruit obtained from seedling trees (Misra *et al.*, 2007). Studies on the performance of rejuvenated trees of different ber varieties under semi arid conditions of Punjab concluded that after nine years of rejuvenation, the fruit size, yield and fruit quality were observed better in the Sanaur-5 cultivars (Singh and Ball, 2010).

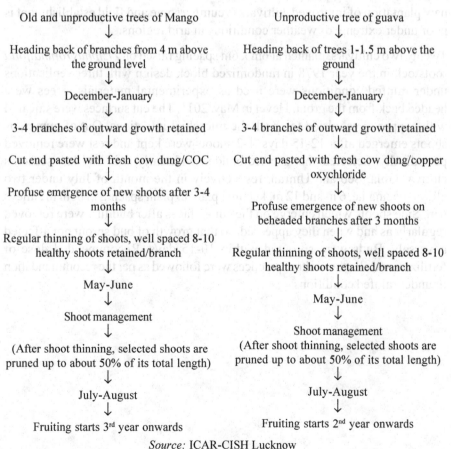

Old and unproductive trees of Mango	Unproductive tree of guava
↓	↓
Heading back of branches from 4 m above the ground level	Heading back of trees 1-1.5 m above the ground
↓	↓
December-January	December-January
↓	↓
3-4 branches of outward growth retained	3-4 branches of outward growth retained
↓	↓
Cut end pasted with fresh cow dung/COC	Cut end pasted with fresh cow dung/copper oxychloride
↓	↓
Profuse emergence of new shoots after 3-4 months	Profuse emergence of new shoots on beheaded branches after 3 months
↓	↓
Regular thinning of shoots, well spaced 8-10 healthy shoots retained/branch	Regular thinning of shoots, well spaced 8-10 healthy shoots retained/branch
↓	↓
May-June	May-June
↓	↓
Shoot management	Shoot management
↓	↓
(After shoot thinning, selected shoots are pruned up to about 50% of its total length)	(After shoot thinning, selected shoots are pruned up to about 50% of its total length)
↓	↓
July-August	July-August
↓	↓
Fruiting starts 3rd year onwards	Fruiting starts 2nd year onwards

Source: ICAR-CISH Lucknow

Ber rejuvenation

The total area under *ber* in India is 90,000 hectares with an annual production of 7,50,000 ton fruits. *Ber* has changed scenario of horticulture in arid regions of Rajsthan and proved as solid solution for drought and famine and offers a sound land use under rainfed situations. But over the time *ber* trees start giving diminished yield and small fruits of inferior quality after bearing normal crops of 25-30 years. In such trees old woods goes on accumulating every year. The other possible reasons of unproductivity in *ber* may be due to aging, general negligence of orchard, poor orchard management, biotic stress and selection of faulty planting material/variety. Fruits obtained from such trees are generally small sized and inferior in fruit quality. Besides, a large number of seedling plantations of *Ziziphus rotundifola* also exist in western Rajasthan and other *ber* growing areas in country. Such established and healthy trees of none descript material and poor genetic potential can be converted into productive once by rejuvenation with improved and region specific commercial cultivars. A large number of seedling plantations in remote areas can also be converted since new plantation of improved cultivars is cum*ber*some and field establishment is poor under extreme of weather conditions in arid regions.

Twenty two cultivars, planted at 6m x 6m spacing raised on *Ziziphus rotundifolia* rootstock in the year 1978 in randomized block design with three replications under rainfed conditions were used as experimental materials. Trees were headed back from the ground level in May, 2011. The cut surfaces were smeared with copper oxycholride to check the microbial infections. Out of numerous shoots emerged after 12-15 days, 1-2 shoots were kept and rest were removed periodically. Selected shoots were budded with early, mid and late season's cultivar Gola, Seb and Umran, respectively in the month of July under two alleys spacing i.e. 6 m and 12 m, keeping plant to plant spacing in both the alleys 6 m. Side shoots which emerge on the cut surfaces after budding were removed regularly as and when they appeared, so that growth of bud sprout not affected adversely. Buds were sprouted within 10-15 days. Recommended dose of fertilizers and all other culture practices were followed as per the recommendation of under rainfed conditions.

Rejuvenation schedule

Old and unproductive trees

↓

Heading back of main trunk from ground level

↓

Emergence of shoots from beheaded trunk

↓

Thinning of newly emerge shoots and select only 2-3

↓

Budding in July-August with Gola, Seb Umran cvs

↓

After sprouting of buds, removal of top portion above the bud portion

↓

Regular thinning and pruning

↓

Fruiting started after second year

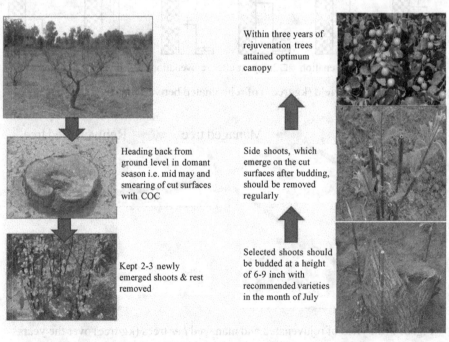

Within three years of rejuvenation trees attained optimum canopy

Heading back from ground level in domant season i.e. mid may and smearing of cut surfaces with COC

Side shoots, which emerge on the cut surfaces after budding, should be removed regularly

Kept 2-3 newly emerged shoots & rest removed

Selected shoots should be budded at a height of 6-9 inch with recommended varieties in the month of July

Fig. 1: Sequential steps of rejuvenation of *ber* orchard

Within three years of rejuvenation trees attained reasonably good canopy and it was slightly higher in 6x12m spacing in the order of Gola, Seb and Umran. Though the rejuvenated trees started fruiting in first year after rejuvenation but the yield was negligible but the yield was remarkably improved during third year after rejuvenation. Gola recorded 3.55 times higher yield (36.4 kg tree^{-1})

with higher per centage of 'A' grade fruit followed by Seb and Umran compared to yield obtained from trees before rejuvenation process. The morphological attributes of fruits obtained from rejuvenated was significantly better than fruits obtained from non-rejuvenated trees of each cultivar. Contrary, chemical attributes i.e. TSS, TSS: acid ratio and total sugars was slighter higher in fruits obtained from non-rejuvenated trees compared to rejuvenated trees but the difference was meager. Hence, rejuvenation technology helps in restoring the production potential of old unproductive and seedling orchards in shortest possible duration.

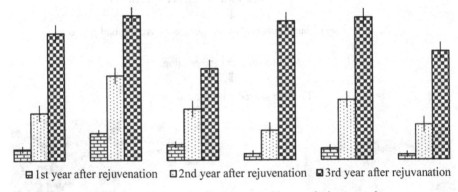

▫ 1st year after rejuvenation   □ 2nd year after rejuvenation   ▪ 3rd year after rejuvanation

Fig. 2: Fruit yield (kg tree^{-1}) of rejuvenated ber varieties over the year

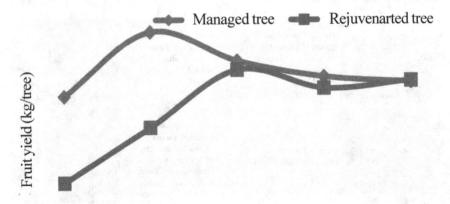

Fig. 3: Fruit yield of rejuvenated and managed *ber* trees (kg/tree) over the years
(*Source:* Singh *et al.*, 2018)

Agri-Horti system in rejuvenated ber

Integration of Jujube in farming system is appropriate option to enhance overall system productivity and livelihood of the poor farmers even in the severe drought. For finding suitable companion crop to be grown in Jujube alleys, tree-crop interaction studies were carried out with nine different intercrops combinations in integrated farming system at CAZRI Jodhpur, India. Greengram was found to perform better in association with *Jujube c.v. Gola* than pearl millet (*Pennisetum glaucum*), dew gram (*Vigna acontifolia*) and horse gram (*Macrotyloma uniflorum*) but system productivity of all the intercrops was higher than their sole crops. Over the time, productivity of *Jujube* trees some time start declining because of aging, poor orchard management, biotic stress, old seedling plantation etc. Vigour and yield in such orchard may be regained by rejuvenation and top working with selected varieties. An attempt has been made to rejuvenate a thirty three years old *Jujube* orchard. Trees were headed back from ground level in summer. Out of the numerous shoots emerged after 12-15 days, 1-2 shoots were budded with early, mid and late season's cultivar. In the next year, inter cropping was carried out in the *Jujube* alleys with greengram (*Vigna radiata*), pearlmillet (*Pennisetum glaucum*) and senna (*Cassia angustifolia*) at two alleys spacing i.e. 6 m and 12 m (plant to plant spacing in both the alleys was 6 m) tested over three *Jujube* varieties viz. Gola, Seb and Umran. Within three years of rejuvenation trees attained reasonably good canopy, though the trees started fruiting in the same year after rejuvenation but it increased remarkably during third year onwards. The variety Gola recorded 3.55 times higher yield (36.4kg tree^{-1}) with higher per centage of 'A' grade fruit (Singh *et al.*, 2015). Interaction of ber varieties revealed that yield of all the three intercrops i.e greengramm (7.93 q/ha), pearl millet (*Pennisetum glaucum*) (17.85 q/ha) and senna (6.0 q/ha) was higher with Gola in 6m x12 m spacing. But system productivity in terms of ber equivalent was found higher in 6m x 6m. Among the intercrops system productivity was found in order of green gram, followed by pearl millet (*Pennisetum glaucum*) and senna. Overall highest system productivity (136.123 q/ha) was recorded in Gola+ greengramm in 6m x 6m spacing.

It has also been observed that when *Jujube* was intercropped with greengram (*Vigna radiata*), SOC stock was higher than the system intercropped with pearl millet (*Pennisetum glaucum*) or senna (*Cassia angustifolia*) or even without intercropping.

Similarly in a 3 hectare silvipasture, the tree component i.e. *Ziziphus rotundifolia* was rejuvenated and converted to horti.- pasture system with *Cenchurus ciliaris*. The plants of *Jujube* started fruiting in second year of rejuvenation with higher TSS. Heading back of *Z. rotundifolia* trees also helped to improve the productivity of pasture (1.82 t/ha).

References

Baba, Jahangeer A., P. Ishfaq Akbar and Vijai Kumar. 2011. Rejuvenation of old and senile orchards: a review. *Annals of Horticulture*, 4(1): 37-44

Bal, J.S. Randhawa, J.S. and Singh, J. 2004. Studies on the rejuvenation of old *ber* trees of different varieties. *J.Res. (PAU)*, 41: 210-13.

Kalloo, G., Reddy, B.M.C., Singh, G and Lal, B. 2005. Rejuvenation of old and senile orchards. Pub. CISH Lucknow, P.40.

Lal, B. and Mishra Dusyant. 2008. Studies on pruning in mango for rejuvenation. *Indian J. Horticulture*. 65(4): 405-408.

Mishra, Dushyant and Lal B. 2007. Performance of commercial mango cultivars during top working on senile trees. *The Hort. J.,* 20(1): 44-45.

Mishra, Dushyant, Pandey, D., Rajneesh Mishra and R.K. Pathak. 2007. Performance of improved aonla cultivars during top working on senile trees. *Indian J. Horticulture*, 64(4): 396-398.

Mistry, P.M. and Patel, B.N. 2009. Impact of heading back plus paclobutrazol on rejuvenation of old and over crowded alphonso orchards. *Indian J. Hort.*, 66 (4): 520-522

Pathak, R. K., Wahid Ali and Dwivedi, R. 1996. Top working in aonla, *Indian Horticulture*, 41: 27.

Singh, Harvinderjeet and Bal, J.S. 2010. Studies on the performance of rejuvenated trees of different *ber* varieties. *Indian J. Horticulture*, 67(3): 315-317.

Singh Akath, S.P.S. Tanwar, P. R. Meghwal, A. Saxena and Mahesh Kumar. 2018. Assessing productivity and profitability of a rejuvenated ber based agri.-horti system under arid conditions. *Indian Journal of Agricultural sciences*. 88(4):573-578

Singh, Harvinderjeet and Bal, J.S. 2010. Studies on the performance of rejuvenated trees of different *ber* varieties. *Indian J. Horticulture*. 67(3): 315-317.

Sree Hari, G. and Sbbi Reddy, G. 1998. Converting seedling mango trees in to high yielding. *Indian Horticuture*, 43: 30.

9

Diversified Farming System for Economic and Ecological Sustainability in Arid Gujarat

Devi Dayal

ICAR-Central Arid Zone Research Institute, Regional Research Station Kukma-Bhuj, Gujarat

The arid zone of India covers about 12% of the country's geographical area and occupies over 31.7 mha of hot desert and about 7 m ha under cold desert. The arid regions of Rajasthan, Gujarat, Punjab and Haryana together constitute the Great Indian Desert known as Thar Desert that accounts for 89.6% of hot arid regions of India. In Gujarat, 6.22 m ha area is under arid zone which constitute 19.6% of the arid area of country (Fig 1). In Gujarat, eight districts fall under arid zone, namely, Kachchh (100% of district area), Jamnagar (80%), Surendranagar (29%), Junagadh (20%), Banaskantha (18%), Mehsana (7%), Ahmedabad (6%) and Rajkot 6%).

Kachchh is the largest district (45, 652 sq km) in the Gujarat state and second largest district after Leh in India. The district experiences arid climate and has unique ecosystem. Agriculture is practiced in about 16% of the total area. Five percent of district area is under cultivable waste land indicating a vast potential for development of agriculture in the region (Fig. 2). Animal husbandry is also an important source of livelihood especially in northern part comprising Banni as the animal population (about 15 lakh) far exceed the human population (Devi Dayal *et al.*, 2009a).

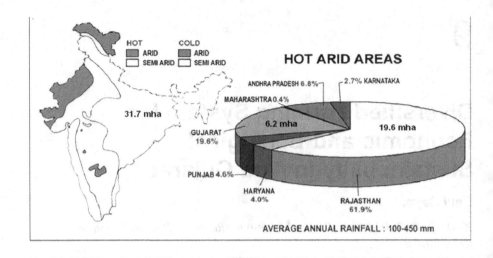

Fig. 1: Distribution of Arid region in the country

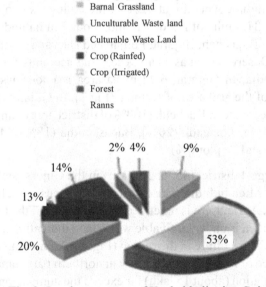

- Barnal Grassland
- Unculturable Waste land
- Culturable Waste Land
- Crop (Rainfed)
- Crop (Irrigated)
- Forest
  Ranns

Fig. 2: Land use pattern of Kachchh district, Gujarat

1. Constraints in crop production under arid ecosystem

i) Soil

The CAZRI workers had grouped soils of Kachchh into five textural groups namely, fine textured-4.07%, moderately fine textured-3.98%, medium textured-3.02%, coarse textured-11.13% and other miscellaneous 63.1%. The soil pH ranges from 8.0 to 9.0 and EC from 0.22 to 29.15 dS/cm. The soils are poor in

OC (0.28-0.62%) and phosphorus (4-18 kg/ha) and medium to high on available potassium (280-360 kg/ha).The soils are deficient in micronutrients, particularly Zn and Mn (Annual Report of CAZRI, Bhuj, 2008). Coarse textured soils have the problem of excessive permeability leading to heavy loss of water and nutrients. The compact surface especially with shallow soil does not allow proper development of root system in several crops and leads to decline in the yield. The soil salinity and alkalinity limits the crop production, especially in Banni and its associated areas.

ii) Climate

The climate of Kachchh region represents arid ecosystem. The average annual rainfall is 326 mm (based on 1972-2014 data) which is highly variable (CV more than 70%) and erratic (eight rainy days with CV more than 73%) leading to protracted droughts. Within a span of 23 years, six severe, three moderate and five mild meteorological droughts (IMD classification) were recorded. Apart from low and erratic rain fall, the region is characterized by high temperature (maximum of 39-45 $^{\circ}$C during May-June), high wind velocity (19-37 km/hr during July) and high potential evaporation (251-266 mm during May-June). The rainfall distribution of Kachchh during 1988-2015 is depicted in Fig 3. In the absence of favourable conditions for agriculture, livestock rearing is the alternative source of livelihood for the majority of the rural population in the arid zone.

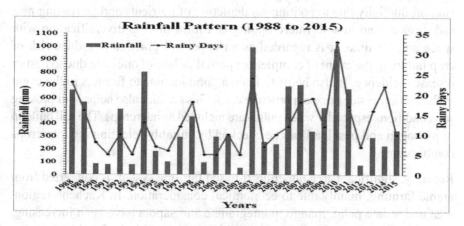

Fig. 3: Rainfall distribution at Bhuj

iii) Water Resources

Water resources in Kachchh region are very limited. The minimum depth to ground water is 10-20 m. The discharge from open well is poor. The excessive groundwater exploitation has led to increased salinity and deepening of ground water. Because of inadequate irrigation facilities from major and medium projects, minor irrigation continues to play a vital role in irrigation. Due to absence of mountain region, the existence of Ranns and scanty rain fall, the region does not have many rivers. In this region, availability of surface water depends on the intensity, duration and amount of rainfall. Out of the total 1482.71 mcm runoff estimated in the region, 847.38 mcm water is being stored in different irrigation storage reservoirs and village tanks, thus , nearly 57.20 % of runoff is surplus and goes a waste into the sea and the Ranns (Singh and Kar, 1966). Under these fragile ecosystems, the stored water need to be efficiently used through micro-irrigation to make the farming system economical and sustainable.

2. Why diversification

The production and life support systems in the hot regions are constrained by low and erratic precipitation (100-420 mm)/year), high evapotranspiration (1500-2000 mm/year) and poor soil physical and fertility conditions. The local inhabitants have evolved suitable land use and management systems of farming, pastoralism and animal husbandry. It has been observed that arable framing system on these fragile ecosystems is neither profitable nor sustainable. However, diversified farming system, including crops, animal husbandry, horticulture, agroforestry and silvi-pasture is more remunerative and eco-sustainable than mere crop cultivation.

The adoption of mono-cropping over same piece of land is reported deteriorate land productivity due to continuous depletion of nutrients and increasing pest and disease load in soil. Intercropping is a form of crop diversification with space and /or time. It is regarded as an important practice to reduce risk of crop failure in the event of complete or partial failure of one crop due to pest or disease incidence. It also helps to fetch a good income to farmers in the event of fluctuation of sale price of particular crop. This practice also helps in improving soil conditions especially when pulses are included as intercrops. The soil buildup of pathogen and pest load can be checked by suitably selecting the intercrops combination.

Recently, significant area under horti-based farming system is converted from arable farming, mainly due to economical consideration. In Kachchh region, area under date palm. mango, pomegranate and sapota have been increasing. Inclusion of crops and or perennial fodder grasses with the horti-system can increase both economic and environmental sustainability.

Animal husbandry had been and will be an important farming system adopted by the arid farmers. The productivity of animal farming is closely dependent on sufficient availability of quality fodder. Adoption of agroforestry or silvi-pasture system and diversify the existing arable cropping is the need of the hour.

Considering the importance of diversification in existing farming system in the arid zones, several studies were undertaken in arid Kachchh region of Gujarat and combinations of arable crop, horticulture and tress based diversified cropping system have been developed, demonstrated and validated on the farmers fields (Devi Dayal *et al.* 2009b).

3) Diversified cropping system with arable crops

The intercropping studies were carried out in different crops for optimizing optimum row ratios and nutrient management for enhanced farm income, while enriching soil health. Castor is an important cash crop of Kachchh region covering an area of about 80,000 ha. Severe attacks of castor semi looper lead to total crop loss to farmers. Various studies were carried out to develop castor based cropping system for the region (Devi Dayal *et al.*, 2009d; Devi Dayal *et al.*, 2010). Sesame and groundnut are other cash crops which are mostly cultivated as monocrop for which diversified cropping system were developed.

i. Castor based cropping system

a. Castor + sesame intercropping

Castor and sesame are important oilseed crops of arid region of Gujarat. Sesame is another very important oil seed crop in Kachchh. The yields of these crops are highly constrained by the climatic variability. Farmers in the Kachchh grow castor as monocrop with wider inter row space. During the initial stages when the growth of castor is slow, the interspaces can be utilized for growing various inter crops to gain additional income and to safe guard against crop failure due to biotic and abiotic constraints.

A demonstration on intercropping of castor with sesame was carried out in Gander during *kharif* 2009 in a row ratio of 2: 3. The sole cropping, which was a usual practice was also undertaken as control. Castor in the intercropping system produced 900 kg/ha and sesame 320 kg/ha. Farmers obtained a net return of Rs. 22500/ ha with a BC ratio of 2.41 under the intercropping system. They observed an income advantage of Rs.9000/ ha over sole cropping of castor and Rs.13350/ha over sole cropping of sesame by adoption of intercropping system. The demonstration was well appreciated by growers of the village and nearby areas.

b. Castor + groundnut intercropping

The studies on castor with groundnut revealed that sole crops of castor and groundnut produced an average yield of 570 and 150 kg/ ha, respectively. The grain yield under intercropping was found to be reduced by 31.9% in castor and

16.6% in groundnut as compared to their sole crop treatments. However, considering the economics of the system, intercropping of castor + groundnut (1:3) gave gross return of Rs 11, 566/ha which was higher by Rs 1451/ha over the sole castor and by Rs 8,026/ha over the sole groundnut.

c. Castor + cluster bean intercropping

Castor being an important cash crop, it is mainly grown as mono crop by the farmers allowing a wider spacing of 180 cm. It is a common practice by farmers who are also rearing livestock, to go for cluster bean separately to meet the fodder requirement apart from economic benefit out of the yield. Under the FPARP project it was advocated to farmers to combine both castor and cluster bean in 1: 3 ratio to fetch higher income as well to meet fodder requirements. The demonstration was conducted at Gander in *kharif* 2009. Under intercropping system castor produced a yield of 950 kg/ha and cluster bean 475 kg/ha. The intercropping system gave a net monetary return of Rs. 21150/ha with a BC ratio of 2.46. The farmers realized an income advantage of Rs.7650/ ha over sole cropping of castor and Rs. 14390/ha over sole cropping of cluster bean by adoption of the intercropping system, which also made it possible to meet their fodder requirements. The farmers were also convinced of the indirect benefit of soil enrichment by inclusion of the leguminous crop in the system (Fig 4).

d. Castor + green gram intercropping

Green gram, a short duration crop is better suited as an intercrop with long duration crops like castor. As castor is grown allowing a wider spacing of 180 cm, growing of green gram as an intercrop would provide an additional income to the farmer, besides providing nutritious fodder to the animals.

A technology demonstration on inter cropping of castor with green gram was conducted at Gander during *kharif* 2009 in 2: 4 ratio. With intercropping, castor produced a yield of 1020 kg/ha and green gram 285 kg/ha. The farmer obtained a net return of Rs.27,520/ha under the inter cropping system, against Rs. 11,140 and Rs.13, 500/ha under sole crop of green gram and castor respectively. The system recorded a BC ratio of 2.57 whereas sole crops of castor and green gram recorded BC ratio of 1.96 and 2.17, respectively. The intercropping system was appreciated by the farmers in the vicinity as well (Fig 5).

ii. Sesame based intercropping system

a. Sesame + cluster bean intercropping

The studies conducted at CAZRI, Regional Research Station, Bhuj on intercropping of sesame + cluster bean indicated that grain yield under intercropping system reduced by 28.8% in cluster bean and 40.4% in sesame as compared to their sole treatments. However, considering the net returns and BCR, intercropping of sesame with cluster bean (1:2) gave the maximum net returns of Rs 7,440/ha along with BCR of 1.80 compared with Rs 5,945/ha and BCR of 1.68 in sole cluster bean and Rs 2,851/ha and BCR of 1.37 in sole sesame. The Sustainable Yield Index (0.74) and Sustainable Value index (0.76) were also higher in intercropping of sesame+ cluster bean (1:2) than that recorded by sole sesame (0.73 and 0.73) and sole cluster bean (0.71 and 0.72). From the studies, it is concluded that intercropping involving cluster bean and sesame was more profitable and sustainable than sole cropping (Meena *et al.*, 2009).

The intercropping technologies developed at the Station and elsewhere were demonstrated on the farmer's field under farmer's participatory action research programme, funded by Ministry of Water Resources, Govt. of India, New Delhi.

b. Cluster bean + sesame intercropping in farmer's field

The field demonstration on intercropping of cluster bean + sesame in 2:1 ratio was conducted in village Gander during 2009. The results indicated that cluster bean produced yield of 490 kg/ha and sesame 140 kg/ha. The intercropping of cluster bean + sesame provided a net return of Rs. 11574/ ha with BC ratio of 2.41. The net return obtained by the intercropping was higher by 41.6 % over sole cropping of cluster bean and by 20.9 % over sole cropping of sesame. The intercropping system recorded a BC ratio of 2.41 as against 1.94 with sole cluster bean and 2.06 with sole sesame.

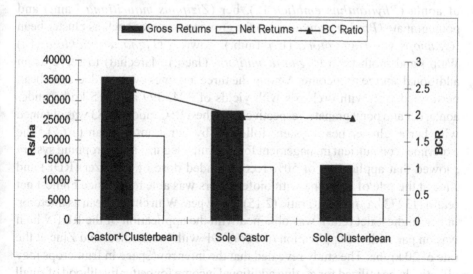

Fig. 4: Yield and economics of Castor + cluster bean

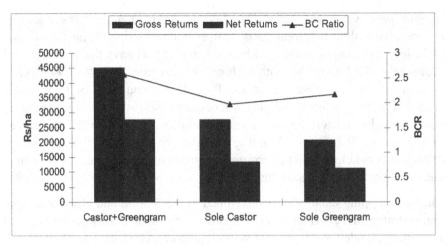

Fig. 5: Yield and economics of Castor + green gram intercropping at Gander

4. Diversified cropping system with horticultural crops

Legumes with fruit crops

Horticulture and fruit crops are one of the main activities in Kachchh region. These crops are widely spaced and have a long pre-bearing stage. Intercrops can be successfully integrated into these crops during the initial years of establishment and in the inter row spaces of bearing trees. A field study was conducted during rainy seasons of 2008 to 2010 at CAZRI, Research Farm, Bhuj to assess the possibilities of utilizing the inter row spaces of fruit orchards of aonla (*Phyllanthus emblica* L.), ber (*Ziziphus mauritiana* Lam.) and pomegranate (*Punica granatum* L.) for growing legumes such as cluster bean (*Cyamopsis tetragonoloba* (L.) Taub.), cowpea (*Vigna unguiculata* (L.) Walp.) and moth bean (*Vigna aconitifolia* (Jacq.) Marechal) to serve as an additional source of income. Among the three legumes evaluated, cluster bean performed well with orchards with yields of 304, 399 and 315 kg/ha under aonla, ber and pomegranate, respectively. Highest B: C ratio of 1.83 was obtained with ber + cluster bean system followed by ber + moth bean (1.65). The experiment on nutrient management for the fruit – legume alley cropping system showed that application of 50% recommended dose of fertilizer (RDF) and Zinc at the rate of 20 kg/ha with biofertilisers was able to produce highest net return (13273 ha^{-1}) and B:C ratio (2.15) in cowpea. With cluster bean as intercrop in ber, highest net return was obtained with the application of the RDF which was on par with the application of 50% RDF with biofertiliser and Zinc at the rate of 20 kg/ha. The study revealed that the inter row space in fruit crops may effectively be utilized for getting additional income for better livelihood of small and marginal farmers in hot arid region of Gujarat (Devi Dayal *et al.*, 2015).

A large area field demonstration on 4 arid legumes intercropped with ber orchard was conducted on the farmer's field during 2010. All the package of practices including INM was adopted. Intercropping of cluster bean cv HG 365 with Ber orchard gave maximum net returns of Rs.13,050.00/ha along with a BCR of 1.71. This was closely followed by intercropping of Moth bean with ber giving net returns of Rs 10,400/ha and BCR of 1.64

5. Fodder crops with orchard

Due to increased animal population the fodder requirement of Kachchh region is very high. To meet the fodder requirement in terms of both quantity and quality, nutritious fodder crops can be introduced into the available inter row spaces. Intercropping of stylo and clitoria with mango orchard was demonstrated on the farmer's field in Kharif during 2011. A strip of 5 rows of stylo was intercropped with two rows of mango. Stylo produced green fodder of 5100 t/ha with a dry fodder of 2800 kg/ha. A Field demonstration on intercropping of clitoria and stylo with established orchard (mango) was also conducted on 3 farmer's field in taluq Bhuj in kharif 2012. Five rows of legumes were intercropped with mango plantation planted at 6m row spacing. A dry fodder yield 2.5 and 3.20 t/ha of clitoria and stylo was harvested. The farmers are convinced with the system as an additional fodder yield of 2.5-3.2 t/ha were recorded without any adverse effect on mango plantation.

6. Silvi-pastoral system

In the dry areas, some grass species like *Cencherus ciliaris, C. setigerus* and *Lasuirus sindicus* etc. are very well adapted to such harsh climate and perform well and make natural rangelands. The climax tree species like *Prosopis cineraria, P. juliflora, Acacia nilotica, Ziziphus numularia* etc come up in these rangelands and make a silipastoral system. Animal husbandry flourishes in such locations and forms an integral part of the prevailing farming system. Silvi-pastoral studies were conducted at CAZRI, Bhuj with combination of tress namely, neem (*Azadirachta indica, Acacia tortilis* and *Leucaena leucocephala* and grasses like *Cencherus ciliaris* and *C. setigerus.* Neem + grass combination was found to be the most adapted silvi-pastoral system for the Kachchh region in terms of both grass yield and tree growth. Among the grasses, *C. ciliaris* was found to be superior to *C. setigerus* in terms of fodder production. Total number of tillers per grass plant and dry fodder yield of grass did not differ significantly due to association of tress with grasses in the silvi-pastoral system (Devi Dayal *et al.,* 2008; Devi Dayal *et al.,* 2009c). Importance of silvipastoral system in improving the fertility status especially organic carbon and potassium of degraded soils have been demonstrated. With 20 years of cultivation under silvipastoral system, the soil organic carbon under grass improved from 0.47 to 0.58% in the surface layers and 0.23 to 0.28% in the lower layers, potassium from 470 to 616 kg/ha in the surface and 197 to 284 kg/ha in the sub-surface layers (Shamsudheen *et al.*, 2009).

Arya *et al.*, (2008) studied establishment of silvipastoral system in highly degraded soil of Kachchh region, Gujarat and found that *Acacia A. jacque montii* performed better than *A, nilotica* and *P. juliflora* in association with grasses (*C. ciliaris* and *C. setigrus*).

7. Impact of research and field demonstration on adoption

The consistent efforts of CAZRI, Regional Research Station through farmers training and demonstrations on farmer's field was beneficial in increasing the adoption rate of diversified cropping system in Kachchh region. The area under intercropping showed a steady increase over the last 5 years from mere 500 ha in 2010 to more than 12,000 ha in 2014.

8. Ecological gains through diversified cropping system

The study was conducted at CAZRI, Bhuj during 2008-10 and carbon sequestration was quantified both in biomass and soil in two pasture systems (*Cenchrus ciliaris* and *Cenchrus setegerus*), two tree systems (*Acacia tortilis* and *Azadirachta indica*) and four silvipastoral systems (combination of one tree and on grass). The silvipastoral system sequestered 36.3% to 60.0% more total soil organic carbon stock compared to the tree system and 27.1–70.8% more in comparison to the pasture system. The soil organic carbon and net carbon sequestered were greater in the silvipastoral system. Thus, silvipastoral system involving trees and grasses can help in better sequestration of atmospheric system compared with systems containing only trees or pasture (Mangalassery *et al.*, 2014).

References

Arya, Ranjana, Kumar, Hemant, Lohara, R, R. and Meena, R, L. 2008. Establishment of silvipastoral system to enhance the productivity of highly degraded aridisol in Bhuj, Kachchh district in Gujarat. Abs. In: National symposium on Agroforestry knowledge for sustainability, climate moderation and challenges ahead (15-17 December, 2008), NRC Agro-forestry, Jhansi (U.P.), P.120.

Devi Dayal, Bhagirath Ram, Shamsudheen, M and Singh, Y. V. 2010. Dissemination of technologies through farmer's participatory approach in Kachchh, Gujarat. Central Arid Zone Research Institute, Kukma- Bhuj, Gujarat, pp. 1-24.

Devi Dayal, Ram, B., Shamsudheen, M., Swami, M.L., Patil, N.V. 2009a. *Twenty years of CAZRI, Regional Research Station, Kukma-Bhuj*, pp. 38. Regional Research Station, Central Arid Zone Research Institute, Kukma-Bhuj, Gujarat.

Devi Dayal, S., Mangalasseri, Meena, S. L. and Bhagirath Ram. 2015. Productivity and profitability of legumes as influenced by integrated nutrient management with fruit crops under hot arid ecology. *Indian J. Agron.* 60 (2): 297-300.

Devi Dayal, Shamsudheen, M. and Ram, B. 2009b. Alternative farming systems suitable for Kachchh region of Gujarat. Regional Research Station, Central Arid Zone Research Institute, Kukma- Bhuj, Gujarat, India, *Technical Bulletin,* pp. 1–22.

Devi Dayal, Shamsudheen. M, Meena, S.L., Bhagirath Ram, Harsh, L. N and Swami, M. L. 2009c. Biomass production of *Prosopis juliflora* and associated changes in soil organic carbon in degraded soils of Kachchh region of Gujarat. National symposium on Prosopis: ecological, economic significance and management challenges, (February 20-21, 2009), Gujarat Institute of Desert Ecology, Bhuj, Gujarat. Pp 17-18.

Devi Dayal, Swami, M. L., Bhagirath Ram, Meena, S. L. and Shamsudheen, M. 2008. Production potential of grasses under silvipastoral system in Kachchh region of arid Gujarat. Abs. In: National symposium on Agroforestry knowledge for sustainability, climate moderation and challenges ahead (15-17 December, 2008), NRC Agro-forestry, Jhansi (U. P.), P. 195.

Devi Dayal, Vyas, S, P., Meena, S. L., Shamsudheen, M and Bhagirath Ram. 2009d. Sustainable production of rainfed oilseeds through intercropping in Kachchh region of Gujarat. *J Oilseeds Res.* 26: 273-274.

Mangalassery, S., Devi Dayal., Meena, S.L. and Ram, B. 2014. Carbon sequestration in agroforestry and pasture systems in arid northwestern India. *Current Science,* 107(8): 1290-1293.

Meena, S.L., Shamsudheen, M. and Devi Dayal. 2009. Productivity of cluster bean *Cyamopsis tetragonoloba*) and sesame (*Sesamum indicum*) intercropping system under different row ratio and nutrient management in arid region. *Indian Journal of Agricultural Sciences,* 79(11): 901–05.

Shamsudheen, M., Devi Dayal, Meena, S.L. and Bhagirath Ram. 2009. Improvement of soil properties under silvi-pastoral system in Kachchh region of arid Gujarat. Fourth World Congress on Conservation agriculture: Innovation for improving Efficiency, equity and environment (Feb. 4-7, 2009), National Academy of Agricultural Sciences, New Delhi, India, pp. 253-254.

Singh, S. and Kar, A. 1996. Integrated natural and human resources appraisal for sustainable devilment of Kachchh district. Report of Central Arid Zone Research Institute (CAZRI), Jodhpur.

10

Intercropping of Seed Spices with Vegetables for Enhancing System Productivity and Profitability

R.S. Mehta, A.K.Shukla, Dipak Kumar Gupta, Kamla K. Chaudhary Keerthika A., S.R. Meena, M.B. Noor mohamed and P.L. Regar

ICAR-Central Arid Zone Research Institute- Regional Research Station Pali, Rajasthan-306401

Since time immemorial India has been leader in respect to spice, area, production and consumption. India is known as 'Land of spices. Out of 109 spices listed by the ISO, India produce as many as 63 owing to its varied agroclimatic regions. Almost all the state and union territory of the country grow one or the other spices. Spices are the sources of livelihood and employment for large number of people in the country. Seed spices are annual herbs, whose dried seeds or fruit are used as spices. Spices are the nature's gift to mankind and since antiquity, spices have been considered indispensable in the culinary art, as they are used to flavour foods and beverages The important seed spices are coriander, cumin, fennel, fenugreek. ajwain, dill,, celery, anise, nigella and caraway. The other seed spices s are poppy seed, rai, yellow mustard, white mustard, parsley and sesame. In India, Rajasthan and Gujarat has emerged as "Seed spice bowl" and together contributes more than 80 percent of total seed spices.

Burgeoning population exerts pressure on limited precious natural resources for enhancing production of food, fodder and fuel. In order to meet out increasing demand for food, fruit, vegetables, oilseeds, fodder and timber wood for burgeoning population, it is necessary to utilize each unit of natural rsesource very efficiently. The productivity of the precious natural resources is to be enhanced per unit time per unit area by agronomic ally managing stresses soil and land ecosystem to harvest more benefit. The cropping system as well as intercropping system practised by using different trees, fruits, vegetables, crops are the very efficient enterprises for realising higher productivity of natural resources available with in specified ecosystem.

The cropping system is one of the important ways to enhance efficiency of the natural resources i.e. land, water, lab our and management. Traditionally cropping system have been defined as the cropping patterns on farm and their interaction with farm resources, other than farm enterprises, and available technology that determine their make-up (Andrews and Kassam, 1976). Okigbo (1980) defined cropping system as the pattern of growing crops in combination and sequences in time and space, in addition to the practices and technologies with which crop are produced. Commonly cropping system refers to temporal and spatial arrangement of crops and management of soil, water and vegetation in order to optimize the biomass production per unit area, per unit time and per unit input. In other words, definitions of cropping systems have been limited to the systematic arrangement of crops as influenced by local factors of crop production.

Intercropping means growing of subsidiary crops between two widely spaced main crop. The main objective of intercropping is to utilize the space left between two rows of main crop and to produce more grain per unit area. The basic concept of intercropping system involves growing together two or more crops with the assumption that two crops can exploit the environment better than one and ultimately produce the higher yield. This practice leads to some benefit like yield advantages as compared to sole cropping, greater stability of yield over different seasons, insurance against aberrant weather conditions, build-up or maintenance of soil fertility, economy of land, production of higher yield and higher economic returns in a given season. Intercropping is an age-old practice and attracted worldwide attention owing to yield advantage if the crops selected are compatible. Intercropping is one of the most important techniques which embodies growing of crop under different plant geometry. With the release of early maturing and diverse crop varieties it is proved beyond doubt that such a system, when based on sound production principles, will provide greater stability, productivity and profitability. Intercropping system has some of the potential benefits such as increased productivity/unit area/unit time, high profitability, improvement in soil fertility, efficient use of resources and reducing damage caused by pests, diseases and weeds (Ghosh *et al.*, 2006). Fennel (*Foeniculum vulgare* Mill.) is an important seed spice crop of Northern parts of the country. Wide spacing and slow growing nature during initial growth period of fennel make possible to raise short duration intercrops in between rows. Inclusion of vegetables like radish (*Raphanus sativus* L.), coriander *(Coriandrum sativum* L.) and fenugreek (*Trigonellafoenum - graecum* L.) as intercrop has been reported to enhance the productivity and profitability per unit area in winter maize (*Zea mays* L.) as compared to its sole cropping (Singh and Kumar, 2002). The system of intercropping not only improves the yield and returns but also reduces the risk of complete crop failure as compared to the sole cropping system (Rao and Singh, 1990).

Types of intercropping

(a) **Mixed Intercropping**: The practice of growing component crops simultaneously with no district row arrangement. It is a traditional method of intercropping.

(b) **Row Intercropping**: It is the system of growing two or more crops in different row arrangement lot of research findings are available on cereal crops based inter cropping. Research work at NRCSS, Ajmer has been initiated to standardize economically viable, environmentally compatible and socially acceptable seed spices based cropping system with vegetable crops like carrot, onion and garlic. Initially the work has been started to standardization fennels, and coriander based inters cropping system with vegetable crops.

(c) **Strip Intercropping**: - It is simultaneous growing of component crops in different strips to permit independent cultivation of each crop. No research information are available on seed spices based strip intercropping. Therefore, it is the need of time to initiate research work for standardizing seed spices based strip inter cropping

When two crops are to be grown together, they are chosen in such away that there is variation in their growth duration. The peak periods of growth of the two crops species should not coincide. In such arrangements, a quick maturing crop completes its life cycle before the other crop starts. Willey (1979) described the concept as temporal complementary. Greater differences in maturity and growth demands of the crop components, more opportunity is provided for greater exploitation of growth factors and over yielding. This will be achieved either by generic difference in crop species or manipulation of planting dates. Normally short and long duration crops are grown together. Based on the per cent of plant population used for each crop in intercropping system, It is divided in to two category *viz*, additive series and replacement series.

Additive Series

It is mostly adopted in India, base crop is sown having 100% of its recommended population in pure stand, another crop known as intercrop is introduced into the base crop by adjusting or changing geometry. The population of intercrop is less than its recommended population in pure stand LER of additive series is greater than replacement series. Additive series is more efficient than replacement series in intercropping system. In additive series the crop duration of both base and intercrop is different. Normally intercrops are of short duration in nature and base crops are of long duration in nature.

Replacement Series

In replacement series both the crops are called component crops. By sacrifising certain proportion of population of one component, another component is introduced. This type of intercropping is practiced in western countries.

Objective of Intercropping

- The main objective of the intercropping is to get more profit per unit area per unit time
- Insurance against total crop failure under adverse weather condition or pest epidemics.
- Increase in total productivity per unit land area
- Judicious utilization of resources such as land, labour and inputs

Characteristics of crops suitable for intercropping

- The component crops should have variation in their growth duration
- The peak periods of growth of the two-crop species should not coincide. In such arrangement quick maturing crop complete its life cycle before the other crop start. Willey (1979) described the concept as temporal complementarily
- Food crops should be mixed with cash crops to ensure both sustenance and cash income
- In intercropping system main crop should deep rooted and bonus crop should have shallow roots for taking moisture and nutrients from different lays of soil.
- At least one crop should be leguminous for maintenance of fertility of soil.

Intercropping system developed at ICAR-NRCSS, Ajmer

Different cropping systems comprising fruits, vegetables and seed spices have been developed at ICAR-NRCSS, Ajmer for enhancing resource use efficiency and profitability to double the income of farmers. Long duration seed spices such as fennel, dill and ajwain takes 180-240 days and their initial growth is slow. Long duration seed spice crops are more branched with higher height, therefore they are grown at row to row spacingof 45-60 cm. Germination of these crops takes 8-10 days and they grow upto 10-15 cm within 70-80 days after sowing. The space left unoccupied in the initial ontogeny of these crops is covered by weeds and irrigation and nutrient applied are utilized by these weeds.

Intercropping of fennel and vegetables under varying fertility situation

Fennel is long duration major seed spices which require wider crop geometry of 50 to 75 cm row to row spacing and 20-25 cm plant to plant distance. In the initial 70-80 days fennel grows upto 10-15 cm height resulting in leaving unutilized spaces which invites growth of unwanted plants which exerts unnecessary competition with crop for water, nutrients and light. Moreover, farmers have to spent money for removal of these unwanted plants in between the rows of fennel resulting in enhancing cost of production. Therefore, in order to address these issues an experiment on effect of intercropping system and fertility levels on yield and profitability of fennel was conducted at NRCSS, Ajmer during rabi season of 2003-04. In the study Mehta *et. al* (2012) reported that application of increasing levels of fertilizers significantly increased fennel yield, carrot/ radish yield, fennel equivalent yield, gross return, net return and BCR and highest yield and economic return was obtained with the application of 120 kg N + 50 kg P_2O_5 per ha. They, further reported that fennel based intercropping system in association of carrot with all ratios resulted higher yield and profitability as compared with radish at respective intercropping ratio. The highest fennel yield (17.35 q ha[-1]), fennel equivalent yield (26.42 q ha[-1]) gross return (Rs. 66050/- ha[-1]), net return (Rs 38584/- ha[-1]) and BCR (1.40) was obtained in 1:1 intercropping ratio of fennel and carrot followed by with 1:2 ratio. Thus, sowing of fennel and carrot in 1:1 ratio with application of 120 kg N +50 kg P_2O_5 is best for realizing the higher system productivity and profitability.

Table 1: Effect of fertility levels and intercropping system on profitability

Treatments	FEY(q/ha)	Cost of cultivation (Rs/ha)	Gross return (Rs/ha)	Net return (Rs/ha)	B: C ratio
75 30	18.96	27050	47400	20350	0.75
90 40	20.92	27327	52300	24973	0.91
105 50	22.42	27604	56050	28446	1.03
120 50	23.52	27881	58800	30919	1.11
CD (P=0.05)	0.49	-	-	-	-
Fennel + Radish (1:1)	17.60	26916	44000	16534	0.60
Fennel + Radish (1:2)	15.82	28958	39550	12084	0.44
Fennel + Radish (2:2)	18.67	26524	46675	19209	0.70
Fennel + Carrot (1:1)	26.42	28016	66050	38584	1.40
Fennel + Radish (1:2)	24.96	28958	62400	34934	1.27
Fennel + Radish (2:2)	25.04	26524	62600	35134	1.28
CD (P=0.05)	0.55	-	-	-	-

Intercropping of fennel with carrot, onion and garlic for higher return

Onion, garlic and carrot are important crops. Among seed spices, fennel is an important major seed spices which require long duration for realizing higher yield. Profitability of fennel production can be enhanced with growing of carrot, onion and garlic in between the space in of rows of fennel. In order to study feasibility and profitability of intercropping of fennel and component crops, a field experiment on effect of different intercropping ratios on growth, yield and system productivity was conducted at NRCSS, Ajmer (Raj.) during rabi season of 2004-05, 2005-06 and 2006-07. In the study Mehta *et al.* (2015) reported that the highest equivalent yield, net return and BCR was obtained in inter cropping of fennel + carrot with 1:1 ratio followed by 2:2 ratio. Fennel and carrot at all the ratios resulted higher equivalent yield, net return, BCR and LER over fennel intercropped by onion and garlic. Intercropping of fennel and carrot in 1:1 ratio is better for realizing higher system productivity, net return and BCR.

Table 2: Effect of different intercropping system on yield of fennel, component crops fennel equivalent, return. BCR and LER.

Treatments	FEY(q/ha)	Cost of cultivation (Rs/ha)	Net return (Rs/ha)	B: C ratio	LER
Sole fennel	17.25	26750	42250	1.58	1.00
Sole Onion	9.26	18950	18090	0.95	1.00
Sole carrot	11.27	22840	22240	0.97	1.00
Sole garlic	15.06	45250	14990	0.33	1.00
Fennel + Onion (1:1)	21.42	28238	57442	2.03	1.55
Fennel + Onion (1:2)	20.09	29532	50828	1.72	1.52
Fennel + Onion (2:2)	20.28	27436	53684	1.96	1.45
Fennel + Carrot (1:1)	23.76	27435	67605	2.46	1.59
Fennel + Carrot (1:2)	20.81	28240	55000	1.95	1.43
Fennel + Carrot (2:2)	21.31	27250	57990	2.13	1.42
Fennel + Garlic (1:1)	21.31	34850	50390	1.45	1.29
Fennel + Garlic (1:2)	17.87	38250	33230	0.87	1.08
Fennel + Garlic (2:2)	20.42	34150	47530	1.39	1.24
CD (P=0.05)	1.91	.	4592	0.16	0.13

Intercropping of coriander with carrot, onion and garlic for higher return

Coriander is second most important major seed spice crop grown at spacing. Efforts have been made at NRCSS for enhancing productivity and enhancing resource use efficiency through intercropping. Hence a field experiment on effect of intercropping system on growth, yield and system productivity was conducted at National Research Centre on Seed Spices, Ajmer, Rajasthan during rabi 2004-05, 2005-06 and 2006-07. The highest coriander equivalent yield (2.11 t/ha), net return (Rs 50,701 / ha) and B:C ratio (2.16) was exhibited by 1:1 ratio followed by 2:2 ratio. Coriander + carrot with all ratios gave higher coriander equivalent yield, net return, benefit cost ratio and land equivalent ratio over coriander intercropped with onion/garlic. Thus coriander + carrot in 1:1 ratio is best for realizing higher system productivity and profitability (Mehta *et.al.*, 2010).

Table 3: Effect of different intercropping system on yield of coriander, component crops fennel equivalent, return. BCR and LER

Treatments	CEY of system (kg/ha)	Cost of cultivation (Rs/ha)	Net return (Rs/ha)	B: C ratio	LER
Sole fennel	15.25	19625	33750	1.72	1.00
Sole Onion	10.03	18830	16262	0.86	1.00
Sole carrot	12.32	22745	20391	0.90	1.00
Sole garlic	14.39	45375	4975	0.11	1.00
Coriander + Onion (1:1)	18.72	24240	41285	1.70	1.45
Coriander + Onion (1:2)	17.01	26648	32876	1.23	1.36
Coriander + Onion (2:2)	17.50	24240	37000	1.53	1.34
Coriander + Carrot (1:1)	21.19	23450	50701	2.16	1.50
Coriander + Carrot (1:2)	19.25	25345	42020	1.66	1.38
Coriander + Carrot (2:2)	20.01	23470	46581	1.98	1.41
Coriander + Garlic (1:1)	17.97	31755	31135	0.98	1.20
Coriander + Garlic (1:2)	16.79	33350	25415	0.76	1.13
Coriander + Garlic (2:2)	17.16	31348	28697	0.92	1.15
CD (P=0.05)	0.90	3203	4038	0.13	0.12

Intercropping of seed spices with vegetables for higher system productivity

A field study on effect of intercropping of seed spices with vegetables for enhancing system profitability was conducted at research farm of ICAR-NRCSS, Ajmer (Rajasthan) during *rabi* season of 2011-12 and 2012-13 The experiment comprising of nine treatments viz., pea + fennel, pea + ajwain, pea + coriander, cabbage + fennel, cabbage + ajwain, cabbage + coriander, carrot + fennel, carrot + ajwain and carrot + coriander was laid in randomized block design with three replications. Based on two-year study results exhibited that that all the vegetables viz. carrot, cabbage and pea performed better in intercropping of fennel compared to ajwain and coriander and accordingly exhibited higher vegetable yield during 2011-12, 2012-13 and in pool. Intercropping of seed spices with carrot resulted higher grain yield of fennel, coriander and ajwain compared with cabbage and pea during both the years and in pool. Among all the intercropping pattern, fennel with cabbage in 1:1 row ratio resulted the highest fennel equivalent yield (30.59 q/ha) and net return (Rs.151160/- per ha) followed by intercropping of carrot with fennel in 1:1 row ratio during 2011-12,2012-13 and in pool. The highest benefit cost ratio of 4.66 was obtained in intercropping of cabbage with fennel followed by intercropping of cabbage with ajwain (4.16). Thus, it is inferred that intercropping of fennels with cabbage in 1:1 row ratio is better for getting higher system productivity, net return and profit followed by inter-cropping of ajwain with cabbage in 1:1 ratio.

Table 4: Effect of intercropping of seed spices with vegetables on gross return, net return and BCR

Treatment	FEY (q/ha)	Cost of cultivation	Gross return (Rs/ha)	Net return (Rs/ha)	BCR
Pea+ Fennel	21.22	28353	127295	98942	3.52
Pea+ Coriander	11.31	26660	67865	41205	1.56
Pea +Ajwain	15.47	24161	92790	68630	2.86
Cabbage+ Fennel	30.59	32331	183490	151160	4.66
Cabbage +Coriander	18.75	30638	112489	81852	2.68
Cabbage +Ajwain	24.66	28138	147950	119812	4.24
Carrot+ Fennel	25.27	29643	151634	121991	4.16
Carrot +Coriander	15.86	27950	95123	67173	2.42
Carrot +Ajwain	20.15	25451	120918	95467	3.77
CD(P=0.05)	2.22	-	13296	10333	0.34

Inter-cropping of fennel with vegetables under low pressure drip irrigation

Under low pressure drip irrigation intercropping study of fennel with different vegetables was done at ICAR-NRCSS, Ajmer during 2013-14, 2014-15 and 2015-16. Intercropping of fennel with five vegetables viz., Cabbage, knolkhol, lettuce, french-bean and fenugreek) was done in three intercropping ratios of 1:1, 1:2 and 2:2. Two crops of *knolkhol* were taken within 80 days after sowing of base crop but only one crop of other vegetable was taken. Based on three-year study it has been confirmed that Intercropping of fennel with *knolkhol* in 1:2 ratio exhibited the highest fennel equivalent yield (3464kg/ha), net return (Rs. 197707/ha ha) and BCR (3.95) followed by 1.1 intercropping ratio of fennel with *knolkhol*. Intercropping of fennel with *knolkhol* in 1:2 and 1:1 ratio produced 85 and 76 percent higher fennel equivalent yield, respectively over sole fennel. Intercropping of fennel with fenugreek and french bean gave higher land equivalent ratio as compared with other vegetables like knolkhol and cabbage Thus, it is inferred that intercropping of fennel with knolkhol in 1:2 or 1:1 ratio is better for realizing higher yield, net return, benefit and system productivity.

Table 5: Effect of fennel based intercropping systems on yield of fennel, vegetables and fennel equivalent yield

Treatment	Fennel Grain yield(kg/ha)	Vegetable Veg. yield (kg/ha)	Fennel equivalent yield(kg/ha)	Net return Rs./ha	BCR
Sole Fennel	1868	0	1868	93571	2.46
Sole cabbage	0	13461	2425	131862	3.18
Sole Knolkhol	0	12499	2541	138729	3.18
Sole Lettuce	0	3336	1302	59610	1.58
Sole French-bean	0	954	807	24552	0.75
Sole fenugreek	225	7170	930	34391	1.17
Fennel+ Cabbage (1:1)	1496	9267	3188	179766	3.70
Fennel+ Cabbage (2:2)	1574	7979	3049	168411	3.46
Fennel+ Cabbage (1:2)	1228	9486	3003	165325	3.40
Fennel+ Knolkhol (1:1)	1557	8435	3291	185754	3.71
Fennel+ Knolkhol (2:2)	1638	7650	3204	180397	3.61
Fennel+ Knolkhol (1:2)	1508	9550	3464	197707	3.95
Fennel+ Lettuce (1:1)	1475	2218	2324	124606	2.77
Fennel+ Lettuce(2:2)	1527	1908	2269	120253	2.66
Fennel+ Lettuce (1:2)	1378	2104	2211	115464	2.56
fennel+ French-bean (1:1)	1659	668	2243	113867	2.59
Fennel+ French-bean (2:2)	1631	597	2152	105568	2.40
Fennel+ French-bean (1:2)	1469	825	2171	110120	2.52
Fennel+ Fenugreek (1:1)	1595	5376	2261	114539	2.67
Fennel+ Fenugreek (2:2)	1596	4963	2216	109950	2.55
Fennel+ Fenugreek (1:2)	1437	5936	2161	107827	2.51
CD(P=0.05)	174	761	293	**15680**	**0.34**

Intercropping of dill with vegetables under low pressure drip irrigation

Under low pressure drip irrigation intercropping study of fennel with different vegetables was done at ICAR-NRCSS, Ajmer during 2013-14, 2014-15 and 2015-16. Intercropping of fennel with five vegetables (Cabbage, knolkhol, lettuce, french-bean and fenugreek) with fennel and cabbage, knolkhol, beetroot, french-bean, and fenugreek with dill) in three intercropping ratio of 1:1, 1:2 and 2:2 were conducted with three replications in randomized block design with low pressure drip irrigation system. Two crops of *knolkhol* were taken within 80 days after sowing of base crop but only one crop of other vegetable was taken. Based on three-year study it has been confirmed that intercropping of fennel with knolkhol in 1:2 ratio exhibited the highest fennel equivalent yield(3464kg/ ha), net return (Rs197707/ha) and BCR (3.95) followed by 1.1 intercropping ratio of fennel with *knolkhol*. Intercropping of fennel with *knolkhol* in 1:2 and 1:1 ratio produced 85 and 76 percent higher fennel equivalent yield, respectively over sole fennel. Similarly, the highest dill equivalent yield (4148/ kg /ha), net return (Rs157173 / ha) and BCR (3.40) was obtained in intercropping of dill and *knolkhol* in 1:2 ratio followed with 1:1 intercropping ratio of dill +cabbage/ knolkhol. Dill + knolkhol in 1:2 ratio and dill + knolkhol in 1:1 ratio produced 158 and 144 percent higher dill equivalent yield, respectively over sole dill. Intercropping of fennel with fenugreek and french bean gave higher land equivalent ratio as compared with other vegetables like knolkhol and cabbage. Intercropping of dill with fenugreek and french bean gave higher land equivalent ratio as compared with other vegetables like knolkhol and cabbage. Thus, it is inferred that intercropping of fennel/ dill with knolkhol in 1:2 or 1:1 ratio is better for realizing higher yield, net return, benefit and system productivity.

Table 6: Effect of dill based intercropping systems on yield of dill, vegetables and dill equivalent yield

Treatment	Grain yield (kg/ha)	Vegetable yield(kg/ha)	Dill equivalent yield(kg/ha)	Net return (Rs/ha)	BCR
Sole dill	1603	0	1603	44128	1.30
Sole cabbage	0	12369	3143	112324	2.70
Sole Knolkhol	0	11695	3360	122349	2.82
Sole Beetroot	0	5740	2323	70810	1.91
Sole French-bean	0	865	1050	18428	0.56
Sole fenugreek	185	6792	1290	27229	0.86
Dill +Cabbage (1:1)	1303	9402	3820	143458	3.14
Dill +Cabbage (2:2)	1396	8267	3627	132476	2.89
Dill +Cabbage (1:2)	1202	9402	3700	134794	2.94
Dill +Knolkhol (1:1)	1354	8402	3924	145950	3.15
Dill +Knolkhol (2:2)	1460	7644	3742	137568	2.98
Dill +Knolkhol (1:2)	1350	9429	4148	157173	3.40
Dill + Beetroot (1:1)	1366	3187	2667	85585	2.18
Dill +Beetroot (2:2)	1401	2971	2621	83919	2.14
Dill Beetroot (1:2)	1283	3625	2757	90674	2.30
Dill +French-bean (1:1)	1455	572	2200	67742	1.74
Dill +French-bean (2:2)	1423	523	2106	60755	1.56
Dill +French-bean (1:2)	1402	686	2272	71925	1.85
Dill +Fenugreek (1:1)	1452	4914	2369	74701	1.97
Dill+ fenugreek (2:2)	1469	4309	2287	69074	1.82
Dill+ fenugreek (1:2)	1333	5209	2310	70931	1.87
CD(P=0.05)	158.11	734.70	342.81	11809	0.28

Management of seed spices based cropping system

The management of cropping system differs from that of single cropping. This is due to inclusion of more than one crop as in intercropping or planting of crops in succession as in sequential cropping. The water, nutrient, weed and diseases pest management in cropping system to be done in organized way.

a)   **Water management of cropping System**: The techniques of water management is the same for sole cropping and inter or sequential cropping. However, the presence of an additional crop may have an important effect on evapo- transpiration. Scheduling of irrigation and method of water application have to be carefully done under intercropping situation since one of component may be sensitive for excess irrigation. Crop yield can be increased without significant increase in water use by selecting suitable cropping system and adopting appropriate crop and soil management practices. The water use efficiency in a cropping system can be increased by identification of appropriate crop combination. The scheduling of irrigation for the component crops may very widely and the critical period of water requirement for the crops may not coincide. In the crops where water requirement vary, skip furrow method of irrigation is advocated. Low pressure drip irrigation is very effective for water management in intercropping system through which irrigation water can be applied simultaneously based on requirement of both the crops. Water productivity in low pressure drip irrigation is higher compared to surface and sprinkler irrigation.

b)   **Weed Management in Inter cropping system**: More complete crop cover and high density available in inter cropping cause severe competition with weeds and reduce weed growth. The weed suppressing ability of intercrop is dependent upon the component crop selected the genotype used and plant density adopted, proportion of component crops, their spatial arrangement and fertility and moisture status of the soil. The intercropping may reduce weed infestation and growth sill there is need for some weeding in most cases. Normally two-hand weeding are required, however, it may be restricted to one weeding under intercropping. Since the canopy coverage in almost complete, weed growth in minimal after first weeding.The critical period of weed free condition may have to be extended a little longer in intercropping than in sole cropping. Chemical weed control is a great problem in inters cropping than in sole cropping. Most herbicide is crop specific. Hence it is difficult to choose chemical that will control a broad spectrum of a weed without causing damage to the component crop.

c)   **Nutrient Management in cropping system:** Nutrient Management become more complex in intensive cropping because of additional factors such as residual factor of nutrients applied to the previous crops, possible effect of legumes in the system complementary and competitive interference from the component crops and influence of the crop residue left in the soil. The nitrogen should be applied at recommended rates and phases to each crop of the cropping system. Phosphorus management in the cropping system needs careful adjustment of P fertilizer doses taking into account the type of fertilizer soil characteristics and their yield level extent of P removal and growing environment. Removal of K in proportion to N is very high in cropping system. Therefore, it is important to apply K fertilizer at recommended rate to maintain soil fertility status. Among micronutrient Zn deficiency in the most common as nearly 50% soils of intensively cultivated area suffering from Zn defiantly. Application of Zn is important for enhancing the importance of other nutrient ion and productivity level of crops.

(d)   **Disease and pest management:** In seed spice based cropping system, the management of disease and pest is very complex. The research information on this aspect is very less. In order to tackle the problem in seed spice based cropping system, integrated disease and pest management approach should be adopted

Conclusion

Seed spice based intercropping system is a better option for realising higher system productivity and profitability along with ensuring sustainability maintenance of soil fertility and land ecosystem.

References

Anonymous 2009: Annual report, NRCSS, Ajmer

Anonymous 2012: Annual report, NRCSS, Ajmer

Andrews, D.J. and A.H. Kassam 1976.The importance of multiple cropping to increasing world food suppliers. In multiple cropping ed. M. Stelly, ASA Special publication 27, Madison, W.I., pp 1-10.

Ghosh, P.K., Mohanty, M., Bandyopadhyaya, K.K., Painuli, D.K. and Misra, A.K. (2006). Growth competition, yields advantage and economics in soybean + pigeanpea intercropping system in semiarid tropics of India II. Effect of nutrient management. *Field Crops Research*, 96: 90-97.

Okigbo, B.N.1980.The importance of mixed stands in tropical agriculture. In opportunities for increasing crop yield, ed.,R.G. Hurd, P.V., Biscoe, and C. Dennis, London, UK: Pitman, Publishing pp 233-244.

Mehta,R.S, Meena,S.S.and Anwer,M.M.2010. Performance of coriander (*Coriandrum sativum*L) based intercropping system. *Indian Journal of Agronomy* 55(4): 37-41

Mehta, R. S. Malhotra, S.K. Lal, G., Meena, S.S. Singh, R., Aishwath, O.P., Sharma, Y. K. ,Kant, K and Khan, M.A.2012.Influence of intercropping system with varying fertility levels on yield and profitability of fennel (*Foeniculumv ulgare* Mill).*International J of Seed Spices* 1(2):24-27.

Mehta R.S., Meena S.S., Lal G., Singh Ravindra and Maheria, S.P. 2013.Higher system Productivity through seed spice with vegetables. In proceeding of National Conference of Plant Physiology-2013 on "Current Trends in Plant Biology Research" held on 13-16 December 2013 at Directorate of Groundnut Research, Junagadh. PP 864-865

Mehta R.S. Singh B. Meena, S.S, Lal, G.,Singh, R. and Aishwath,O.P.2015.Fennel (*Foeniculum vulgare* Mill.) based intercropping for higher system productivity. *International Journal of Seed Spices* .5 (1):56-62

Mehta, R.S., Singh B., Meena S.S., Singh R., Ranjan J.K., Meena N.K. and Maheria, S.P. 2015.Intercropping of dill (*Anethum sowa* Roxb) with vegetable for enhancing system productivity. In National Seminar on"Hi-tech Horticulture for Enhancing Productivity, Quality and Rural Prosperity" January 19-20, 2015, ICAR-National Research Centre on Seed Spices, Tabiji, Ajmer.

Mehta, R.S, Meena S.S., Singh, B. Lal, G., Singh, R..& Aishwath, O.P.2015. Intercropping of seed spices with vegetables for enhancing system profitability. XII Agricultural Science Congress, ICAR-National Dairy Research Institute, Karnal, Haryana

Rao, M.R.and Singh, M. 1990. Productivity and risk evaluation in contrasting intercropping system. *Field Crop Research,* 23: 279-293.

Singh, S.N. and Kumar, A.2002. Production potential and economics of winter maize based intercropping systems. *Annals of Agricultural Research,* 23 (4): 532-534.

Vashishtha, B.B and Malhotra S.K.2003. Seed Spice Crops:Status and production potential under low rain areas. Abstract in International Symposium for sustainable dry land Agriculture systems held at ICRISAT Sahelian Centre, Niamey, Niger, December, 2-5,2003.

11

Important Horticultural and Fodder Crops for Horti-Pastoral System in Arid Region

Vikas Khandelwal[1], Dheeraj Singh[1], Hansraj Mahla[2], S.P.S. Tanwar[2] A.K. Shukla[1]

[1]ICAR-Central Arid Zone Research Institute, Regional Research Station Pali, Rajasthan-306401
[2]ICAR-Central Arid Zone Research Institute, Jodhpur, Rajasthan-342003

The arid zone eco-system of Rajasthan is very fragile and is prone to serious imbalance even with the slightest disturbance due to mismanagement of resources. Among the eastern margin of the mean annual rainfall is 500 mm, while in the western most part it is 100 mm. The rainfall is monsoon driven. It comes between June and September (in 9-21 short spells). Large coefficient of variation (40-60%) and erratic distribution during the monsoon are characteristics features of the rainfall, leading often to prolong drought and failure of rainfed crops (More and Samadia, 2007). Temperature as low as 4.4° C, and as high as 50° C has been recorded in Western Rajasthan. Farming system and cropping system approach for sustainable use of farm resources and reduced risks have been successfully demonstrated in perennial horticulture. There is great pressure of livestock on available feed and fodder, as land available for fodder production has been decreasing. The requirement of dry and green fodder by 2025 would be 1170 mt and 650 mt while supply is expected around 411.3 mt and 488 mt deficit will be 64.87 % green fodder and 24.92 % dry fodder (Table 1).

Table 1: Scenario of Feed and Fodder Availability and Future Requirement (in million tones)

Year	Supply (In million tones)		Demand (In million tones)		Deficit as % of demand (actual demand)	
	Green	Dry	Green	Dry	Green	Dry
1995	379.3	421	947	526	59.95 (568)	19.95 (105)
2000	384.5	428	988	549	61.10 (604)	21.93 (121)
2005	389.9	443	1025	569	61.96 (635)	22.08 (126)
2010	395.2	451	1061	589	62.76 (666)	23.46 (138)
2015	400.6	466	1097	609	63.50 (696)	23.56 (143)
2020	405.9	473	1134	630	64.21 (728)	24.81 (157)
2025	411.3	488	1170	650	64.87 (759)	24.92 (162)

Source: Draft Report of Working Group on Animal Husbandary and Dairying for Five- Year Plan (2002-2007, Govt. of India, Planning Commission, August-2001).

The pressure is bound to increase in future. Therefore, there is need of vertical growth in terms of productivity, diversification, farming systems to meet the gap between demand and supply. In horti-pastoral system forage are grown in wide inter rows spaces of fruit trees for economic utilization of orchard lands. Therefore, the solution lies in maximizing forage production in space and time, through integration of forage crops in established or grown up plantations as well as establishing orchard in the rainfed (Kumar *et al.*, 2009). Under such situation incorporation of fruit trees along with animal husbandry in common farming system is advisable to improve lively hood, income and nutritional security of the farmers. Therefore, it is need to utilize rainfed areas through horti-pastoral system and husbandry to sustain nutritional security and livelihood improvement. Fruit crops chosen must have flowering and fruit development synchronous with moisture availability, tolerant to salinity in soil and saline irrigation water, deep and strong root system to pierce the hard pan and to extract moisture from deeper layers of soils are the criteria for selecting fruit crops in arid region (Panda, 2014). Fruit crops like ber, pomegranate, aonla, datepalm, tamarind and bael are suitable for arid region.

The yield depends upon type of soil, its fertility and availability of adequate water in time. The yield also depends upon quality of seeds, seed rate, fertilisers used, weeding done and general agroclimatic conditions.

Strategies to increase green fodder production and availability

- Use quality seeds of improved varieties/hybrids of fodder crops
- Follow recommended agronomical practices of cultivation
- Follow suitable crop rotation

- Select short duration fodder crops
- Sow legume as an inter-crop or as a mixed crop with non-legume crop to enhance the nutritional value of fodder and improve soil fertility
- Plant fodder trees/shrubs on farm boundaries to get green fodder during the lean period
- Harvest fodder at the appropriate stage to get the maximum nutrients

A. The important horticultural crops and their cultivars are as follows

Ber (*Zizipus mauritiana*)

Ber (*Zizipus mauritiana* Lamk.) or the *jujube* is the hardiest cultivated fruit tree grown in north Indian plains. It has been truly called as poor man's fruit. Its fruits are eaten fresh as well as dried and processed into delicious candy. The major growing states are Maharashtra, Rajasthan, Gujarat, Haryana, Madhya Pradesh, Andhra Pradesh, Uttar Pradesh and Punjab. Agri-Technique:
- Soil and climate: It can be grown in a wide range of soil. It grows well in deep sandy loam soils with neutral or slightly alkaline reaction. It prefers hot and dry climate for quality fruits and production.
- Nutrition: FYM (30-40 kg), N (600 g), P (250 g) & K (250 g) per plant per year.
- Flowering: Time and duration of flowering varies with cultivar and location. Flowering starts from July and extended upto November in different regions of the country.
- Yield: 50 to 80 kg per tree from 5 to 6 years old tree.
- Important cultivars: Gola, Seb, Umran, Mundia, Banarasi, Kaithli, Goma Kirti, Thar Sevika, Thar Bhubhraj

Aonla (*Emblica officinalis*)

Aonla (*Emblica officinalis*) is a very promising fruit crop for arid region. It is also known as Indian gooseberry. Besides being hardy in nature, it is salt tolerant and can grown successfully in saline soil and irrigated conditions. Agri-Technique:
- Soils and climate: Sandy loam soils are the best suited for its cultivation. Soil should be at least 2 m deep, otherwise, plant will not prolong beyond 12 years in fruiting. Its cultivation is more successful in subtropical regions.
- Nutrition: 15 kg FYM, 180-360 N, 540-1080 P and 180-360 potash g/tree/yr.
- Flowering and fruit set: Flower bud differentiation in Banarasi aonla was observed during the first week of March and flowers started opening from the last week of March.
- Yield: Desi aonla starts bearing fruits after 6 years of plantation, whereas budded or grafted trees start bearing after 3 years. From 10-15 years aged trees, approximately 200 kg fruit yield can be obtained /year/tree.
- Important cultivars: Krishna, Kanchan, Hathijhool, Chakiya, Banarasi, Neelum, Amrit, Anand 2, Laxmi 52, Goma Aishavarya

Pomegranate
(*Punica granatum* L.)

Pomegranate (*Punica garanatum* L.) is one of the important commercial crops with fine table and therapeutic values, extensively cultivated in India. It is commercially grown for its sweet-acidic taste. Its commercial cultivation is feasible in Rajasthan with the introduction of promising pomegranate cultivar such as Jalore Seedless. Agri-Technique:

- Soils and climate: In deep heavy, loamy and well drained soil with pH range of 5.5 – 7.5. The best quality fruits can be produced in areas of cool winters and hot & dry summers where rainfall is low.
- Nutrition: FYM (20 kg), N (600 g), P (250 g) & K (250 g) per plant per year.
- Flowering and fruit set: Ambe bahar (February –March), Mrig-bahar (June-July), Hastha Bahar (September – October). In arid condition Mrig bahar bears better quality fruits.
- Yield: The fully grown up tree of about 10 years old produces 80-120 fruits (16-20 kgs).
- Important cultivars: Dholka, Musket Red, Jodhpur Red, Ganesh, Jalore Seedless, G 137, P 26, P 23, Mridula, Arakta, Rubi Red, Bhagava, Phule Arakta

Bael(*Aegle marmelos* (Linn.) Correa)

Bael (*Aegle marmelos* (Linn.) Correa) is a hardy fruit tree and can be grown successfully in dry areas. The importance of bael fruit lies in its curative properties, which make the tree one of the most useful medicinal plants of India. Agri-Technique:

- Soil and climate: It can be grown in acidic, alkaline and stony soils having pH range from 5-10. It can withstand maxmum temperature of 46.6 °C and 6.6 °C.
- Nutrition: 60 g N, 40 g P and 60 g K and 10 kg FYM for one year plant. This dose should be doubled as the years advance.
- Flowering and Fruit set: The flower buds appear in the month of May and flowers start opening from the middle of May.
- Yield: Commercial bearing starts after 7 to 8 years of plantation in seedling bael trees as compared to 4 to 5 years in budded trees. The yield increases with the age and size of the tree. 10 to 15 years tree gives 200 to 400 fruits/year.
- Important cultivars: CISH 2, NB 9, Pant Urvashi, Pant Sujata

Date palm
(*Phoenix dactylifera*)

Date palm (*Phoenix dactylifera*) is a crop of desert oases. The datepalm is a monocotyledons plant and its stem is unbranched and it does not grow in thickness with age. It is a diocious species, i.e. male (staminate) and female (pistilate) flowers are produced on separate palms. Agri-Technique:

- Soil: Thrives well in light as well as in heavy soils which have at least 2.5 to 3.0 m depth. Deep sandy loam soils are the best.
- Climate: (1) A long summer with high day as well as night temperature. (2) A mild winter (up to a fall of 2.2 C for short periods without frosts). (3) Absence of rain, at the time of flowering and fruit setting with low relative humidity and plenty of sunshine.

- Nutrition: FYM 40-50 kg, N (262 g), P (138 g) & K (540 g) per plant per year. Fruiting: Under optimum management they start bearing at the age of 4 to 5 years. One palm can produce 10-12 bunches. It produces most when it is between 20-50 years old.
- Yield: The yield varies according to age of the tree, cultivar and location. 40 to 50 kg fruits can be harvested from full grown tree.
- Important cultivars: Halawy, Barhee, Khuneji, Medjool, Khadrawi

Tamarind (*Tamarindus indica*)

Tamarind (*Tamarindus indica*) is a leguminous tree in the family Fabaceae indigenous to tropical Africa. It is a a slow-growing, long-lived, massive tree reaches, under favorable conditions, a height of 80 or even 100 ft (24-30 m), and may attain a spread of 40 ft (12 m) and a trunk circumference of 25 ft (7.5 m). Seedling plants start yielding in 8-10 years after planting. Grafts and budded plants start yielding in 4-5 years after planting. Harvesting is done during January-April. Average yield is 26 tones of pods/ha.

Agri-Technique:

- Soil: Deep loamy soils with adequate moisture would be the best for its growth.
- Climate: Tamarind is a hardy tree, which grows well under warm climatic conditions of tropics and subtropics, wherein summers are hot and dry and winters are mild.
- Nutrition: Application of compost/FYM at 40-50 kg/tree /year would suffice the most need of nutrients. N 200 g, P 150 g and K 250 g per plant per year.
- Fruiting: Seedling tree of tamarind comes to bearing in 10-14 years after planting. Flowering starts from April - June and pods ripen from Feb - April.
- Yield: A fully developed tree can give about 200 - 250 kg fruits / year.
- Important cultivars: PKM 1, No. 263, Yogeshwari, Pratisthan, DTH-1, DTH-2

Khejri (*Prosopis cineraria* (Linn.) Druce)

The khejari tree (*Prosopis cineraria*) is referred as the golden tree of Indian deserts, it plays a vital role in preserving the ecosystem of arid and semi-arid areas. Basically, Khejri tree is a symbol of socio-economic development of the arid regions. Since all the parts of the tree are useful, it is called kalp-taru. It is also known as the 'king of desert', and the 'wonder tree'. This multipurpose tree has played an important role in the rural economy of Western Rajasthan as it provides fodder, fuel, vegetable fruits etc. and widely cultivated as agroforestry tree under which the grain yield and forage biomass production of the field crops improved. The pods are eaten as vegetable and values as fodder.

Propagation and planting

Fruit crop	Propagation method	Planting space (m)	Planting time
Aonla	Patch budding during July-August	6x6	July-August
Ber	I-budding during June-August	6x6	July-August
Pomegranate	Hard wood cutting, air layering	5x5	July-August
Bael	Patch budding during May-July.	6x6	July-August
Datepalm	Offshoot/Tissue culture	8x8	July-August
Tamarind	Seeds, budding, cuttings, layering, approach grafting	6 x 6 m (Dwarf varieties) 12 x 12 (Tall varieties)	July –August

B. Important fodder crops and their varieties are as follows

Sorghum (*Sorghum bicolor* L. Moench) Sorghum is the most important cereal fodder crop grown in summer/rainy season. Covering the maximum cultivated area among fodder crops, sorghum is grown in all parts of the country except the cool hilly areas. It has high tolerance to drought and excessive rainfall. To avoid prussic acid or cyanide toxicity to livestock, the crop should be harvested at 50% of flowering or after irrigation at the pre-flowering stage. Agro technique:

- Sowing Time: March April June to August upto Nov.
- Distance: 25 cm
- Seed rate seedlings per Hectare: 20 Kg (By machine), 50 kg Broad Casting
- Fertiliser (Hectare): N-90-120 kg, P-30 kg, K-10 kg
- Harvesting: 1st cut after 70 days. Then each cut after every 45-50 days.
- Yield per Hectare: 1st cutting 25-30 tonnes Then 10-15 tonnes each cut
- Important cultivars: Raj chari-1, Raj chari-2, M P chari, HC-171, PSC-1, CSH-22 MF, CSH-24 MF

Bajra (*Pennisetum glaucum*) Bajra (*Pennisetum glaucum*) is the most widely grown type of millet. It is well adapted to production systems characterized by drought, low soil fertility and high temperature. It performs well in soils with high salinity or low pH. It is an important forage crop of the arid and semi arid regions of the country. Agro technique:

- Sowing Time: Mar-April June-Aug
- Distance: 25 Cm
- Seed rate seedlings per Hectare: 10 kgs
- Fertiliser (Hectare): N 70-120 kg, P 20-30 kg, K 5 kg
- Harvesting: 3 months
- Yield per Hectare: 450 to 550 quintal total
- Important cultivars: Raj Bajra Chari-2, GFB-1, Raj-171, CZP 9802, Pusa-383, Avika Bajra Chari (AVKB-19), CO-2

Maize (*Zea mays*) Maize (*Zea mays*) is one of the best cereal fodder crops grown during summer, rainy and/ or early winter season. It produces rich and nutritious green fodder which is a good source of carbohydrates. It is susceptible to water logging. In early stage up to 35 days after sowing, the crop is drought tolerant. Agro technique:

- Sowing Time: March April June to Aug Oct-Nov
- Distance: 25 cm or 30 to 45 cm
- Seed rate seedlings per Hectare: 45 kg, 70kg Broadcasting
- Fertiliser (Hectare): N 80-120 kg P30-40 kg
- Harvesting: 1st cut after 60-65 days
- Yield per Hectare: 30-40 tonnes
- Important cultivars: Pratap Makka Chari-6, African tall composite, JS-1006, Vijay composite

Napier/Elephant Grass (*Pennisetum purpureum*)

Napier/Elephant Grass (*Pennisetum purpureum*) is supposed to be native of tropical Africa. The name 'Elephant Grass' derives from it being a favorite food of elephants. It is most susceptible to frost. The crop becomes ready for first cutting about three months after its planting and thereafter each subsequent cutting may be taken at about two months interval. However, it depends upon the vegetative growth of the crop. On an average 6-8 cuttings can be taken which gives about 400-600 quintal of green fodder per hectare. Agro technique:

- Sowing Time: March-July June-July
- Distance: 90 x 90 cm or 90 x 60 cm
- Seed rate seedlings per Hectare: 25000
- Fertiliser (Hectare): 250-300 quintal/hectare F.Y.M After cutting 25 kg N
- Harvesting: 7 – 8 weeks 1st cut thereafter 4-6 weeks
- Yield per Hectare: 30-35 tonnes from each cut
- Important cultivars: Pusa Giant Napier, Pusa Napier-1, Pusa Napier-2, Gajraj

Motha Dhaman/Bird wood Grass (*Cenchrus setigerus*)

Motha Dhaman/Bird wood Grass (*Cenchrus setigerus*) is commonly known as, cow sanbur (English), anjan, mode dhaman grass, kata-dhaman, kala-dhaman in India. It is native to Africa (Kenya, Tanzania, Eritrea, Ethopia, Somalia, Sudan, Egypt), Asia (Southern Iran, Yeman, India, Pakistan) occurring in open dry dush and grassland, usually on alkaline soils, sometimes on heavy black clays with impeded drainage. It prefers light-textured, sandy soils, adapted to a wider range of soils than is *Cenchrus ciliaris*. It is adapted to arid and semi arid climates (annual rainfall as low as 200 mm) with a long dry season. Also it is more drought tolerant than *Cenchrus ciliaris*. It can be cut every 30 days at 10 cm. Once established, can stand heavy grazing even by sheep. It is moderately palatable and readily accepted by stock.Important cultivars: CAZTI-76, Marwar Dhaman (CAZRI-175)

Anjan Ghas/Buffel Grass (*Cenchrus ciliaris*)

Cenchrus ciliaris is commonly known as 'Dhaman' in rajasthan and 'Anjan' in other parts of India, is considered as very drought resistant species. It is a native of tropical and subtropical Africa, India and Indonesia. It is widely distributed in hotter and drier parts of India and is found in open bush and grassland in its natural habitat. It is widely distributed in the plains of Rajasthan, Gujarat, Punjab and Western UP extending up to foot hills of Jammu (an altitude 400 mm). It is polymorphic, perennial and warm season bunch grass with extensive native range in the form of various ecotypes and cytotypes.Agro technique:

- Sowing Time: June-July
- Distance: 50 x 30 cm
- Seed rate seedlings per Hectare: 5kg (30,000 seedlings)
- Fertiliser (Hectare): N 60kg P 20kg (for one month crop) & N 40kg P 20kg (subsequent years)
- Yield per Hectare: 9–10 tonnes/ha and DMY 4–5 tonnes/ha
- Important cultivars: Marwar Anjan (CAZRI-75), Bundel Anjn-1 Bundel Anjan-3

Cow pea
(Vigna unguiculata)

Cow pea (*Vigna unguiculata*) legume crop is grown under both irrigated and rainfed conditions. It is widely cultivated across the country excluding the temperate hilly areas. It has a great potential for sustainable agriculture in marginal lands and semi-arid regions of the country. It has great potential as a mixed crop when sown with maize, sorghum and millets to produce an ideal 'legume and cereal' fodder mixture. It is more tolerant to heavy rainfall than any other pulse crop. It suffers from water stagnation and heavy drought. It grows quickly and can yield 25 to 45 tones/hectare of green fodder. It also use as green manure crop. Agro technique:·
Sowing Time: Feb-May June-Aug.

- Distance: 30 to 45 cm
- Seed rate seedlings per hectare: 35 to 40 kg, 50 kg Broadcasting
- Fertiliser (hectare): N 15 kg, P 90 kg, K 30 kg
- Harvesting: 1st cut 40-50 days their after 30-35 days
- Yield per Hectare: 2 to 3 cuttings 35 tonnes
- Important cultivars: Cowpea-74, HFC-42-1 (Hara lobia), EC-4216, GFC-1, GFC-2, GFC-4, Type-21, UPC-5286, Sweta (No. 998), Charodi, UPC-625

Oats (*Avena sativa*)

Oats (*Avena sativa*) is an important grass fodder of winter season. This crop is well suited to limited water availability situations. It has excellent growth and shows quick regeneration capacity after cutting. The green fodder is succulent, rich in carbohydrates and palatable and contains 10-12% protein and 30-35 % dry matter. The yield ranges from 30 to 50 tones /hectare. Agro technique:

- Sowing Time: Oct-Dec
- Distance: 25 cm
- Seed rate seedlings per Hectare: 100 kgs
- Fertiliser (hectare): N-90-120 kg P-30 kg K – 15 kg
- Harvesting: 60-65 days first cutting and second after about 90 days of sowing.
- Yield per Hectare: 450 to 500 quintal
- Important cultivars: Kent, HFO-114, UPO-94, UPO-212, Weston-11, JHO-822

Berseem or Egyptian clover
(Trifolium alexandrum)

Berseem or Egyptian clover (*Trifolium alexandrum*) is a legume crop of the winter season grown mainly in Bihar, Haryana, Madhya Pradesh, Punjab, Rajasthan and Uttar Pradesh. It provides a nutritious and succulent fodder during the winter/early summer season. It is rightly called as 'king of fodders'. It has 20% protein and 70% dry matter digestibility. It gives six to seven cuts between November to May and produces 70 to 80 tones /hectare of extremely palatable and nutritious green fodder. The fodder of Berseem is known as 'milk multiplier'. Being a leguminous crop it also fixes atmospheric nitrogen in the soil and improves soil fertility. Agro technique:

- Sowing Time: Oct-Dec
- Distance: 20 cm
- Seed rate seedlings per Hectare: 30 kgs
- Fertiliser (Hectare): 15-20 tonnes well decomposed organic manure. N- 15 kg P-120 kg K – 40 kg
- Harvesting: 40-45 after sowing, subsequent cutting thereafter done at an interval of 20-25 days.3-4 cuttings/yr
- Yield per Hectare: 600 to 800 quintal
- Important cultivars:Diploid- Mescavi, Fahali, Wardan, BL-1, BL-10, BL-22, BL-30, BL-42, JB-3, IGFRI-S-99-1, UPB-104, UPB-1905, and KhdrabiTetraploid: Pusa Giant, T-526, T-529, T-724, T-674, T-730

Lucerne/Alfaalfa (*Medicago sativa*)

Lucerne/Alfaalfa (*Medicago sativa*) is also known as the 'queen of fodder' and it is the most popular fodder crop in the country after berseem and sorghum. The crop can be grown both in irrigated and rainfed areas. The crop can give seven to eight cuts from November to June with an average green fodder yield of 60 to 80 tones /hectare. The green fodder containing about 15-20% protein, has a digestibility of about 72%. In some areas, it is cultivated as a perennial crop. Agro technique:

- Sowing Time: Oct-Dec
- Distance: 20 cm in line
- Seed rate seedlings per hectare: 25 to 30 kg
- Fertiliser (hectare): N-15 kg P-160-200 kg K – 40-80 kg
- Harvesting: 45-65 days
- Yield per Hectare: 100 to 600 quintal
- Important cultivars: Sirsa-8, Sirsa-9, NDRI Selection 1, Anand-2 (GAUL-1), Co-1, LL Composite-3, LL Composite-5, Chetak (S-244), T-9, GAUL-2 (SS-627), Lucern no. 9-L, Anand-3, RL-88, Anand Lucern-3 (AL-3).

References

Kumar, S., Satyapria, Singh H.V. and Singh K.A. 2009. Horti-pasture for nutritional security and economic stability in rainfed area. *Progressive Horticulture,* 41 (2): 187-195.

More, T.A. and Samadia, D.K. 2007. Prospects of horticulture in arid zone. In Dryland Ecosystems: Indian Perspective, CAZRI and AFRI, Jodhpur. Pp. 149-167.

Panda, S.C. 2014. Integrated farming systems. In Forage Crops and Grasses. pp 126. Agrobios (India) Publication, Jodhpur.

12

Henna - A Potential Crop for Hot Arid and Semi-Arid Region

M.B. Noor mohamed[1*], Shukla, A.K.[1], Keerthika, A.[1], Dipak Kumar Gupta[1], P.L. Regar[1] and P.K. Roy[2]

[1]ICAR-Central Arid Zone Research Institute, Regional Research Station-306401 Pali, Rajasthan
[2]ICAR-Central Arid Zone Research Institute, Jodhpur, Rajasthan-342003

Henna or Mehandi (*Lawsonia inermis* L.) a perennial shrub belongs to family Lythraceae. It is commonly cultivated as a commercial dye crop possess natural dyeing properties and are used for hair dyeing and for staining of palm, feet and other body parts since times immemorial. The plant finds their importance due to presence of orange red dye in its leaves and essential oil in its flower. Oil extracted from leaves and flower called 'Otto of henna' which is utilized as perfume (Jaimini *et al.*, 2005). It can grow well in harsh climate of arid and semiarid regions. In India, over 90 percent of the henna production comes from Sojat region of Pali district of Rajasthan and Sojat is only center for its processing and trading in India. In India it occupies about 40,000 ha area out of which 35,000 ha alone is in Pali district (Sojat and adjoining tehsils).

Geographical distribution

It occurs naturally in North Africa mainly in countries of Algeria, Cyprus, Egypt, Eritrea, Ethiopia, Indonesia, Iraq, Iran, Jordan, Kenya, Kuwait, Lebanon, Libya, Malaysia, Morocco, Oman, Philippines, Qatar, Saudi Arabia, Syrian Arab Republic, Tanzania, Tunisia, Turkey, Sahara and Yemen. It is naturally occurring in many of these countries along the water courses and semi-arid regions where it can with stand dryness as well as drought. It has been introduced into countries like Australia, Benin, Burkina Faso, Cameroon, Central Africa, Chad, Congo, China, Gabon, Gambia, Ghana, Guinea, India, Liberia, Mali, Mauritania, Niger, Nigeria, Pakistan, Senegal, Sierra Leone, Spain, Sudan, Togo and Zanzibar (Suresh Kumar *et al.*, 2005).

Distribution in India

In India, it was once grown extensively in Punjab and Gujarat. However, gradually the major cultivation has shifted to the arid fringes of Rajasthan, with Pali district having the maximum area under henna. It is taken as an annual ratoon crop under rainfed conditions. Besides this, leaf cuttings are also collected from existing farms and field boundary plantations in Gujarat and Madhya Pradesh. In Uttar Pradesh henna is grown for its scented flowers used in the essence industry. Kanauji, Lucknow, and Varanasi in Uttar Pradesh are reputed centers for the commercial extraction of mehndi oil from flowers and its use in traditional essence. Since henna is grown as a dryland crop under limited moisture conditions greater use of available moisture by the plants is possible.

Henna is commonly propagated in some parts of Rajasthan and Gujarat by using cuttings for hedges and by seed for commercial plantations. Its cultivation is concentrated in Sojat and surrounding areas in Pali district in Rajasthan (Kavia *et al.,* 2004). In Pali district, henna is mainly cultivated as a rainfed crop and cultivation area is concentrated around the ancient town Sojat city. Sojat and Marwar Junction tehsils of Pali district constitute around 95 percent area under henna cultivation (Khem Chand *et al.,* 2002), the remaining part is constituted by other tehsils of district and adjacent tehsils of Jodhpur and Naguar.

Botanical description

Lawsonia inermis is tall shrub to small tree, 2-6 m high. It is glabrous, multibranched with spiny tipped branchlets which are green when young but turn red with age. Bark greyish-brown, unarmed when young, older plants with spine-tipped branchlets. Young branches quadrangular, green but turn red with age. Leaves opposite, entire, subsessile, elliptic to broadly lanceolate, 1.5-5 x 0.5-2 cm, glabrous, acuminate; veins on the upper surface depressed.

Floral biology and phenology

Flowers small, white, numerous; in large pyramidal terminal cymes, fragrant, 1 cm across, 4 petals crumpled in the bud. Calyx with 2-mm tube and 3-mm spread lobes; petals orbicular to obovate, white or red; stamens 8, inserted in pairs on the rim of the calyx tube; ovary 4 celled, style up to 5 mm long, erect. Number of days required to complete anthesis is 9 to 25 days with a mean of 16.25 days. Flowering started in July and continue till mid-September in different plants. Flowering was asynchronous (flowers were developing at different times). The length of individual inflorescence is 2.40 cm to 9 cm. Number of flowers per individual inflorescence had a range of 27 to 142. Percentage of viable pollen grains is ranged from 62.5 to 99.1 percent at the time of full bloom (Manjit Singh *et al.,* 2005).

Fruits are small, brown, globose capsules 4-8 mm in diameter, many-seeded, opening irregularly, split into 4 sections, with a persistent style. Berries are green when young and later it turns blueish black in colour. Fruits are non-edible in nature. Seeds 3 mm across, angular, with thick seed coat.

Types and Variability of Henna

There are two botanical varieties of henna viz., *alba* with pale or light-yellow petals, and *rubra* with rose (red) petals and light green sepals. However, in semi-arid regions of Rajasthan the yellow flowered henna is cultivated as dye crop perhaps due to the better expression of leaf dye under such environment. There are two distinct phenotypic variants or ecotypes have been observed among the yellow flowered henna in Rajasthan. They are locally called *desi* and *muraliya* (also *muraili, mureli, mooli*) types.

Muraliya type: It has woody canopy with small leaves of distinctly greyish green and hard pointed branchlets. It regenerates into tall, erect and lax canopied plants having conspicuous terminal clusters of small branchlets, small and thick greyish green leaves. The muraliya plants bears late flowering after several years.

Desi type: Generally desi type plants are favored for the large-scale cultivation. Because it has big leaves, less woody stem and easy to harvest compared to muraliya types. It has higher leaf yield potential and being easier to handle due to the absence of pointed thorn-like branchlets.

Ecological requirement

Henna is a hardy shrub capable of growing under diverse soil and climatic conditions. It prefers well drained sandy to sandy clay loam soils in younger and older alluvial plains. It thrives well in deep, fine sandy or medium textured well drained soil and are neutral to moderately alkaline (pH 7.7 to 9.9) is considered best for henna cultivation. The henna tolerates moderate salinity of 8 to 12 dS cm^{-1} in subsoil. The henna could be successfully cultivated in saline soil and its quality of leaf also better in saline soils. Although the plant thrives under arid to tropical and warm temperate climatic conditions, it needs moderate rainfall and a generally hot, dry and open climate for good harvest of quality leaves. In Rajasthan, most of the henna growing area receives around 400 mm of annual rainfall and gives an average productivity of 1.0t ha^{-1} dry leaves.

As a dye crop it requires hot, dry and sunny weather condition with RH <50% and also it thrives well in moderate temperature of about 30-350C during active growth period for higher dye content. The distribution of rainfall is important since the crop requires intervening dry sunny periods along with cool nights for proper ripening and expression of high dye content of leaves.

Propagation Techniques

Propagation by Seed

The seeds have a xeromorphic structure. Henna seeds have hard seed coat and it takes more time to germinate with less than 20% germination. Presowing treatment is prerequisite before sowing to break hard seed coat. Seeds are sown in nursery beds during March when temperature is optimal for germination (25°C -30°C). They are first soaked in water for 7 -10 days, with frequent changing of water that gives 20% germination. Soaking of seeds in 3% salt solution for 24 hours results in 60-70% germination. Nursery beds are prepared about 30m² size and 1 kg of henna seeds mixed with fine sand are broadcasted. About 10 kg of seed is needed to raise seedlings for transplanting 1 hectare. Seed germination of henna up to one month after sowing ranged 44 percent to 100 percent with mean of 73.6 percent.

Propagation by cuttings

In general, seed germination is not difficult to raise nursery, even though initial soaking, proper sowing depth, maintenance of moisture during initial germination stages are some of the critical points. Henna is easy to root species and vegetative propagation is also not a problem. Generally, plant growth hormones are used to stimulate the rooting in henna cuttings viz., IBA, IAA. The cuttings collected from henna plants with three to four nodes and 10 to 15 cm length are suitable. It should be soaked with fungicide for 10 minutes. After that the cuttings are dipped into either IAA or IBA solution in 3000ppm or 5000ppm. The treated cuttings are planted in nursery bed. Best time for vegetative propagation in henna is July -September.

Transplanting

Transplantation is carried out immediately after monsoon rainfall in the month of July and August. Before planting, land should be prepared by one or two deep ploughing using mould board or disc plough followed by disc harrowing and planking. Before levelling, one final ploughing with country plough is considered desirable. The seedlings, at least 3-4-month-old, after uprooting from the nursery beds are cut back to leave about 10 cm of both main stem and tap-root, and the root portion drenched with suitable pesticide against termite attack. Thereafter, they are placed in matching peg holes made in the field. Rains immediately after the field transplantation lead to better seedling establishment.

The seedlings are transplanted at a spacing of 30 x 30 cm. However, trials at Central Arid Zone Research Institute showed that a spacing of 45 x 30 cm gave higher dry leaf yield followed by 60 x 30 cm and mechanical intercultural operation for hoeing and weeding possible at this spacing. Henna can be planted under agro-forestry system (Table 1).

Table 1: Effect of crop spacing on leaf yield (kg ha^{-1}) of henna under rainfed conditions at Pali Marwar

Spacing	1999 (354 mm)	2000 (245 mm)	2001 (500 mm)	Pooled
30 X30 cm	401.2	403.1	1242.3	6872.2
45 X 15 cm	295.4	386.7	1118.8	600.3
45 X 30 cm	509.6	803.4	1399.7	904.2
60 X 15 cm	338.9	376.2	1138.2	617.8
60 X30 cm	426.2	778.1	1398.2	867.5

(Rao *et al.*, 2003)

Nutrition and Fertilizer management

Very little or no fertilizer is being used for henna cultivation in our country. But there was a better response of henna crop to nutrition in trails conducted at Central Arid Zone Research Institute, RRS, Pali. Application of farmyard manure at the rate of 5 t ha^{-1} proved beneficial in terms of seedling establishment and dry leaf yield as well as quality (Table 2). It should be mixed well in soil during field preparation before transplanting. It is essential to apply nitrogen and phosphorous. Nitrogen fertilization stimulates tiller growth and leaf development and phosphorous plays important role in formation of roots, their proliferation and improvement in their functional ability. Application of 60 kg ha^{-1} N has recorded higher yields and is on par with 90 kg ha^{-1} in arid regions (Table 3). The input levels may vary enormously in different locations depending upon soil type.

Table 2: Effect of Farmyard manure on dry leaf yield (kg ha^{-1}) of henna under rainfed condition at Pali-Marwar

Farm yard manure	Leaf yield (kg ha^{-1})
No FYM	970
FYM 5t ha-1	1099

(Rao *et al.*, 2004)

Table 3: Effect of nitrogen levels on leaf yield (kg ha^{-1}) of henna under rainfed conditions at Pali-Marwar

Nitrogen (kg ha^{-1})	1999 (354 mm)	2000 (245 mm)	2001 (500 mm)	Pooled
0	382.2	510.9	1145.7	679.6
30	395.8	543.7	1226.7	722.1
60	390.4	567.9	1326.3	761.6
90	408.6	575.6	1339.3	774.5

(Rao *et al.*, 2004)

Irrigation and Intercultural Operations

Henna is usually grown as a rainfed crop during Kharif. Even though henna is rainfed-crop irrigation is beneficial in absence of adequate rainfall or prolonged drought spell during rainy season. Second harvest also possible where irrigation facilities are available and the crop is usually irrigated at an interval of 15-20 days leading 2-3 times higher leaf yield. Before the crop reaches physiological maturity, it is left unirrigated for about a fortnight to achieve high dye content in the leaves. However, frequent irrigation is believed to decrease the leaf quality by lowering dye content.

At least one hoeing-cum-weeding is required after a month of transplantation or growth in ratoon crop. One weeding is necessary during rainy season. Preemergence application of Atrazine @ 1.5 kg a.i. ha^{-1} will control most of the annual grasses and broad-leaved weeds in the henna plantation without any adverse effect on henna leaf yield.

Crop protection

There are no major diseases associated with henna in the arid regions. Termite is the major pest of henna in the field. It can be controlled by soil application of Chlorpyriphos (10%) or Furadon 10 G 25 kg ha^{-1} during field preparation and by dusting of standing crop with chlorpyriphos or furadon. Under prolonged moist and cloudy conditions there may be high incidence of semi loopers that can be controlled by foliar spray of Quinalphos 30 EC at 1.25 litre ha^{-1}.

Harvesting and Processing

The branches of the henna are cut near the base (6-8 inches). The cutting is done mainly by women who use leather gloves in the left hand and sickle (non-serrated half circle) in the right hand. Most of the farmers harvest in September-October. After harvesting, the branches with leaves are collected on any clean place near the crop field for sun drying. These are turned upside down daily for 3-5 days using wooden implement 'Bewla" (having about 6 ft. handle and a fork in the front). Thereafter these are thrashed using the same implement. The general view was that if the plants are harvested early, the quality of leaves will not be up to the mark. The time of harvest is most crucial because if farmer delays harvesting, he may lose produce in the form of shedding of leaves. And also, there is a risk of spoilage of harvested leaves in the field due to even a very little shower of rain or moist (humid) weather, if untimely or early harvesting occurs. After harvesting of the crop should be dried properly, otherwise the leaves loses its colour and there will be fast deteriotion of quality.

Quality parameters of Henna

The quality of henna powder is determined by its colour, purity, its dyeing property and fineness (BIS, 1985). The powder colour of henna is based on the colour of the dried leaves and it varies from olive green to brown colour (Table 4). In case of leaves getting moist while drying the leaves turn on a dark rusty brown colour due to the leaching out of the dye and it produce brown coloured powder. And also, the processing technique may affect the powder colour. Higher temperature during the grinding process will lead to increase in the browning of powder and loss of pigments presumly due to more of oxidation of pigments. Generally, consumers are preferring more for green colour because this colour is sensed as proof of the herbal origin, freshness and purity of the powder.

Table 4: Effect of harvesting time of henna on powder colour and its dye content

Month of harvest and curing	Powder colour	Lawsone content (%)
June	Rusty brown	2.38
July	Greenish brown	2.68
August	Brown	2.75
September	Yellowish brown	2.60
October	Green	2.82

(Roy et al., 2005)

Yield of Henna

Dried leaves of henna are stored in gunny bags in a dry place for marketing. For storage and transport big bags prepared by stitching two gunny bags are commonly used by the farmers and these can accommodate about 40kg leaves. The average dry leaf yield is about one ton per hectare under rainfed condition of Pali region. Higher yield up to 2.5 t per hectare is obtained under irrigation. A separate regulated market for exclusive trading of dry henna leaves is functioning in Sojat city. Henna farmers generally sell their produce in this market and get remunerative price varying from about Rs.20 to 30 per kg. Trading also takes place at the Bhawani mandi of Kota district in Rajasthan.

Economic Uses of Henna

Leaves are used as a natural dye to colour skin, nails, hand, body and hairs and also getting popular for tattooing in all over world. Henna is used as drug for medicinal as well as veterinary purposes (Ghosh, 1999). The henna roots are used in Jaundice and enlargement of spleen. The leaves are externally applied for headache and rubbed over the soles of feet in the burning of the feet. The essential oil is used to keep the body cool. The leaves are used as a prophylactic against skin diseases, jaundice, leprosy and small pox (Kirtikar and Basu, 1981).

The dye is extensively used in silk and wool industries. Its flower is used to make scent. Scent manufacturing is very common in India at Kanauji (Uttar Pradesh) and Ujjain (Madhya Pradesh).

Prospects of Henna

Henna cultivation is profitable under low rainfall conditions and give assured income returns at low cost investment in drought prone arid and semi-arid regions. Due to its drought hardiness, it can be cultivated on land that are drought prone, marginal or unsuitable for arable cropping. Due to its deep root system and perennial nature, it reduces the soil erosion in arid regions. Economic production of leaves starts from the third year onwards that continues for the next 15-30 years. Low use of input and maintenance, easy to regenerate, high rate of success in establishment, no or very little usage of fertilizers, less or no pest and disease attack etc., leads to adequate benefits for promoting the henna cultivation for the economic and ecological development of hot dry tracts of the country.

Future thrust

There are no released varieties of henna at present and only populations raised from seeds are existing at farmers field in India. After the harvesting of henna, there is no use of crop residues including branches and stem. There is a need of proper marketing channel and processing techniques to improve the henna cultivation. The rain during or just after the harvesting deteriote the quality of leaves and reduces the market value of leaves. So proper storage techniques should be evolved. The cost of henna harvesting and weeding is very high due to the higher wages of skilled labour. So that, mechanization is necessary to harvest henna for large and marginal plantations.

References

Ghosh, A. 1999. Herbal veterinary medicine from the tribal areas of Bankura district, West Bengal. *Journal of Economic and Taxonomic Botany*, 23: 557-560.

Jaimini, S.S., Tikka, S.B.S., Prajapati, N.N. and Vyas, S.P. 2005. Present status and scope of henna cultivation in Gujarat. In: Henna: Cultivation, Improvement and Trade, Eds. Manjit Singh, Y.V. Singh, S. K. Jindal and P. Narain, Jodhpur: Central Arid Zone Research Institute. Pp. 5-7.

Kavia, Z.D., Verma, S.K. and Singh, H. 2004. Traditional methods of cultivation of Henna (in hindi). *Bhartiya Krishi Anusandhan Patrika*, 19: 168-174.

Khemchand, Jangid, B.L. and Gajja, B.L. 2002. Economics of Henna cultivation in semi-arid Rajasthan. *Annals of Arid Zone*, 41(2): 175-181.

Kirtikar, K.R. and Basu, B.D. 1981. Lythraceae. In: K.R. Kirtikar and B.D. Basu (Eds.). Indian medicinal plants. Vol.II, International book distributors, Raipur Road, Dehradun, India, Pp. 1076-1080.

Manjit Singh, Jindal, S.K. and Deepa Singh. 2005. Natural variability, propagation, phenology and reproductive biology of Henna. In: Henna: Cultivation, Improvement and Trade, Eds.

Manjit Singh, Y.V. Singh, S. K. Jindal and P. Narain, Jodhpur: Central Arid Zone Research Institute. Pp. 13-18.

Rao, S.S., Roy, P.K. and Regar, P.L. 2003. Effect of crop geometry and nitrogen on Henna (*Lawsonia inermis* L.) leaf production in arid fringes. *Indian Journal of Agricultural Sciences*, 13(5): 283-285.

Rao, S.S., Regar, P.L. and Roy, P.K. 2004. Crop weather interaction on henna (*Lawsonia inermis* L.) leaf production in arid fringes). Abstracts: In National symposium on geo informatics applications for sustainable development. February 17-19, 2004, WTC, IARI, New Delhi.

Roy, P.K., Manjit Singh and Pratibha Tewari. 2005. Composition of Henna powder, quality parameters and changing trends in its usage. In: Henna: Cultivation, Improvement and Trade, Eds. Manjit Singh, Y.V. Singh, S. K. Jindal and P. Narain, Jodhpur: Central Arid Zone Research Institute. Pp. 39-43.

Suresh Kumar, Singh, Y.V. and Manjit Singh. 2005 Agri-history, uses, Ecology and distribution of Henna (Lawsonia inermis L. syn. alba Lam.). In: Henna: Cultivation, Improvement and Trade, Eds. Manjit Singh, Y.V. Singh, S. K. Jindal and P. Narain, Jodhpur: Central Arid Zone Research Institute. Pp. 11-18.

13

Henna Based Production System for Arid Region

P.L.Regar, B.L.Jangid, S.P.S. Tanwar, Keerthika A.
M.B. Noor mohamed, A.K. Shukla and D.K. Gupta

ICAR-Central Arid Zone Research Institute
Regional Research Station, Pali-Marwar, Rajasthan-306 401

Distribution

Several countries grow henna on a commercial scale for its dye-bearing leaves, notably India, Egypt, Iran, Niger, Sudan and Pakistan. In India, it was once grown extensively in Punjab and Gujarat. However, gradually the major cultivation has shifted to the arid fringes of Rajasthan, with Pali district having the maximum area under henna. It is taken as an annual Ratoon crop under rainfed conditions. Presently over 39,000 ha of henna plantations in the district with nearly 70 per cent occurring in the Sojat area produce around 20,000 t dry leaves worth Rs. 40 crores. Besides this, leaf cuttings are also collected from existing farms and field boundary plantations in Gujarat and Madhya Pradesh. In Uttar Pradesh henna is grown for its scented flowers used in the essence industry. Kannauj, Lucknow, and Varanasi in Uttar Pradesh are reputed centres for the commercial extraction of mehndi oil from flowers and its use in traditional essence. Since henna is grown as a dryland crop under limited moisture conditions greater use of available moisture by the plants is possible.

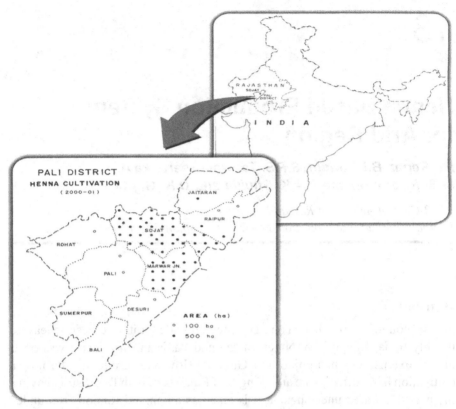

Fig. 1: Distribution of henna growing area in Rajasthan

Plant Type

There are two botanical varieties of henna viz., *alba* with pale or light yellow petals, and *rubra* with rose (red) petals and light green sepals. However, in semi-arid regions of Rajasthan the yellow flowered henna is cultivated as dye crop perhaps due to the better expression of leaf dye under such environment.

Among the yellow flowered henna in Rajasthan, two distinct phenotypic variants or ecotypes have been observed. They are locally called *desi* and *muraliya* (also *muraili, mureli, mooli*) types. The *muraliya* type regenerate into tall, erect and lax canopied plants having conspicuous terminal clusters of small branchlets, small and thick greyish green leaves. Comparatively, *desi* type is favoured for cultivation as it is leafier and easy to harvest compared to the woody and thorny shrubs of *muraliya* type. Plant height range from about 30 cm to over 150 cm and form canopies ranging from spreading dense bush to erect lax open canopy in the field. Both bear pale yellow coloured flowers.

Fig. 2: Botanical varieties of henna: *alba* and *rubra*

Soil and Climatic requirement

Being a hardy shrub plant, henna can withstand diverse soil conditions from sand to wastelands. However, medium textured soils that are retentive of moisture in subsoil and are neutral to moderately alkaline are considered better suited for henna cultivation. In Pali district of Rajasthan, the henna fields have sandy loam to clay loam 40-60 cm deep soils with weakly calcareous substratum and soil pH ranging from 7.7 to 9.9 in the profile. It is also seen that this crop tolerates moderate salinity of 8 to 12 dS cm^{-1} in subsoil.

Climate seems to play a dominant role in the production of quality leaf crop. Although the plant thrives under arid to tropical and warm temperate climatic conditions, it needs moderate rainfall and a generally hot, dry and open climate for good harvest of quality leaves. Such conditions commonly occur in the semi-arid or dry subtropical regions. In Rajasthan, most of the henna growing area receives around 400 mm of annual rainfall and gives an average productivity of 1.0t ha^{-1} dry leaves. The distribution of rainfall is important since the crop requires intervening dry sunny periods along with cool nights for proper ripening and expression of high dye content of leaves.

Propagation and Nursery Practices

It is propagated by both reproductive and vegetative methods. Seedlings are transplanted for commercial cultivation due to their higher survival rate in field but cuttings are used for growing hedges. At least one lakh nursery raised seedlings are needed for transplanting one hectare.

Field Preparation and Transplantation

Land selected for crop establishment should be prepared well by one deep ploughing during summer using mould board or disc plough followed by cross-cultivation with disc harrow or cultivator. Finally, good planking for proper levelling and line marking with tractor drawn cultivator according to selected line spacing is necessary of the field.

Transplantation is carried out immediately after monsoon rainfall in the month of July and August. The seedlings, at least 3-4 month old, after uprooting from the nursery beds are cut back to leave about 10 cm of both main stem and tap-root, and the root portion drenched with suitable pesticide against termite attack. Thereafter, they are placed in matching peg holes made in the field. Rains immediately after the field transplantation lead to better seedling establishment.

The seedlings are transplanted at a spacing of 30 x 30 cm. However, trials at Central Arid Zone Research Institute showed that a spacing of 45 x 30 cm gave higher dry leaf yield and mechanical intercultural operation for hoeing and weeding possible at this spacing. Henna can be planted under agro-forestry system.

Manure and Fertilizer

Henna responds to the application of manure and fertilizer. Farm yard manure at the rate of 10-20 t ha^{-1} is mixed well in soil during field preparation in May-June, and 60 kg P_2O_5 and 60 kg N per hectare applied at planting. DAP is prefer over the use of urea fertilizer by farmers for meeting the N requirement. Thereafter, every subsequent year similar quantity of N in two equal split doses, once after 40 days of growth and then a month later, are given to increase leaf yield from the ratoon crop. Henna also shows response to the application of secondary and micro-nutrient or even growth hormones. However, nutrient x environment interactions play significant role in the overall response of the crop to applied nutrients.

Weeding and hoeing/Inter-culture

In the main field one hoeing-cum-weeding after a month of transplantation/ rains helps to check weed growth, during weeding damage to fresh transplants must be avoided. However, in subsequent years of re-growth bullock or tractor drawn implements are employed for inter-cultural operations to control weeds and to conserve soil moisture.

Fig. 3: Intercultural operation in Henna

In-situ water harvesting and Planting Configuration

Under rainfed farming conditions, storage of rainfall in the potential root zone either through water harvesting or timely weed control to avoid competition for moisture play key role. Planting configurations and in-situ moisture conservation practices has an important role in optimising the production of henna. Dry leaf yield was optimum at 45 X 30cm spacing as compared to other spacing and farmers practice.

Inter-row water harvesting in this planting configuration brought about 27 % increases over farmers practice. The significance of compressed planting geometry into water harvesting (ITrWH) for minimizing evaporation losses through better canopy spread on plating zone had been reflected by obtaining 12.8 % higher yield over no water harvesting. Field bund prevents runoff in the event of heavy and intense rainfall events and increase soil moisture storage in the plant root zone. Bunds tops stacked with harvested henna woody left over acts good fencing.

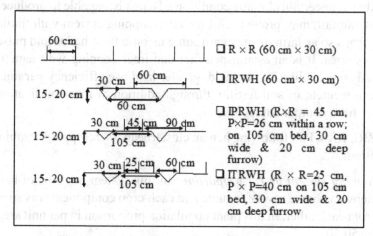

Fig. 4: Tractor operated water harvesting systems

Fig. 5: In-situ water harvesting in henna planted field

Henna- legume intercropping system

In Rajasthan, henna is grown mostly as pure crop under rainfed conditions. Generally farmers grow henna without any kind of fertilizer application in the region. Growing of crops in mixed stands is regarded as more productive then growing them separately for better use of land and to avoid the risk of crop failure due to uncertain climatic conditions. Henna being able to produce even with low rainfall, may prove useful for intercropping system with traditional legume crops so as farmers may earn some income from henna and pulses for their daily uses. It is an established fact that intercropping with legume has certain advantages like higher yield, greater land use efficiency per unit land area, improvement of soil fertility through addition of nitrogen fixation and excretion from component legume.

At CAZRI, RRS, Pali following henna: clusterbean intercropping combination were tested:

Intercropping system in 1:1 proportion: In this system one row of henna is to be altered with one row of legume and each crop component was spaced at 60 cm from each other. In this plant population proportion in per unit area land would be 50:50.

Intercrop system in 1:2 proportions: Two rows of legumes at 40 cm distance are to be planted in between two rows of henna (spaced at 1.20 meter). In this, population of legume would be twice to that of henna on same piece of land.

Intercrop system in 1:3 proportions: In this system three rows of legumes in between two rows of henna (1.20 meter) are incorporated. Legume populations would be thrice to that of henna.

Intercrop system 1:5 proportions: growing legume components with henna in 5:1 proportion, where in henna is spaced at 3.6 meter and five rows of legume planted in three meter space. The proportion, legume to henna would be 83:17 in this system. Each row whether of henna or legume are spaced at 60 cm from the other one.

Strip-crop system: A 2.4-meter wide strip of henna is alternated with 2.4-meter wide strip of legume. In this system population percentage proportion would be in a ratio of 50:50. Both the components are spaced at 60 cm apart so as to facilitate intercultural with tractor to control weeds and conserve soil moisture. This system may have an advantage over replacement series described about that crops may not compete for light interception as both are raised on separate strips. Secondly both crops may have different zone for soil moisture and nutrient extraction. However, the advantage of legume in fixing nitrogen may not be there as are with replacement series of inter-crop system described above.

Alley crop system in 2:4 proportions: In this system two rows of henna planted in isosceles configuration in 60 cm rows at 3-meter distance. Four rows of legume were arranged in the open strip. Legume and henna would be in the proportion of 66:33 in this system. Henna would be maintained in the form hedge from third year onwards after plantation.

Alley crop system in 2:10 proportion: Paired row of henna in isosceles configuration at 60 cm are planted at 6-meter distance. The 6-meter space was used to plant 10 rows of legume component. Population of legume and henna would be in 10.2 proportions in this system, which means population of legume and henna would be in 83:17 proportions. The photograph plates 1- 5 depict the view of different intercrop systems of henna.

Criteria for inter cropping

Relative Yield Total (RYT): The sum of the relative yields of component crops is called Relative Yield Total and is denoted by RYT. The mixture yield of a component crop expressed as a proportion of its yield as a sole crop from the same replacement series is the relative yield of the crop. However, Mead and

Riley (1981) have argued against its use because the objectives of intercropping are essentially agronomic, that is to find the best ways of growing crops together. When the RYT is equal to or less than 1, there is a no advantage to intercropping.

In case of henna dry leaves are major component for marketing, hence, the yields of component crops are to be transformed into relative henna dry leaf value.

Land Equivalent Ratio (LER): LER is defined as the total land area required under sole cropping to give the yields obtained in the intercropping mixture. Land Equivalent Ratio (LER) is an index of combined yield for evaluating the effectiveness of all forms of intercropping. LER is the sum of the two partial land equivalent ratios so that

$$LER = L_i + L_j$$

When LER = 1, there is no advantages to inter cropping in comparison with sole cropping. When LER >1, a large area of land is needed to produce the same yield of sole crop of each component than with an intercropping mixture. For example, when LER = 1.25, 25% more land is needed to produce the same yield form the components as sole crop.

Factors influencing intercropping system

a. *Light factor*: the rate of dry matter production in crops depends on the efficiency of interception of photosynthetic active radiation (PAR).

b. *Importance of water*: water is a most important soil factor in arid/ semi-arid and sub tropical regions for any intercropping system because inadequate rainfall may frequently limit crop production. Depth of rooting, lateral root spread and root-density are factors that affect competition for water between component crops.

c. *Role of nutrients*: Balance nutrition of component crop is most important factor. However, farmers do not apply any fertilizer to henna and legume in this region. But these crops may compete for N, P and K and other micronutrients. Since henna is perennial crop it may exhaust these nutrients in long course of time. So balance application of nutrients is essential to attain the sustainable production of any intercropping system.

Table 1: Performance of Henna: cluster bean inter-cropping system (2004) Rainfall 325.9mm

Treatments	Henna DLY (kg /ha)	Guar yield (kg /ha)	Henna leaf equivalent yield (kg /ha)	Net profit (Rs. ha⁻¹)
H: C (1:1)	316	31	342	-519
H: C (1:2)	310	63	363	-88
H: C (1:3)	290	71	348	-433
H: C (1:5)	40	80	107	-3447
Alley 3m	93	58	143	-3007
Alley 6m	47	75	110	-3371
Strip 2.4m	360	60	411	2922
Sole henna	517	-	517	5142
Sole crop	-	122	95	-4429

Table 2: Performance of Henna: cluster bean inter-cropping system (2005) Rainfall 239.1mm

Treatments	Henna DLY (kg /ha)	Guar yield (kg /ha)	Henna leaf equivalent yield (kg /ha)	Water use (mm)	Water use efficiency (kg/ hamm)
H: C (1:1)	422.5	69.3	491.8	367.6	1.3
H: C (1:2)	427.7	126.0	553.7	368.7	1.5
H: C (1:3)	389.3	166.4	555.7	367.1	1.5
H: C (1:5)	145.7	201.1	346.8	368.3	0.9
Alley 3m	180.1	210.0	390.1	387.4	1.0
Alley 6m	121.8	267.9	389.7	379.9	1.0
Strip 2.4m	491.3	213.4	704.7	385.0	1.8
Sole henna	655.7	-	**655.7**	382.2	1.7
Sole crop	-	354.0	354.0	382.2	0.9

Note: H-henna, C- cluster bean

Table 3: Dry leaf yield, Guar seed yield and yield attributes of Henna - Guar intercrop

| | Henna | Guar | | Henna | Guar | |
Treatment	DLY (kg/ha)	Seed yield (kg/ha)	DLY PER (kg/ha)	DLY (kg/ha)	Seed yield (kg/ha)	DLY PER (kg/ha)
Year	2011 (654.9mm)			2013 (559mm)		
H:C (1:1)	1271.33	48.7	1329.7	529.5	26.7	561.5
H:C (1:2)	1067.64	37.7	1112.9	460.1	12.8	475.5
H:C (1:3)	1004.85	15.3	1023.2	439.8	5.4	446.3
H:C (1:5)	781.55	120.7	926.4	529.5	81.2	626.9
Alley 3m	833.26	94.9	947.1	616.3	60.4	688.8
Alley 6m	735.14	112.4	870.0	494.8	89.6	602.3
Strip 2.4m	1419.62	126.3	1571.2	1197.9	45.0	1251.9
Sole henna	1741.00	-	1741.0	1345.5	-	1345.5
Sole guar	-	139.6	167.5	-	117.0	140.4

Highest DLY (655.7 kg/ha) was recorded in sole henna. Maximum henna equivalent yield (704.7 kg/ha) recorded in strip cropping followed by sole henna and lowest (346.8 kg/ha) in 6m wide henna alley. WUE of 1.8 kg/hamm in strip cropping system was followed by sole henna 1.7 kg/hamm. Rainfall received 645.6mm. DLY of henna with 1:1, 1:2, 1:3 and strip cropping was higher than 3m and 6m wide henna alley. Seed yield of clusterbean declined in 1:1, 1:2 and 1:3 inter-cropping system owing to moisture competition.

Pests and Diseases management

Termite is the most important pest of henna in the field followed by ants and mites. Soil application of Chlorpyriphos (10%) or Furadon 10G at 25 kg ha[-1] before final field preparation is essential during crop establishment. Further, the standing crop is dusted with Chlorpyriphos or Furadon before interculture. Under prolonged moist and cloudy conditions there may be high incidence of semi-loopers on the young and green leaves of henna. Foliar spray of Endosulphon 30EC at 1.25 l ha[-1] is recommended for its control. Among the important field diseases, this crop is affected by leaf blight that can be controlled by spraying any copper fungicide.

Economics and Marketing

Henna cultivation is considered an economical enterprise under erratic and low rainfall conditions. In Sojat area of Rajasthan it involves an establishment cost of about Rs. 16,000 per hectare and gives an average net return of around Rs. 14,000 per hectare per annum. Similarly, Singh and Gupta (1998) reported an average gross return of Rs. 21,000 per hectare from the henna raised on stony gravelly arid wasteland without any fertilizer application.

Colour Plates

Fig. 2: Botanical varieties of henna: *alba* and *rubra*

Fig. 5: *In-situ* water harvesting in henna planted field

Fig. 6: Intercropping in henna

A separate regulated market for exclusive trading of dry henna leaves is functioning in Sojat city. Henna farmers generally sell their produce in this market and get remunerative price varying from about Rs. 20 to 30 per kg. Trading also takes place at the Bhawani mandi of Kota district in Rajasthan. Dry henna leaves after processing at Sojat and Faridabad (Haryana) are marketed not only inside the country but exported to other countries too. Reportedly, about 40-50 per cent of the total leaf production is exported to France, UK, Syria, Algeria, Jordan and many other countries including USA. During 1993-94 alone, henna leaf powder worth Rs. 18.4 lakh was exported from Rajasthan (Agarwal, 1999).

Processing

It principally involves size reduction of the dried leaves followed by blending into various dyeing products. Processing, in general, leads to about 80-100 per cent value addition in the raw produce of henna. More than 25 major and 100 small-scale grinding units located in Sojat city of Rajasthan are engaged in processing henna leaves. The leafy produce is first shredded into smaller pieces in a thresher and then ground to fine powder form by passing through the pulveriser, stone-burr or hammer mill. Graded powder obtained is finally blended in many ways before packaging and marketing. Since henna dye is an acid mordant dye, the leaf powder is commonly blended with Indian gooseberry (aonla) for improving dyeing capacity of the product. Admixture with jujube, khejri, or *Cassia auriculata* (anwal) leaves and even fine green sand is also practiced to get different quality grades.

Conclusion

- Henna cultivation is considered an economical enterprise under erratic and low rainfall conditions.

- The planting configuration and water harvesting only conserved moisture but also minimized the infestation of weeds and enhanced the yield of henna.

- Yield of henna and intercrop clusterbean is depends on the rainfall received during the growing season. Henna plantation takes four years to attend effective dry leaf yield production. Intercrop clusterbean yields compliment during this establishment and intercultural results in better moisture conservation. After establishment henna plantation intercrop is economically not advisable. Farmer opts to have intercrops with henna strip cropping of 2.4m width or operational width suitable for mechanical intercultural operations is advised.

14

Tree Borne Oilseeds (TBO'S) of Arid Region

Keerthika A., Dipak Kumar Gupta, M.B. Noor mohamed, B.L. Jangid and A.K. Shukla

ICAR-Central Arid Zone Research Institute, Regional Research Station Pali, Rajasthan-306401

India is a huge importer of crude oil and spends about Rs. 1,200 billion of foreign exchange every year to meet 75% of its oil needs (Anand, 2006). This has affected its balance of payment adversely, especially after the unprecedented rise incrude oil prices. Being an agricultural country endowed with varied climates, nutrient-rich soil and ability to grow many different crops, India offers a great promise as a producer of surplus raw material for biodiesel and bioethanol production. India's biofuel production currently accounts for only 1 percent of global production (Shinoj *et al.,* 2012). Although all oils can be used, edible oils as a source for biodiesel production have to be ruled out because they are required for cooking and food purposes. Therefore, non-edible oils are the premier raw material for biodiesel production in India. Seeds rich in these oils are mostly produced by perennial species. Hence these trees are referred to as tree-borne oilseed species (TBOS) and they produce TBOs (Aurora and Kumar, 2015). TBOs are cultivated/grown in the country under different agro-climatic conditions in a scattered form in forest and non-forest areas as well as in waste land/deserts/hilly areas. The country has enormous potential of oilseeds of tree origin like mahua (*Madhuca indica*),neem (*Azadirachta indica*), simarouba (*Simarouba glauca*), karanja (*Pongamia pinnata*), ratanjyot (*Jatropha curcas*), jojoba (*Simmondsia chinesis*), cheura (*Diploknema butyracea*), kokum (*Garcinia indica*),wild apricot (*Prunus armeniaca*), wild walnut (*Aleurites molucana*), kusum (*Schleichera oleosa*), tung (*Vernicia fordii*), etc. The best characteristic is that it can be grown and established in the wasteland and have varied agro-climatic conditions. (Aurora and Kumar, 2015)

The country has an estimated potential of more than 5 million tonnes of tree borne oilseeds. However, only 8-10 lakh tonnes are being collected and less than 1.5 lakh tonnes of oil are being extracted out of an exploitable potential of more than 1 million tonnes of oil from tree origin sources (Hedge, 2012). In addition to the existing potential of TBO, of the total land area of India (328.7 million ha), about 16.8% is wasteland. Of this total, approximately 32.3 million ha would be suitable for growing these oilseeds. In addition to wastelands, other unused lands, including agricultural border fences, hedgerows, and land along roads, railways, and canals, might total a further 8 million ha. (Planning commission, 2003)

National level initiatives on Biodiesel

India introduced its biodiesel program in 2003 with the formulation of the National Mission on Biodiesel. The program focused on producing biodiesel from *Jatropha curcas*, a small shrub that grows on degraded land or wasteland producing non-edible oilseeds. The December 2009 National Policy on Biofuels called for an indicative blending target of 20% by 2017 for both bioethanol and biodiesel.Currently under the 12th plan, a Mini Mission-component III on tree borne oilseeds has been setup. Two National Networks viz., 'National Network on Jatropha and Karanja' and 'National Network on Wild Apricot and Cheura' were constituted by involving State Agricultural Universities, national research institutions to address different researchable issues under the banner of National Oilseed and Vegetable Development Board, Gurgaon. Both National Biofuel policy (2009) and National Agroforestry Policy (2014) being launched in the country, the first policy targets the replacement of fossil fuels by biofuel to the extent of 5% by 2012, 10% by 2017 and above 10% beyond 2017 and the second policy aims at integrated land use option for livelihood, environment and energy security. (Dhyani *et al.*, 2015)

Agro-techniques of different TBO'S of Rajasthan

Jojoba (*Simmondsia chinensis*)

Jojoba (*Simmondsia chinensis*) pronounced as Ho-ho-ba which is a scraggy looking desert bush belonging to the family simmondaceae. It is also recognized as "Desert Gold". Jojoba is called by many names viz., Bushnut, Bucknut, Pig nut, Goat Berry, Wild hazel or Quinine plant.it is evergreen long living short bush, 1- 1.5m in height with good canopy. It has been introduced in India in recent past. Jojoba plant is an evergreen, long living bush with a life span upto 200 years. Initially, Jojoba was introduced in Rajasthan state at the Central Arid Zone Research Institute (CAZRI) Jodhpur, in the year 1965 from Israel. Looking to the immense potential of Jojoba in Rajasthan, a society called "Association of Rajasthan Jojoba plantation and research project (AJORP)", was set up in Jaipur in the year 1995 with hundred per cent funding from the Department of

Distribution of Tree Borne Oilseeds (TBO's) in India

S.No	Botanical Nmae	Distribution	Oil (%)	Uses
1.	*Jatropha curcas*	Throughout India	30-40	Industrial, medicinal and soap
2.	*Simarouba glauca*	Central America, Mexico, southern states of India	55-70	Biodiesel, Edible oil, Pharmaceuticals, Cosmetics, Soaps, Lubricants, Paints, warnishes, Wax and candles
3.	*Calophyllum inophyllum*	Along sea coasts	50-73	Illumination and soap
4.	*Madhuca indica*	Central India, plains of N. India	35	Cocoa butter substitute, vanaspati, soap
5.	*Shorea robusta*	Central Himalayas and foothills of Himalayas in sub temperate regions	13	Cocoa butter substitute, soap, vanaspati
6.	*Mesua ferrea*	Forests of N.E. India,Karnataka, Kerala	45	Soaps, lubricants, illumination, medicinal
7.	*Mallotus phillippinensis*	Himalayan foothills, Maharasthra Bengal, Orissa, M.P.		Paints, varnish
8.	*Garcinia indica*	Forests of W. ghats in S. Maharashtra and Slopes of Nilgiri hills	33-44	Cocoa butter substitute, soap, candles

(Neelakandan, 2004)

Distribution of Tree Borne Oilseeds (TBO'S) in Rajasthan

S.No	Scientific name	Native range	Distribution in Rajasthan	Oil (%)	uses
1.	*Olea europaea* (Olive)	Indo-Israel	Bikaner, Ganganagar, Hanumangarh, Jaipur and Jhunjhunu	60-70	Edible oil, cosmetics and Pharmaceuticals.
2.	*Simmondsia chinesis* (Jojoba)	Native to sonoran desert of mexico, California and Arizona.	Jhunjhunu, Sikar, Bikaner, Jaipur, Jodhpur, Sriganganaga and Churu	50	Lubricants, waxes, sources of acids and alcohols, cosmetics and Pharmaceuticals
3.	*Salvadora oleoides* (Pilu)	Arid regions of Punjab and W. India		33	Soaps and candles
4.	*Azadirachta indica* (Neem)	Throughout India		30-35	Insect repellant and pesticide properties, soaps and cosmetics
5.	*Pongamia pinnata* (Karanja)	Throughout India	Southern and semi-arid parts of Rajasthan	27-39	Soap, lubricants, Ilumination and industrial purposes
6.	*Balanites aegyptica* (Hingot)	Tropical and Northern Africa, Syria, W. Asia and Sudan.		44 - 50	Soaps and detergents.

Land Resources, Ministry of Rural Development, Government of India, New Delhi. Ministry also established two national level models of Jojoba plantation and Research farms in 100 hectares. (Dangi *et al.,* 2009 and 2010).

The plant is deciduous in nature, bearing male and female flowers on separate plants. In Indian conditions at Rajasthan, the flower buds generally appear during December – February and the fruits ripen in April. Occasionally, it has also been observed that flower appears in June and fruit ripens in April. Thus two flowering and fruiting seasons has been reported.

Jojoba can be grown on any type of soil including gravel and rocky soil except heavy soil and soil prone to flooding. Its pH requirement ranges from 5 to 8 indicating its tolerance to acidity as well as alkalinity. It can tolerate extreme temperature ranging from 5 to 54 °C. One kg seed contains about 800-1200 seeds depending on their size and shape. About 5 to 6 kgs of fresh seeds are required for raising seedlings of one hectare. The germination starts after 8–10 days of sowing and continue up to 25 – 28 days depending upon moisture of soil. Generally, 25–30 °C temperature found most suitable for germination. The seedlings attain height of 20–30 cm after 5 months of sowing in favourable soil moisture and temperature conditions. Five months old seedlings are ready for plantation. Planting is done at spacing of 4 X2 m (1250 plants/hec) and 2x2 m (2500 plants/hec). Best ratio of 1 male: 10 female flowers are recommended. Male plants started flowering only after one and half to two years, while female plants generally flowers after two years. Irrigations from flowering to seed setting stage *i.e.,* from December to April are required for higher seed harvest. On an average , the seed yield at 5 years reported to be 400-500g/plant which increased to 1.0 kg /plant by 10th year and sometimes 2.0kg/plant by 13-15th year. (Kureel, 2008/ NOVOD, 2008)

The crops like mung bean, moth, gram, groundnut, vegetables of cucurbitaceae family can be successfully grown as intercrops.

Olive (*Olea europaea*)

Olive is a commercial plant of the Mediterranean since ages and in India it was imported for use in salads and other culinary preparations. In 2007, with Indo-Israel collaboration, its cultivation was tried in Rajasthan and first fruits were harvested in 2011-12. The oil content of the seeds from Indian cultivation ranged from 9-14 % in comparison to 12-16 % of the other olive growing countries. At present, olives are grown in an area of 182 ha in several parts of Rajasthan with 14000 MT.

Improved varieties/accessions: Arbequina Olive; Barnea Olive; Coratina Olive; Frantoio Olive; Koroneiki Olive; Picholine Olive and Picual Olive.

Olive requires deep, rich and well drained loamy or clay loam soils for best growth and yield. To provide good aeration for root development, soil should be about 5 ft deep are well suited. The soil pH range 6 to 7.5 is best and ideal for good quality and higher yields. During the month of September to October fruits are collected. The stones from fruit pulp should be separated by dipping in caustic soda containing 10 percent NaOH. The washed stones should be sown immediately on raised nursery beds. The distance of row to row should be 15cm and seed to seed should be 5 cm. for better germination. Nursery beds should be irrigated and mulched regularly.

Planting during November-December or February-March in holes that can be dug manually or mechanically, in dimensions of about 60 cm x 40 cm (manual digging) or 20 cm x 30 cm (mechanical digging). Planting depth should be same as in the nursery. In dry areas, planting holes must be 5-10 cm deeper. After planting, the surrounding earth should be covered with straw to minimize water loss from the soil. Fertilizers @ of 1000-1500 kg ha^{-1} P and 500-800 kg ha^{-1} K at the time of planting and 3-4 fertilizations with ammonium nitrate (20-30 g tree^{-1}) every year followed by irrigation. Young trees are irrigated in the initials 2-3 years. First bearing starts at 5-10 years, with flowering in Mar-April and fruit harvesting begins from October. (Dhyani *et al.*, 2015)

Seedlings of olives can be grafted by tongue grafting. The ideal time for grafting is March-April. Patch budding and i- budding are most effective methods.

In the initial year's cotton, tomato, potato, pumpkins, etc. can be grown as per the region and it should be restricted among the rows of the olive trees to minimize competition among the plants.

Pilu (*Salvadora oleoides*)

Salvadora oleoides, popularly known as pilu, is a small, multipurpose agroforestry tree commonly grown in western Rajasthan and Gujarat states of India (Kaul 1963). *S. oleoides* is also called jhal, baha, pilu and khakan. The tree generally flowers in March-April. It grows well in the sand dunes of deserts to heavy soils, non-saline to highly saline soils and dry regions. *S. oleoides* can be found growing in groups of up to 50-500 trees in its natural population, and can age up to 150 years old (Bhansali, 2011).

It is much branched, evergreen shrub or a small olive like tree with short twisted, hollow trunk. Seeds are dispersed by birds. *S. oleoides* can be regenerated by seeds, coppices and root suckers. However, very little information is available regarding seed germination. Natural regeneration of plants from seeds is rare, probably because seeds mature at the onset of the monsoon season. The seeds of *S. oleoides* are not stored because viability is not retained. De-pulping the fruits and pre-treating the seeds promote early germination. Freshly harvested seeds with 26 percent moisture content showed 90 percent germination. Viability of seeds is reduced by 50 percent 15 days after harvest, and 100 percent after 24 days of storage.

Fruits are collected in May –June, immediately depulped by rubbing and then washed and dried. Fresh seeds sown in may shoed germination up to 87% within 15 days of sowing. Nursery can be raised using saline water of 15 dsm⁻¹ by placing bags in 20cm deep trenches and the saplings can be grown for 90 days in polybags. The saplings of 3 months old are ready for transplantation. Best period for transplanting is rainy season (July- September). Pit of 30X30X30 cm is dug and planted at spacing of 4x4m on saline black soil. Farm Yard Manure at 1 kg/pit and DAP 50 g per pit may be applied prior to planting. Ammonium sulphate at 100g/plant at the time of flowering enhance berry formation and seed yield. A good tree yields fruit about 30-40 kg.

According to local people, *S. oleoides* trees have undergone profuse flowering but negligible fruit setting in the last 10-12 years. The reasons cited by the local people were changing environmental conditions which caused severe incidences of pests and diseases.

Karanj (*Pongamia pinnata*)

Pongamia is a drought tolerant, semi-deciduous medium sized tree with short bole and spreading crown. It is widely grown from tropical dry to subtropical dry forest zones. It is a good shade bearer, suitable for planting in pastures, for afforestation in watershed areas and drier part of the country. It grows under a wide range of climate and soil conditions and can grow even in dry areas with poor, marginal, sandy and rocky soils. In addition to drought, it can tolerate saline conditions. Flowering takes place during April to June. Pods ripen from March to May of the following year. Pods are generally collected during December to May and the time of collection in different parts of the country differs according to the climate.

Pod collection is done by beating the branches or from ground. The pods are dried in the sun and thrashed to extract the seed, which is dried in shade before storage. Seed yield per tree varies from 9 to 90 kg. Soaking of seeds in cold water for 24 hr, hastens and improves germination. About 1200 seeds weigh one kg. Bold and healthy seeds are either sown in polybags or directly in field during July August. The soaking of seeds for 24 hours in IBA (30 ppm) or GA3 (20 ppm) enhances germination and vigour. Vegetative propagules developed through semi-hardwood cutting, air layering or cleft grafting can also be used. (Dhyani, 2015)

Plantation: Pits of 60cm x 60cm x 60cm are dug for planting at the spacing 5m x 4m to accommodate 500 plants ha-1. Addition of 5kg FYM per-pit is recommended. One year old seedlings are transplanted at the onset of monsoon.

The dwarf and short duration oilseed and pulse crops i.e., mustard,groundnut, sesamum, chickpea, blackgram, cowpea, horsegram, soybean and millets like maize, bajra etc can be grown successfully as intercrops upto 4-5 yrs after planting without affecting karanja plants to increase the economic feasibility.

Desert Date (*Balanites aegyptica*)

Balanites aegyptiaca (L.) Del. belongs to family balanitaceae. It is an evergreen woody plant of fragile ecosystem of the Great Indian Thar Desert. Flowering and fruiting occurs during October (Bhandari, 1990). One tree produces 100 - 150 kg/ year (Rathore et al, 2004). Best selected trees can yield date fruits up to 52 kg/trees. Plant attains maturity within 5-7 years

The oil is consisted on four major fatty acids: palmitic (16:0), stearic (18:0), oleic (18:1), and linoleic (18:2), constituting 98-100% of the total fatty acids in the oil Plantation of this species should be promoted in arid regions as shelterbelts and in saline soil for reclamation of soil, it will be helpful in conservation of biodiversity and desert development programme.

Neem *(Azadirachta indica)*

Neem is one of the most valuable tree species found in India. Neem tree is noted for its drought resistance. It thrives in areas with sub-arid to sub humid conditions with an annual rainfall between 400 and 1200mm. It can grow on wide range of soils up to pH 10. They can grow on all type of soils including clay, saline and alkaline soils but does well on black cotton soils. It thrives best on well drained deep and sandy soils.

Fruiting begins in 3 to 5 years. The tree bears flowers in February – March in southern parts and March – April in northern parts of the country. Fruits mature during May- June in southern parts and June – July in northern parts. Fruit yield is 10- 25 kg per tree per year in the initial years. A mature tree produces 30-100 kg fruit/year per tree depending on rainfall, soil type and neem ecotype or genotype.

Propagated both by seeds or vegetative propagation (cuttings/air-layering/grafting or tissue culture). Although neem is a prolific seed producer, yet seed supply is a frequent problem. The viability of fresh seed decreases rapidly after two weeks. Seedlings can be sown at a distance of 15 to 20cm and a depth of 1 to 1.5cm and then watered. Germination takes place in a week's time. After 5 to 6 weeks, the seedlings are removed from the nursery and planted in polythene bags. 5-6 months old seedlings can be transplanted in a pit size of 30cm x 30cm x 30cm at a spacing of 5m x 5m during monsoon season. About 5-6 kg green manure, 20-25 gm endosulphan dust, 10 g urea, 20 g single super phosphate, 20 g MOP and 1-2 kg neem cake plant-1 is recommended during transplanting. (Dhyani *et al.*, 2015)

Intercropping: In the gestation period of 5 years, groundnut, mustard, chickpea, cowpea, horse gram, soybean etc can be successfully grown.

Constraints in exploitation of TBOs: Under the existing situation of tree borne oilseeds being of forest origin, the problems encountered are: collection from scattered locations, high dormancy and problems in picking and harvesting in avenue and forest plantations, non-availability of quality planting material or seed, limited period of availability, unreliable and improper marketing channels, lack of post-harvest technologies and their processing, non-remunerative prices, wide gap between potential and actual production, absence of state incentives promoting bio-diesel as fuel, and economics and cost-benefit ratio (Babu *et al.*, 2007)

Conclusion

The TBO's are found in the wild condition. Several research institutes are carrying out research to identify suitable genotypes and accessions. With the recent in renewable resources and introduction of policies, value of oil in TBOs has been highlighted. Extensive cultivation in marginal and waste lands seems viable, and the development of improved planting materials and better management of nursery and block plantations may result in a sustainable biofuel business.

References

Anand, N. 2006. Biofuel – an alternative fuel for India. *The Bull on Energy Effi.*, 6: 32-35

Bhandari, M.M. 1990. Balanites, 88: In Flora of the Indian Desert, MPS Repros, Jodhpur, India.

Bhansali, R. 2011. Promoting the use of *Salvadora oleoides* as a multipurpose agroforestry species. *Agroforestry Newsletter,* 38. Pp.2-6

Dhyani, S. K., Vimala Devi, S. and Handa A. K. 2015. Tree Borne Oilseeds for Oil and Biofuel. Government of India, Ministry of Rural Development. 2005. Wasteland Atlas of India. Delhi.

Hegde, D.M. 2012. Carrying capacity of Indian Agriculture: Oilseeds. *Curr Sci,* 102(6): 867-73.

Neelakantan, K.S. 2004. Tree Borne Oilseeds - an Over view. Strategies for Improvement and Utilization of Tree Borne Oilseeds. ICAR - Winter School (Eds M.Paramathma, K.T. Parthiban, K.S. Neelakantan). Forest College and Research Institute, Mettupalayam 641 301. pp. 44-51

Rathore, M. and Meena, R.K. 2004. Nutritional evaluation of some famine foods of Rajasthan Desert. *Indian Forester*, 130 (3): 304-312.

Sudhakara Babu, S.N., Sujatha, M. and Rajeshwar Rao, G. 2008. Non-Edible Oil-Seed Crops For Biofuel Production: Prospects and Challenges. In: Proceedings of the International Technical Workshop on the "Feasibility of Non-edible Oil Seed Crops for Biofuel Production". Pp.23-38
Technical Bulletin 2/2015. ICAR-CAFRI, Jhansi. pp: 50.

15

Traditional and Improved Agroforestry Systems in Arid Western Rajasthan with Special Reference to Climate Change Adaptation and Livelihood Security

J.C. Tewari

ICAR-Central Arid Zone Research Institute, Jodhpur, Rajasthan-342003

Of the total land area of the world, arid zones account for 18.8%. Throughout the arid zones, there is no dearth of problems, but rapidly increasing desertification (some call it land degradation) is a problem of worldwide dimension. Water is a scare commodity in arid zones. Much of rainfall is lost by evapo-transpiration and as a result, ground water recharged only by seepage through soil profile. However, it is a common phenomenon in arid zones of the worlds that ground water is frequently used at the rate that exceeds recharge. The situation in arid tropics of India, which is spread over an area of 31.7 million ha, is no more different.

Agriculture is the human enterprise that is most vulnerable to climate change. Tropical agriculture including agriculture in arid tropics, particularly subsistence agriculture is vulnerable, as farmers do not have adequate resources to adapt to climate change. Land-use options that increase livelihood security and reduce vulnerability to climate and environmental change are necessary. Rural folk in arid western Rajasthan, the principal hot arid region of the country have been practising arable cropping in association with these scattered trees on crop fields since time immemorial. This combined protective-productive system of integration of trees into farming system rooted in the principles of ecology, productivity, economics and sustainability, is generally referred as agroforestry. In fragile ecosystems of hot arid regions of India, agriculture alone cannot be a dependable enterprise, hence desert dwellers with their traditional wisdom are integrating forestry into farming since ages in order to confer stability and generate assured income (Tewari *et al.*, 1999; Narain and Tewari, 2001). Traditional resource management adaptations, such as agroforestry systems, may potentially provide options for improvement in livelihoods through simultaneous production of food, fodder and firewood as well as mitigation of the impact of climate change.

The Indian Hot Arid Regions

Location and Distribution

The hot arid regions of India lie between 24° and 29° N latitude, and 70° and 76° E longitude, covering an area of 31.70 million hectares, and involving seven states: Rajasthan, Gujarat, Punjab, Haryana, Andhra Pradesh, Karnataka, and Maharashtra. An area-wise break up of hot arid regions is presented in Table 1. In total, 11.8% of the country is under a hot arid environment. The arid regions of Rajasthan, Gujarat, Punjab, and Haryana together constitute the Great Indian Desert, better known as the Thar Desert. As arid western Rajasthan accounts for 61% of hot arid region of the country, therefore it is considered principal hot arid region.

Table 1: Distribution of arid regions in different states of India

State(s)	Area (million hectares)	Percent of total
Rajasthan	19.61	61.00
Gujarat	06.22	19.60
Punjab and Haryana	02.73	09.00
Andhra Pradesh	02.15	07.00
Karnataka	00.86	03.00
Maharastra	00.13	00.4
Total	31.70	-

Climate and Soils

The production and life support systems in this part of the hot Indian arid zone are constrained by climatic limitations including: low annual precipitation (100–300 mm); very high temperature during the summer season (mean maximum temperature = 41°C) touching a maximum of 48° to 50°C; short (December to mid-February) cool and dry winters (the mean winter season temperature varies from 10° to 14°C); high wind speed (30–40 km/hour); high evapotranspiration; and general low humidity (an aridity index of 0.045–0.19) (Sharma and Tewari, 2005).

Sand dunes are a dominant land formation of the region. More than 58% of the area is sandy and intensities of dunes vary from place to place. In general, soils contain 1.8–4.5% clay, 0.4–1.3% silt, 63.7–87.3% fine sand, and 11.3–30.3% coarse sand. They are poor in organic matter (0.04–0.12%), and low to medium in phosphorus content (0.05–0.10%). The nitrogen content is mostly low, ranging between 0.20 and 0.07% and infiltration rate is very high, at 7–15 cm/hour (Dhir, 1997). Because of the complete absence of any aggregation, the soils are highly erodible.

Population and Livelihood Resources

Since the vast arid expanse of Ghaggar is blank on Chalcolithic and Iron Age maps of the Indian subcontinent, it seems that human occupation of the region was thin, and depended upon hunting and limited pastoralism (Dhir, 1982). From the analysis of available authentic records, the population of the region registered an increase of 490% from 1901 to 1991. Decennial variations in population from 1901 to 1991 in the region showed a growth rate of 186% (between 1901 and 1971) as compared with 132% for the whole country. The population growth rate for the decades 1971–1981 and 1981–1991 were 36.7% and 30.7%, respectively. The changes that occurred in the density of population in the Thar between 1971 and 2011 were:

1971   48 persons/km^2

1981   69 persons/km^2

1991   89 persons/km^2

2001   101 persons/km^2

2011   127 persons/km^2

The population density is quite high compared with the global average of 6–8 persons/km^2 for arid zones. The major factor responsible for such phenomenal growth of population in the region is a wide gap between birth and death rates. Moreover, social values are predisposed to having more children and positive sanctions for fertility outnumber negative ones.

In spite of erratic and unevenly distributed rainfall, agriculture is the mainstay of rural populations, and mixed crop-livestock, mixed livestock-crop farming, and livestock farming, forms the spectrum of economic activities (Tewari *et al.*, 1999). The distribution of agricultural land holdings by size is fairly large at 4–6 ha. Next to land, livestock constitutes the most important asset of the cultivators. The number of livestock in the region is estimated to be 29.6 million in the year 2000, which mainly includes cattle, sheep, goats, camels and buffaloes.

On average, a household has 1.32 buffaloes, 2.28 cows, 1.48 cow young stock, 0.62 buffaloes, 0.38 buffalo young stock, 6.48 sheep, 6.50 goats and 0.49 camels (Sharma and Tewari 2005). The other sources of rural livelihood are caste occupations such as carpentry, black smithy, oil pressing, pottery making, tanning, leather work, dyeing, gold smithy, etc.

Agroforestry in Arid Region of Western Rajasthan

The image of the desert of little but vegetation-less dense unbroken stretches of sand and inaccessible terrain conditions is shattered upon entering in arid western Rajasthan. There are sparsely distributed trees, and underneath growth of arable crops and or grasses, and other herbaceous flora (especially during the *kharif* season, as agriculture is predominantly rainfed) in long stretches interspersed with distantly distributed village settlements and *dhanis* (a unique settlement pattern of the arid western Rajasthan intended for the life of agriculturists families during the active cropping period away from the village, but nearby their fields) are features of the region (Tewari *et al.*, 1999). In fact, desert of arid western Rajasthan is a vegetated desert having a very rich floral diversity of about 682 plant species, out of which 131 are known for their economic uses.

Major Traditional Agroforestry Systems

In general rainfall appears to be governing factor for evolution of traditional agroforestry systems (Sharma and Tewari, 2005). On the basis of rainfall, four types of major traditional agroforestry systems have been identified (Table 2). Data indicated that with decrease in rainfall, the density of woody component of the system decreased substantially and as well as the productivity of arable crops and grasses were declined.

Table 2: Traditional agroforestry systems in arid region of western Rajasthan, the principal hot arid region

Rainfall zone (mm)	Agroforestry system	Trees/shrubs (Nos/ha)	% Density of prominent species
>400	*P. cineraria - A. nilotica* based	31.4	80.5
300-400	*P. cineraria* based	14.2	80.0
200-300	*Zizyphus* spp. - *P. cineraria* based	91.7	100.0 (91.7% *Zizyphus* spp.)
<200	*Zizyphus* spp. - *P. cineraria* – *Salvadora* spp. based	17.2	87.2 (65% *Zizyphus* spp.)

Rainfall seems to play determinant role in crop production. *P. cineraria - A. nilotica* based agroforestry system was found to be highly productive and *Zizyphus* spp. - *P. cineraria–Salvadora* spp. system was least productive

(Table 3). It was very interesting that yield of pearl millet, main cereal crop of the region, below the canopy of woody components in any system was not affected at all, rather in *Zizyphus* spp. - *P. cineraria* and *Zizyphus* spp. - *P. cineraria* - *Salvadora* spp. systems it was substantially increased below the tree canopies in comparison to open spaces. However, legumes such as moth bean, cluster bean and mung bean exhibited slightly yield reduction below the tree canopies, 5.0% *(Zizyphus* spp. - *P. cineraria* system) to 19.5% *(P. cineraria* system); 12.5% *(Zizyphus* spp. - *P. cineraria* - *Salvadora* spp. system) to *18.2% (Zizyphus* spp. - *P. cineraria* system); and 10.0% *(P. cineraria* system) to 33.7% *(P. cineraria* - *A. nilotica* system), respectively. Yield reduction under tree canopies for sesamum ranged from 26.3% *(P. cineraria* -*A. nilotica* system) to 46.1% *(P. cineraria* system).

Table 3: Crop production in traditional Agroforestry system in arid region of western Rajasthan

AF Systems Crops Systems	*P. cineraria* - *A.* *nilotica* based	*P. cineraria* *based*	*Zizyphus* spp. - *P. cineraria* *based*	*Zizyphus* spp. – other spp. based
Pearl millet				
Below the canopy	10.5	7.9	2.0	2.0
Away from canopy	10.8	7.9	1.8	1.2
Moth bean				
Below the canopy	2.8	2.1	1.9	1.4
Away from canopy	3.0	2.6	2.0	1.6
Cluster bean				
Below the canopy	2.5	1.8	0.9	0.8
Away from canopy	2.9	2.1	1.1	1.0
Sesamum				
Below the canopy	1.4	2.1	1.9	—
Away from canopy	1.9	3.9	2.6	—
Mung bean				
Below the canopy	1.4	0.9	—	—
Away from canopy	2.1	0.1		

Traditional silvi-pastoral systems are integral part of traditional agroforestry of desert region of Rajasthan. Forage production under agroforestry tree species is given in Table 4. Under *P. cineraria* and *Tecomella undulata* trees, it was estimated to be 1.5 t/ha. Maximum herbage biomass in peak growing season (1st week of September) was found in *P. cineraria,* stands (Table 5) and minimum in *A. senegal* stands (Tewari *et al.,* 1999).

Table 4: Estimates forage production with agroforestry tree species in arid region of western Rajasthan

Tree species	Forage yield (kg/ha)	Palatable grass (%)
P. cineraria	1500	74.8
Zizyphus spp.	1000	87.5
Tecomella undulata	1500	56.7
Albizzia lebbek	1400	57.9
A. senegal	600	52.5

Table 5: Herbage production in protected natural Silvipasture systems in arid region of western Rajasthan

Silvipasture	Herbage (t/ha) biomass	Herb spp. richness (nos)	Herb spp. diversity	Palatable herbs (%)	Carrying capacity (sheep/ha/yr)
P. cineraria based	3.9	2	0.41	50.00	4.5
Zizyphus spp. based	2.3	7	2.31	42.8	2.3
Acacia senegal based	1.2	2	0.59	50.0	1.4
Acacia tortilis based	1.5	1	0.00	0.00	0.00

Fuelwood and leaf fodder production potential of *P. cineraria - A. nilotica* system maximum, while it was minimum for *Zizyphus* spp. - *P. cineraria - Salvadora* spp. system (Table 6). Thus productions from woody components of traditional agroforestry systems follow the rainfall gradient like that of arable crops.

Table 6: Fuelwood and top feed production from traditional agroforestry systems in arid region of western Rajasthan

Agroforestry systems	Plant density (nos/ha)	Fuel wood (t/ha/yr)	Leaf fodder (t/ha/yr)
P. cineraria-A. nilotica based	31.4	1.23	0.29
P. cineraria based	14.2	0.60	0.20
Zizyphus spp. - P. cineraria based	91.7	0.62	0.90
Zizyphus spp. - P. cineraria - Salvadora spp	17.2	0.18	0.14

Improved Agroforestry Practices

In the light of increasing pressure on land resources, Central Arid Zone Research Institute (CAZRI), Jodhpur initiated systematic studies on agroforestry systems in late 1970s. Since then a number of improved agroforestry practices in order to enhance overall productivity and economic returns of the farming communities have been developed and standardized. Following improved practices have been found promising and remunerative, and easy to fit in existing traditional agroforestry systems.

(a) Agri-Horticulture

Leguminous crop sown under *ber (Zizyphus mauritiana* cv. Seb) plantation produced 0.2 t/ha of grain and 0.8 t/ha quality *ber* fruits from same land unit even when seasonal rainfall is 200 mm (Table 7), thus rendering a drought proofing mechanism to the system. The density of *ber* plants were kept 400 individuals/ha (Gupta, 1997). The economics of this improved system indicated that in case of sole leguminous crop (mung bean) farming, the net profit per hectare was Rs. 4800/-, however, in case of *ber* intercropping, the profit was to a tune of Rs. 8000/- per ha.

Table 7: Improved agroforestry practices: agri-horticultural *(Ber* + Mung bean)

Treatment	Annual rainfall (mm)	Fruit yield (kg/ha)	Grain yield (kg/ha)	Net profit (Rs/ha)
Sole crop	200	-	520	4800
Intercropped *with Ber*	200	800	200	8000

(b) Horti-Pasture

Ber based horti-pasture have proved highly remunerative in *Thar* desert on farmers' field. *Ber* trees were planted at spacement of 6x6 m and grass *Cenchrus ciliaris* was introduced between tree rows after third year. On an average, the dry grass production was 1.55t/ha/year (Tewari *et al.,* 1999) (Table 8). The fruit, leaf fodder and fuel wood production from *ber* was 2.77, 1.87, 2.64 t/ha/year, respectively.

Table 8: Improved hortipasture *(Ber* + *Cenchrus ciliaris)* on farmers' field

Year	Dry grass (t/ha)	Tree products		
		Fuelwood (t/ha)	Leaf fodder (t/ha)	Fruit yield (t/ha)
1	1.50	3.10	2.13	3.10
2	1.67	2.74	1.80	2.74
3	1.42	2.46	1.71	2.49
4	1.66	2.49	1.63	2.85
5	1.48	2.43	2.01	2.80
6	1.57	2.63	1.94	2.67
Average	1.55	2.64	1.87	2.77

(c) Silvi-Pasture

Arid region of Rajasthan is reported to have 89 species of grasses. The species like *Lasiurus sindicus* is efficient builder of biomass with energy use efficiency of 1.4-2.0% (Harsh *et al.,* 1992). In an improved silvi-pasture, *Hardwickia binata* was taken as tree component at 3m x 3 m spacing with *Cenchrus ciliaris* grass. Results of 9 years study (Table 9) revealed that average carrying capacity of the practice was 4.1 sheep/ha/year against 3.7 for sole C. *ciliaris* pasture and 1.6 for sole *H. binata* plantation. In this type of improved silvipasture practice, in addition to grass + top feed production to the tune of 3.06 t/ha/year, 0.26 t/ha/year of fuelwood is also obtained.

Table 9: Production potential of *Harwickia binata* + *Cenchrus ciliaris* based improved silvipasture.

Year after planting	Grass yield (t/ha)	Leaf fodder yield (t/ha)	Total fodder yield (t/ha)	Carrying capacity (sheep/ha/yr)
1	0.82	-	0.82	1.9
2	1.23	-	1.23	2.9
3	1.64	-	1.64	3.9
4	2.05	-	2.05	4.9
5	1.64	-	1.64	3.9
6	1.64	-	1.64	3.9
7	1.24	2.31	3.55	8.5
8	0.84	0.66	1.50	3.6
9	0.84	0.66	1.50	3.6
Average	1.33	0.40	1.73	4.1

Role of Agroforestry in Adapting to Climate Resilient Agriculture in Arid Regions

Considering the role of agriculture in the social and economic progress of developing countries, the vulnerability of agricultural systems to the impacts of climate change has received considerable attention from the scientific community (Fischer *et al.*, 2002; IISD, 2003; Kurukulasuriya and Rosenthal, 2003). Much of the available literature suggests that the overall impacts of climate change on agriculture especially in the tropics will be highly negative, although in a few areas there may be minor increases in crop yields in the short term (Maddison *et al.*, 2007). Table 10 presents some of the projected changes in climate and their potential impacts on agriculture as summarised in the IPCC report (Parry *et al.*, 2007).

Table 10: Projected changes in climate and their impact on agriculture

Phenomena and direction of change	Likelihood of occurrence	Major projected impacts on agriculture
Warmer and fewer cold days and nights; warmer/more frequent hot days and nights over most land areas	Virtually certain (>99% chance)	Increased yields in colder environments decreased yields in warmer environments increased insect outbreaks
Warm spells/heat waves: frequency increases over most land areas	Very likely (90-99% chance)	Reduced yields in warmer regions due to heat stress; wild fire danger increase
Heavy precipitation events: frequency increases over most areas	Very likely	Damage to crops; soil erosion, inability to cultivate land due to water logging of soils
Area affected by drought: increases	Likely (66- 90% chance)	Land degradation, lower yields/crop damage and failure; increased livestock deaths; increased risk of wildfire
Intense tropical cyclone activity increases	Likely	Damage to crops; wind throw (uprooting) of trees; damage to coral reefs
Increased incidence of extreme high sea level (excludes tsunamis)	Likely	Salinisation of irrigation water, estuaries and freshwater systems

Agroforestry, the integration of trees and shrubs with annual crops production, is an age old management system practiced by farmers in arid regions of India provide shade, a steady supply of food and/or income throughout the year, arrest degradation and maintain soil fertility, diversify income sources, increase and stabilize income, enhance use efficiency of soil nutrients, water and radiation, and provide regular employment.

Carbon sequestration potential under different land use options are given in figure 1. Recognizing the ability of agroforestry systems to address multiple problems and deliver multiple benefits, the IPCC Third Assessment Report on Climate Change (IPCC, 2001) states that "Agroforestry can both sequester carbon and produce a range of economic, environmental, and socioeconomic benefits. For example, trees in agroforestry farms improve soil fertility through control of erosion, maintenance of soil organic matter and physical properties, increased N, extraction of nutrients from deep soil horizons, and promotion of more closed nutrient cycling." We believe that agroforestry interventions provide the best "no regrets" adaptation measures in making communities resilient to the impacts of climate change and do discuss the same in relation to the challenges posed by the changing and variable climate.

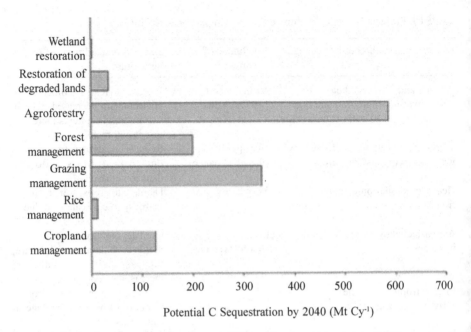

Potential C Sequestration by 2040 (Mt Cy^{-1})

Fig. 1: Carbon sequestration of different land management options
(adopted from IPCC 2000)

Agroforestry systems can be useful in maintaining production during drier years, a common phenomenon in arid regions of India. During drier years or in complete drought situations, deep root systems of trees are able to explore a larger volume for water and nutrients, which helps to maintain depleting soil moisture conditions to some extent. In drought prone environment of arid western Rajasthan, as a risk aversion and cooping strategy, the traditional agroforestry systems avoid long term vulnerability as trees act as an insurance against drought, insect-pests outbreaks and other threats, instead of a yield-maximising strategy aiming at short term monetary benefits (Rathore, 2004).

Adaptation to climate change is now inevitable. Research on agroforestry as an adaptation to climate change and as a buffer against climate variability is in the process of evolving. Many pathways through which agroforestry qualify as an adaptation to climate change, especially in arid regions are discussed below.

Agroforestry systems play a critical role in controlling soil erosion

There are several ways by which climate change manifests soil degradation. Higher temperatures and drier conditions lead to lower organic matter accumulation in the soil resulting in poor soil structure, reduction in infiltration of rain water and increase in runoff and erosion (Rao *et al.*, 1998) while the expected increase in the occurrence of extreme rainfall events will adversely impact on the severity, frequency, and extent of erosion (WMO, 2005). These changes will further exacerbate an already serious problem is being faced by arid regions of the country.

The woody component in agroforestry systems helps in reducing soil erosion which is the most harmful abiotic stress in arid region. It was observed if rows of trees are planted right angle to wind direction a tremendous amount of soil drop can be checked from agricultural fields. The shelterbelts on agricultural fields form a type of agroforestry practice (Gupta *et al.*, 1984). Following data (Table 11) depict how effective are agroforestry practices for controlling the soil erosion in arid regions.

Table 11: Effect of different type of shelterbelts on soil erosion

Type of shelterbelt	Total amount of soil loss (kg/ha)		
	Year I	Year II	Mean
Prosopis juliflora	93.2	609.3	351.2
Cassia siamea	91.5	277.1	184.2
Acacia tortilis	106.0	494.1	300.0
Agricultural field without trees	262.7	831.0	546.8

Agroforestry systems have inbuilt capacity to enhance the use efficiency of rain water

Water is already a scarce resource and climate change is expected to make the situation worse. Climate change has both direct and indirect impacts on water availability. The direct impacts include changes in precipitation patterns, while the indirect ones are increases in losses through runoff and evapotranspiration.

There are several mechanisms whereby agroforestry may use available water more effectively than the annual crops. Firstly, unlike in annual systems where the land lies bare for extended periods, agroforestry systems with a perennial tree component can make use of the water remaining in the soil after harvest and the rainfall received outside the crop season. Secondly, agro-forests increase the productivity of rain water by capturing a larger proportion of the annual rainfall by reducing the runoff and by using the water stored in deep layers. Thirdly, the changes in microclimate (lower air temperature, wind speed and saturation deficit of crops) reduce the evaporative demand and make more water available for transpiration.

The tree canopies in agroforestry systems intercept the rain and reduce runoff (Khan *et al.*, 1995). In a study at CAZRI, Jodhpur it was found that in *Acacia tortilis* silvi-pasture stand canopy interception was 21.4%, whereas in *Colophospermum mopane* silvi-pasture it was 13.1%. Rainfall interception was positively related with canopy cover and negatively related with throughfall. Average surface run off in *A. tortilis* based silvi-pasture stand was 53% higher than in *C. mopane* based silvi-pasture. This indicated that hydraulic response to rain is dominated by plant species character however; the percent annual runoff and soil erosion was very low in situations with trees on agricultural fields in comparison to bare soil condition. Thus, the enhanced use efficiency of rain water by woody species in agroforestry systems improves agricultural productivity.

Agroforestry systems provide economically viable and environment friendly means to improve soil fertility

Nutrient mining from continuous cropping without adequately fertilizing or fallowing the land is often cited as the main constraint to increase in productivity in arid regions. It is observed that on average soils in Indian arid regions have poor nitrogen and phosphorus content. Organic carbon status in certain parts of arid regions has been depleted to an alarming state. While fertilisers offer an easy way to replenish the soil fertility, at the current prices it is very unlikely that there will be any change in the investments made by farmers of arid region in fertilizers. In this context, agroforestry systems have attracted considerable attention as an attractive and sustainable pathway to improve soil fertility. CAZRI, Jodhpur made substantial progress in the identification and promotion of agroforestry systems aimed at improving soil fertility.

Soil fertility under the tree species (*Prosopis cineraria* and *Acacia nilotica*) and in open field conditions was studied intensively (Singh and Lal, 1969). In general, organic carbon content decreases with depth but the same was slightly higher under the canopy of both the tree species than in open field condition. The total nitrogen, available P and K are maximum under the canopy of *P. cineraria* followed by *A. nilotica*. The values for the same were minimum in open field condition.

The soil fertility build up under the fourteen year old stands of *P. cineraria* and *P. juliflora* were found to be of higher order as compared to open field condition. Similarly, status of micro-nutrients was also higher under tree canopies than in the open field (Table 12).

Table 12. Available macro and micro-nutrients under different tree based agroforestry systems

Tree species	Macronutrients (kg/ha)			Micronutrients (ppm)			
	N	P	K	Zn	Mn	Cu	Fe
Prosopis juliflora	231	7	333	0.89	9.3	0.58	3.3
Prosopis cineraria	221	11	479	1.44	10.8	0..89	2.8
Open field	199	6	3.2	0.19	7.0	0.38	3.5

Agroforestry systems play a very important role in moderating the micro-climate in arid regions

The full genetic potential of many crops and verities can only be realized when environmental conditions are close to optimum. Trees on farm bring about favourable changes in the microclimatic conditions by influencing radiation flux, air temperature, wind speed, saturation deficit of understorey crops all of which will have a significant impact on modifying the rate and duration of photosynthesis and subsequent plant growth, transpiration, and soil water use (Monteith *et al.*, 1991). Following examples present microclimatic moderation effect in harsh climatic conditions of arid regions of western Rajasthan.

Air temperature exhibited appreciable variation under the canopy of *A. tortilis* during monsoon period (Tewari *et al.*, 1989). A decline of 0.1°C to 0.7°C was evident beneath the canopy cover of *A. tortilis* than that of surrounding open areas at 07 hrs. Around 14 hrs, this decline was of the order of 0.6°C to 2.0°C. Similarly, it was found during monsoon season that the soil temperature just beneath the tree cover was lower by as much as 10°C to 16°C in top soil zone and 4°C to 5°C at 30 cm depth when compared to open field conditions, thereby indicating better soil-thermal regime. Seven year old *A. tortilis* trees provided more humidity on cluster bean grown with the trees. This finding proves the efficiency of agroforestry practices for maintaining favourable moisture status on arable lands (Ramakrishna and Shastri, 1977).

Agroforestry systems improve biodiversity

Continued deforestation is major challenge for forests and livelihoods. Although agroforestry may not entirely reduce deforestations, particularly in arid regions, it may acts as an effective buffer to deforestation. Trees in agro-ecosystems in Rajasthan have been found to support threatened cavity-nesting birds and offer forage and habitat to many other bird species. Biodiversity conservation may not be a primary goal of agroforestry systems. Nevertheless in some cases traditional agroforestry systems in arid western Rajasthan found to support very good species diversity and also act as a buffer to parks and protected areas (Pandey, 2007).

In a study at Govardhanpura and Gokulpura cluster of villages in Bundi district having relatively little better rainfall regime, where a highly degraded silvi-pasture (45 ha) was re-vegetated and fenced to minimize abiotic and biotic disturbance exhibited marked difference in biodiversity of plant species (Dixit *et al.*, 2005). The area was divided into 6 different blocks. Though initially only five woody and some grass species were introduced with some water conservation structures in the area, after five years the fortified area with physical and social fencing exhibited the woody species richness to the tune of 20 species with a Beta (between habitat) diversity of 2.2. Maximum number of woody species (25%) belonged to family Leguminosae/Mimosoideae. Rest of woody species were distributed among 11 other families/sub-families. The herbaceous vegetation richness increased to 36 (number of species) of which 80% were palatable. These 36 herbaceous species were distributed among 24 families. Cyperaceae family was represented by maximum percentage of the species. The Beta diversity of herbaceous species was 2.32 with index of general diversity of 3.437. The value of general diversity for woody and herbaceous species was very-very low in adjoining unprotected area.

A dominance-diversity (d-d) curves were developed for herbaceous vegetation on the basis of relative biomass values for all the blocks and also for open degraded land. With exception of open degraded land site, the d-d curves on all the blocks exhibited similar log normal distribution. This indicated that several herbaceous species were sharing relatively low importance values (biomass) or similar range of importance values. The log normal or curves approaching to log normal, in fact, indicate relatively stable population. The d-d curve of open degraded land site was closer to geometric series. Whittakar (1972) opined that communities having low diversity exhibit geometric series. In open degraded land ruthless exploitation of herbaceous vegetation has resulted in tremendous loss of biodiversity.

Agroforestry for Enhancing Livelihood Options in Arid Regions of India

Agroforestry contributes to livelihood improvement in arid regions of India, where people have a long history of accumulated local knowledge. Arid regions are particularly notable for ethano-forestry practices and indigenous knowledge system on growing trees on farm lands. The economics of previously discussed traditional agroforestry systems of arid western Rajasthan indicated that net B:C ratio of such systems is generally on positive sides (Table 13).

Table 13: Economics of traditional agroforestry systems of arid western Rajasthan

AF system	Expenditure (Rs/ha)	Returns (Rs/ha)			Gross Returns (Rs/ha)	Net Returns (Rs/ha)	Net B: C ratio
		Crops	Fuel wood	Leaf fodder			
P. cineraria- A. nilotica based	1850	4103	1230	870	6203	4353	2.3
P. cineraria based	1550	3670	600	420	4690	3140	2.0
Ziziphus spp. - P. cineraria based	1550	1506	620	600	2726	1176	0.7
Ziziphus spp. - P. cineraria - Salvadora spp. based	1500	1400	500	500	2400	900	0.6

In fact, multipurpose trees in agroforestry systems provide a rational formula for drylands. In hot arid regions trees on farmlands are the organisms with an ecological investment strategy (Oldeman, 1983). Matter and energy are immobilised in long lived trees and this majority of accumulated production is harvested at longer interval than in case of arable crops.

Jujube (Ber) is a very useful tree for agroforestry of arid regions. The leaves make excellent fodder, crude protein 13-17% and fiber 15%. Leaves are also food for silkworms. The branches and twigs are excellent fuel wood having calorific value 4878 K cal/kg. Root bark and leaves have 7% and 2% tannin, respectively and are sometimes mixed with other sources for tanning leather. The fruits provided by all the improved cultivars are delicious and rich source of vitamin A, C and B complex. Thus cultivation of jujube in arid regions offers solution to many problems. Tewari *et al.* (2001) demonstrated that agroforestry, particularly improved jujube based horti-pasture system can be major source of livelihood in many parts of arid regions (Table 14).

Table 14: Economic returns (in terms of gross income) from a Jujube based horti-pastoral system

Particular	Production (t/ha)	Returns (Rs.)
Dry grass	1.55	3,100
Fuel wood	2.64	2,640
Leaf fodder	1.87	7,480
Fruit	2.77	33,240
Total	-	46,640

This system provides production in terms of human food, livestock feed and cooking energy needs to communities. Such multifunctional agroforestry systems provide social wellbeing to the people in addition to goods and services.

The most spectacular evidence in context of livelihood improvement potential of agroforestry in arid region came through *Acacia senegal* based localized traditional agroforestry system in parts of Barmer and Jodhpur districts (15 villages). *A. senegal* is source of gum Arabic which has very high commercial value. In addition to harvesting crop grain for food and crop straw for fodder, the farmers harvested substantial quantity of gum Arabic through small intervention by CAZRI, Jodhpur. The data regarding production of gum and revenue earned by the farmers in last three years are given in Fig. 2. CAZRI gum inducer has been so popularised in the villages that farmers have made their own cooperatives for purchasing of gum inducer and sale of produce in the market.

During the year 2008-09, 12,000 trees were treated through CAZRI gum inducer which resulted in exudation of 5.4 tonnes of gum Arabic and farmers earned Rs. 27 Lakhs from the sale of gum.

Number of *Acacia senegal* trees (in thousands) treated with CAZRI gum inducer by farmer of western Rajasthan, production of gum arabic (in tons), total income earned by the farmers (in lac rupees) and revenue generated by CAZRI through sale of gum inducer

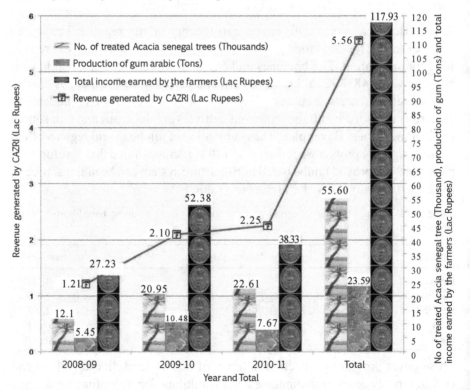

Fig. 2: Gum arabic production in certain villages of Barmer and Jodhpur districts

Subsequently in the year 2009-10 and 2010-11, 21,000 and 22,600 trees were treated, respectively resulting in production of 10.5 t and 7.6 t gum respectively. In the process farmers earned 52 Lakhs in the years 2009-10 and Rs. 38 Lakhs in the year 2010-11. In this way the farmers of 15 villages earned Rs. 117 Lakhs from the sale of gum Arabic with the average sale price in local market Rs. 500/kg.

These examples demonstrate the potential of agroforestry in arid regions for breaking the poverty and food insecurity circle. In addition to providing food and feed, there are numerous other non-timber tree products (excluding gums and resins) collected from trees on farm lands. *Prosopis cineraria* trees in addition to providing fuel and fodder give nutritious vegetable in form of pod, locally known as '*sangri*'. Likewise seeds of *Acacia senegal* are also used in preparing vegetable. During drought years their value is increased tremendously. In fact, scattered trees on agricultural fields in arid regions of western Rajasthan are multipurpose in nature and they have ability to satisfy the expectations and aspiration of rural folk regarding production of their basic needs. Thus, agroforestry provides unique basis of life support in hot arid regions.

Synthesis

Climate change impacts on agriculture are of the greatest concern to most developing countries, particularly in tropics because of higher dependence on agriculture, subsistence level of operation, low adaptive capacity and limited institutional support. The scenario for arid tropics of India is no more different. In conclusion, for sustainable development agroforestry systems in arid region are essential options in present-time as they are most suitable for optimum production, livelihood improvement and climate change mitigation in inhospitable environmental conditions of drylands. Agroforestry systems act as sink for atmospheric carbon, while helping to attain food security, increase farm income, improve soil health and discourage deforestation.

References

Dhir, R. P. 1982. The human factors in ecological history. In: *Desertification and Development: Dryland Ecology in Social Perspective*. Academic Press, New York, pp. 311-331.

Dhir, R. P. 1997. Characteristics and behavioral aspects of arid and semi-arid zone soils. In: Yadav M. S., Singh Manjit, Sharma S. K., Tewari J. C. and Burman U. (eds.). *Silvipastoral Systems in Arid and Semi-arid Ecosystems*. CAZRI, Jodhpur, India. pp.39-46.

Dixit, S., Tewari, J. C., Wani, S. P., Vineela, C., Shaurasia, A. K. and Panchal, H. B. 2005. *Participatory Biodiversity Assessment: Enabling Rural Poor for Better Natural Resource Management*. ICRISAT, Patancheru, A.P., India, P.16.

Fischer, G., Shah, M and Van Velthuizen, H. 2002. *Climate Change and Agricultural Vulnerability*. International Institute for Applied Systems Analysis. Report prepared under UN Institutional Contract Agreement 1113 for World Summit on Sustainable Development. Luxemberg, Austria.

Gupta, J. P. 1997. Some alternative production systems and their management for sustainability. In: Gupta J. P. and Sharma B. M. (Eds.) *Agroforestry for Sustained Productivity in Arid Regions*, Scientific Publishers, Jodhpur, India, pp. 31-39.

Gupta, J. P., Rao, G. G. S. N., Ramakrishna, Y. S. and Ramana Rao. 1984. Role of shelterbelts in arid zone. *Indian Farming*, 34(7):29-30.

Harsh, L. N., Tewari, J. C., Burman, U. and Sharma, S. K. 1992. Agroforestry in arid regions. *Indian Farming*, 45: 32-37.

IISD. 2003. *Livelihoods and Climate Change: Combining Disaster Risk Reduction, Natural Resource Management and Climate Change. Adaptation in a New Approach to the Reduction of Vulnerability and Poverty*, a conceptual framework paper prepared by the Task Force on Climate Change, Vulnerable Communities and Adaptation.

IPCC. 2000. *Special Report on Land Use, Land Use Change and Forestry*. Summary for Policy Makers. Geneva, Switzerland. P.20.

IPCC. 2001. *Climate Change 2001: The Scientific Basis*. Contribution of the Working Group 1 to the Third Assessment Report of the IPCC. Cambridge, Cambridge University Press.

Khan, M. A., Tewari, J. C. and Issac, V. C. 1995. Hydrology of small forested catchments in the arid region of Rajasthan. *Annals of Arid Zone*, 34(4): 259-262.

Kurukulasuriya, P. and Rosenthal, S. 2003. *Climate Change and Agriculture: A Review of Impacts and Adaptations*. Climate Change Series Paper 91, World Bank, Washington, District of Columbia, P.106.

Maddison, D., Manley, M. and Kurukulasuriya, P. 2007. The Impact of Climate Change on African Agriculture A Ricardian Approach. World Bank Policy Research Working Paper, 4306.

Monteith, J. L., Ong, C. K. and Corlett, J. E. 1991. Microclimatic interactions in agroforestry systems. *For. Ecol. Manage.*, 45, 31– 44.

Narain, P. and Tewari, J.C. 2001. Agroforestry and soil conservation in arid regions of India. In:*Proc. First Regional Workshop on Thematic Programme Network-2 on Agroforestry and Soil Conservation (UNCCD)*. ICRISAT, Hyderabad, India. pp. 3-13.

Narain Pratap and Tewari, J. C. 2005. Trees on agricultural fields: a unique basis of life support in Thar Desert. In: Tewari V. P. and Srivastava R. L. (eds.) *Multipurpose Trees in the Tropics: Management and Improvement Strategies*. Arid Forest Research Institute, Jodhpur, pp. 516-523.

Oldeman, R. A. A. 1983. Thedesign of ecologically sound agroforests. In: Huxley P. A. (ed.) *Plant Research and Agroforestry*. International Council for Research in Agroforestry (ICRAF), Nairobi, Kenya. pp. 173-207.

Pandey, D. N. 2007. Multifunctional agroforestry systems in India. *Current Science*, 92(4): 455-463.

Parry, M.L., Canziani, O.F., Palutikof, J.P., Linden van der, P.J. and Hanson, C.E. 2007. *Climate Change 2007: Impacts, Adaptation and Vulnerability. Contribution of Working Group II to the Fourth Assessment Report of the Intergovernmental Panel on Climate Change*, Cambridge University Press, Cambridge, UK, P.1000.

Ramakrishna, Y. S. and Shastri, A. S. R. A. S. 1977. Microclimate under *Acacia tortilis* plantation. *Annual Progress Report*, CAZRI, Jodhpur, pp. 69-70.

Rao, K. P. C, Steenhuis, T. S., Cogle, A. L., Srinivasan, S. T., Yule, D. F. and Smith, G. D. 1998. Rainfall infiltration and runoff from an Alfisol in semi-arid tropical India. II. Tilled systems. *Soil and Tillage Research*, 48:61-69

Rathore J. S. (2004). Drought and household coping strategies: A case of Rajasthan, *Indian Journal of Agricultural Economics*, 59(4): 689-708

Sharma Arun, K. and Tewari, J. C. 2005. Arid zone forestry with special reference to Indian hot arid zone. In: *Forests and Forest Plants, Encyclopaedia of Life Support Systems (EOLSS)*, Developed under auspices of the UNESCO, EOLSS publishers, Oxford, UK (http://www.eolss.net).

Singh, K. S. and Lal, P. 1969. Effect of *Prosopis cineraria* and *Acacia nilotica* on soil fertility and profile characteristics. *Annals of Arid Zone,* 8:33-36.

Tewari, J. C., Bohra, M. D. and Harsh, L. N. 1999. Structure and production function of traditional extensive agroforestry system and scope of intensive agroforestry in Thar Desert. *Indian Journal of Agroforestry,* 1(1): 81-94.

Tewari, J. C., Harsh, L. N. and Venkateswarlu, J. 1989. Agroforestry research in arid regions: a review. In: Singh R. P., Ahlawat I. P. S. and Saran G. (Eds.) *Agroforestry Systems in India – Research and Development,* Indian Society of Agronomy, New Delhi, pp. 3-17.

Whittakar, R. H. 1972. Evolution and measurement of species diversity. *Taxon,* 28: 213-251.

Tewari, J. C., Tripathi, D. and Narain Pratap (2001). Jujube : A multipurpose tree crop for arid land farming systems. *The Botanica,* 51: 121-126

WMO. 2005. *Climate and Land Degradation.* WMO-No. 989 2005, World Meteorological Organization.

16

Integration of Livestock in Horticulture Based Farming System in Rainfed Regions of India

A.K. Misra

ICAR-Central Arid Zone Research Institute, Jodhpur, Rajasthan-342003

Rainfed farming has a distinct place in Indian Agriculture, occupying 67% of the cultivated area, contributing 44% of the food grains and supporting 40% of the human and 65% of the livestock population. It is characterized by resource poor farmers, poor infrastructure and low investment in technology inputs (Singh, 1999). The land use in rainfed areas is quite diverse with a variety of crops, cropping systems, agroforestry and livestock farming. The productivity levels have remained low over the years and the share of rainfed crops in the total food grain production has been declining. The rainfed ecosystem suffers from the problems of (i) frequent droughts due to high variability in the quantum and distribution of rainfall, (ii) poor soil health due to continued degradation and inadequate replenishment of nutrient exhaustion, (iii) low animal productivity due to an acute scarcity of fodder and (iv) low risk bearing capacity of farmers due to poor socio-economic base, credit availability and infrastructure.

Traditionally, the Indian farmers are rural small holders and have integrated their livestock with crop production. Livestock production could be referred to as backyard of traditional or subsistence production. The small and marginal farmers and landless labourers dominate livestock production in rainfed areas. As crop production on 67% of the agricultural land still depends on rain, it is prone to drought, rendering agricultural income uncertain for most farmers. Shackled to subsistence production as a result of a shortage of finance and credit facilities, these farmers become entangled in a strangling debt cycle. The combination of an unfavorable land: person ratio and fragmented land holdings make it difficult to support large families on crop income alone. Continuing population growth will require further intensification in the crop-livestock-horticulture mixed systems and livestock especially ruminants will continue to play a vital role. Under rainfed conditions, integrating livestock with perennial crops offers a unique opportunity to produce valuable animal products on land that is currently used for other purposes, making the overall system more sustainable and environmentally sound. In most cases, the productivity of main crop may also be improved. The integration of livestock with horticulture based production system is not a simple system. The species of animal need to be well selected and their management carefully considered in order to obtain maximum benefit without negatively affecting the production of main horticultural crops.

There is need to define much better the research methodology in order to evaluate the effects on various system components. The comparative advantages of various species need to be studied for the different systems, for example, identifying a forage species that can be used as cover crop for orange plantation for sheep to graze on. Above all, the protection of the environment and sustainability should be essential objectives of all agricultural systems.

Several on farm trials are conducted under Institute Village Linkage Programme to find out suitable technological alternatives which promote better resource use trough synergies from crop-livestock-horticulture integration. Maintenance of soil fertility and provision of livestock feeds appear to be the main areas of reciprocal benefits, while animal traction may be an option for increasing land and labour productivity. In this paper we assess and discuss the challenges facing rainfed agriculture and the potential contribution of crop-livestock-horticulture integrated systems towards agricultural growth and sustainability.

Livestock Production Systems

The livestock production systems are rather complex and generally based on traditional and socio-economic considerations, mainly guided by available feed resources. The livestock production can be described as low input system. These traditional production systems are designed to be self-sufficient at the household level and are dependent on the low-cost agro-by-products as nutritional input to animals for producing quality food of high biological value and clothing materials of choice. Pastoralists as well as certain group of city dwellers (having scavenging livestock) have been forcing the animals to adapt remarkably to all available, often less/non-conventional feed and garbage in the surrounding environment.

About 2/3rd of the total livestock population of the country is found in rainfed areas, in which small ruminants (sheep and goat) predominate due to availability of forest grazing and wasteland. Livestock rearing is done mostly by small and marginal farmers and landless labourers with holding size of 1-3 cattle/buffalo and 20-50 sheep per household (Misra and Mahipal, 2000). The general characteristics of livestock production of small farmers under village condition are:

- as a complement to crop production,
- utilization of marginal land and non-marketable farm products,
- utilization of readily available/surplus family labour,
- required minimal cash inputs as well as simple and traditional technology,
- non-market-oriented production,
- very low degree of economic risk.

The prevalent livestock production systems in India can broadly be classified in to:

- Small holder livestock production: The traditional, resource-driven and labour intensive ruminant production system (sheep, goat, cattle, buffalo), which produces a multitude of services to subsistence farm. This kind of livestock production system is very common in rural areas and practiced by small and marginal farmers or landless people. Low technology uptake, insufficient market facilities and infrastructure and small economies of scale are common. Small farm animal rearing involves little external inputs such as feeds, medicines or breeding stock, and are relatively well adapted to the surroundings.

- Commercial livestock production: The modern, demand-driven and capital intensive, which is more relevant to poultry and to some extent dairy cattle and buffalo in cities and peri-urban areas. This system is very efficient and has good market access. Increasing intensification and concentration of animal increases pollution and disease risk to humans.

Small holder livestock production is a very important component of rainfed agriculture both in economic and socio-cultural terms, and needs integration within a whole farm system in which crop production generally is the major component. Unfortunately, in India, during the past five decades, research activities that are directly relevant to the development of small farm production, have received little attention both from scientists and policy makers. There has hardly been any sociological and economically relevant research on the integrating existing systems to increase the productivity of small holder systems.

Genesis of Crop –Livestock-Horticure Production System

Crop-animal systems have evolved and developed over many centuries. The type of crop and animal system that has developed at any particular location is a function the agro-ecological condition (Sere and Steeinfeld, 1994, Devendra

2011). Climate, edaphic and biotic factors decide whether cropping is feasible and, if so, the type of crops (grain and horticulture). This, in turn, determines the quantity, quality and distribution of animal feed resources throughout the year. The feed-base and disease challenge determine the potential animal production systems that develop. Feed resources provide a direct link between crops and animals and the interaction between the two largely dictates the development of such systems. Figure 1 provides an illustration of the phases in the development of crop-livestock systems (Devendra and Thomas, 2002; Devendra, 2011).

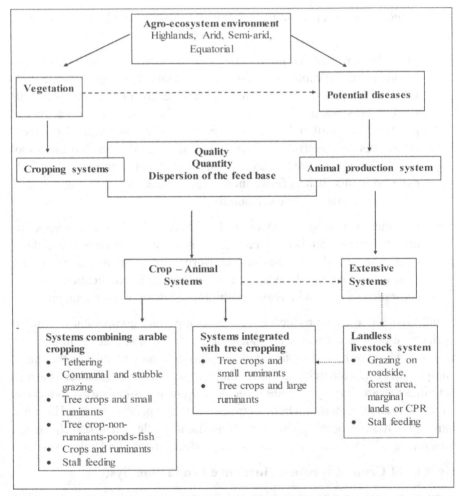

Fig. 1: Genesis and types of animal production systems

Types of Integrated System

Integrated farming systems are the practices of agriculture in numerous ways in different locations. The systems vary according to the type of agro-ecological zones, biophysical environment, extent and quality of the natural resources available, and the level of poverty with resource-poor farmers. It is important to keep in perspective the terms integration and integrated systems. Integration involves various components, namely crops, animals, land and water. Integrated systems refer to approaches that link the components to economic, social and ecological perspectives. The process is holistic, dynamic, interactive, multi-disciplinary and promotes efficiency in natural resource management. The integration of various crops and animals enables synergistic interactions, and result in a greater additive and total contribution than the sum of their individual effects (Edwards *et al.,* 1988).Crop-livestock integration involve the natural resources (crops, animals, land and water) in which these sub-systems and their synergistic interactions have a significant positive and the greater total effect than the sum of their individual effects (Edwards *et al.,* 1988). The management and use of the natural resources in a mutually reinforcing manner, enables ecological and economic sustainability. Two broad categories of integrated farming systems in dryland areas of India are identified:

Systems combining animals and annual cropping

- Systems involving non-ruminants, ponds and fish
- Systems involving ruminants.

Systems combining animals and perennial cropping

- Systems involving ruminants
- Systems involving non-ruminants.

In general, systems combining annual crops are more common in irrigated and lowland areas where water is available and enables intensive crop production. Systems combining animals and perennial crops are more common in rainfed areas. Between the two systems, dairying involving mainly cattle and buffalo is very common in the first system involving annual cropping. These are the areas, which are usually irrigated where intensive cereal cropping is common, and a distinctive market pull exists because of the peri-urban demand for milk. Thus, dairy production is an expanding feature in these situations in which it is integrated with crop cultivation and provides an economic motivation for farming systems.

Although dairy production is more commonly integrated with annual crops, mainly cereals, in some cases it is also practiced in areas where tree crops are cultivated. Good example is dairying in coconut growing areas in the South India. Widely spaced trees allow for intercropping with grasses, and the use of multipurpose trees as fences such as Leucaena and Gliricidia to provide forage for livestock.

Characteristics of Crop-Livestock-Horticulture Integration

Livestock can enhance farm productivity by intensifying nutrient and energy cycle. Stubble fields and other crop residues, e.g. after threshing, are important sources of forage in small holder systems. Weeds from cultivated fields, lower mature leaves stripped from standing crops, plants thinned from cereal stands, and vegetation on fallow fields offer additional fodder resources related to food cropping. When animal consume vegetation and produce dung, nutrients are recycled more quickly than when the vegetation decays naturally. Grazing livestock transfer nutrients from the range to cropland and concentrate them on selected areas of the farm. The livestock themselves can do the work of collecting, transporting and depositing the nutrients and organic matter in the form of urine and dung. The main crop-livestock interactions in integrated farming systems are presented in Table-1 (Devendra, 1997).

Table 1: Potentially important perennial crops and their beneficial effects

Crop	IFS model	Preferred animal species	Estimated profitability
Citrus	Citrus-sheep	Small ruminants	More profitable through addition of manure and urine
Cashew	Cashew-small ruminants	Large and small ruminants	More profitable than coconut alone
Coconuts	Coconut-crop-animals	Large/small ruminants and poultry	More profitable than coconut alone. Increase in nut and copra yield.
	Coconut-dairy-poultry		Increased return with livestock by 59%
Fruit trees	Mango-stylo-shhep	Sheep	More profitable through addition of manure and urine
Oil palm	Oil plam-cattle	Large and small ruminants	Saving in wedding costs

Table 2: Crop-livestock interactions in mixed farming systems

Crop production	Livestock production
• Crops provide by-products in the form residues and also concentrates that can be utilized by the livestock. • Improved forages can be introduced into cropping systems to provide feed for livestock.	• Large ruminants (cattle, buffaloes) provide draught power for land preparation and for soil conservation practices. • Both ruminants and non-ruminants provide manure and urine for the maintenance and improvement of soil fertility. In many

Contd.

- Native pasture, improved pastures and cover crops grown under perennial tree crops can provide grazing for livestock.
- Perennial trees provide valuable shade for livestock, which reduces heat stress.

farming systems it is the only source of nutrients for cropping.

- Animal grazing vegetation under tree crops can control weeds and help to increase yields of the plantation crops. They are reduced weeding costs by 16-35%)· The effective utilization of the feeds from the horticulture crops gives valuable animal products such as milk, meat and eggs.
- The sale of animals and animal products and the hiring out of draught animals can provide cash income.
- Animals provide entry points for the introduction of improved forages into cropping systems. Herbaceous forages can be under sown in annual and perennial crops and shrubs or trees established as hedgerows in integrated farming systems.
- Integration of livestock adds value to the horticulture crops, and can demonstrate total factor productivity.

Relevance and Potential Importance of Integration

The relevance of crop-livestock integration is associated with the characteristics of the systems, and the management and use of the production resources. In particular, the nature and extent of the crop-livestock interactions are especially important, in relation to the negative and positive benefits, effects on the environment and sustainable agriculture. The potential importance of the system from a research and development perspective is directly related to three considerations (Devendra, 1996):

- Increased productivity in the future through strategic interventions will involve the preponderance of small farm systems, which currently pro-duce over 90% of food from livestock.

- Improved efficiency in the use of natural resource, and

- Development of sustainable agriculture that is consistent with poverty alleviation, food security, and environmental integrity.

The stress on the use of natural resources (land, water, crops and animals) will be unprecedented, and will necessitate efficient use in a way that will ensure productivity enhancement and environmental integrity (Figure 2). Among the natural resources, land-use pattern will be paramount. In the past, the major productivity of cereals through the "Green Revolution" came mainly from the irrigated areas. With it came many benefits and prosperity, especially to rice and wheat farmers, but the higher agricultural growth exacerbated poverty and food insecurity among the poor in rainfed areas. Given the fact that these areas are overused, attention now shifted to the rainfed areas. Examples of such demonstrable impact are (Devendra, 2000):

- Sustainable crop-livestock systems that involve nutrient recycling efficiency, with no pollution.

- Development of food-feed systems that provide increased feed for animal throughout the year, increased crop yields, and soil fertility.

- Improved livelihoods, promotion of stable household and surveillance, and increased economic output and rural development.

Diversification into livestock keeping extends the risk reduction strategies of farmers beyond multiple cropping and thus increases the economic stability of the farm system. Spreading risk by practicing both crop and livestock production may lead to lower productivity within each sector than in specialized farms, but total production per unit area may even be increased, as both crop and livestock yields can be gained from the same area of land. Diversification is central to such efforts, and integrated farming systems also have other distinct advantages as follows:

- Reduction in, and spread of, socio-economic risks through diversification

- Promotion of linkages between system components (land, water and crops);

- Generation of value-added products (e.g. Utilization of fibrous crop residues to produce meat, milk and fibre);

- Supply of draught power for crop cultivation, transportation and haulage operations;

- Contributions to soil fertility through nutrient cycling (dung and urine);

- Contribution to sustainable agriculture and environmental protections;

- Prestige, social and recreational values, and development of stable farm households.

The economic importance of animals within the integrated farming systems is considerable and is often underestimated. This value increases with a shift from subsistence agriculture to the more open market economics, to include specialization and intensification of the production systems.

Food production is the primary objective, but the role of animals clearly surpasses this function. Within integrated systems, animals play a particularly vital role, the extent of which is dependent on the type of production system, animal species and scale of the operation. In this context, livestock production is becoming an increasingly important in dryland regions, in which this component generates significant, and more importantly, daily cash income, as well as contributing to the improvement of the livelihoods of very poor people and the stability of farm households. It is for these reasons that livestock, particularly, dairying in India is considered to be an important instrument of social and economic change, and is identified with rural development (Kurien, 1987).

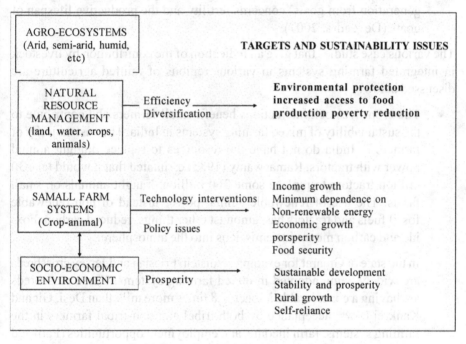

Fig. 2: Sustainable crop-livestock-horticulture integrated system

Economic Benefits of Crop-Livestock-Horticulture Integration

The economic benefits due to positive crop-animal-soil interactions are especially significant. It is relevant to consider the magnitude of economic benefits from livestock in integrated systems. A review of the existing information as well as from more recent studies suggests the following results with reference to the use of livestock:

i)   Increased animal production and income

ii) Increased yield of feed and fodder resources

iii) Savings in weeding costs

iv)   Internal rate of return: The IRR of cattle under integration was 19% based on actual field data. Haji Basir Ismail (2005) calculated the economic returns from four hectares of land under oil palm, inter-cropping as well as fodder cultivation for a seven year period. The beneficial incomes generated as a percentage of total incomes in favour of integration for cattle, sheep and goats were 44.4%, 86.6% and 91.5% respectively. The income from goats was the highest followed by cattle.

v)   An important biological advantage influencing productivity and income generation from goats' concerns fertility and the productive lifespan of goats (Devendra, 2007).

The various case studies that give an indication of the contribution by livestock in integrated farming systems in various regions of rainfed agriculture are discussed below:

- The crop-livestock interactions benefit small farmers and contribute to the sustainability of mixed farming systems in India. The vast majority of farmers in India do not have the resources to replace draught animal power with tractors. Ramaswamy (1985) estimated that it would take 30 million tractors to replace some 300 million draught animals on small farms. The use of renewable animal power instead of non-renewable fossil fuels and tractors has, amongst other things, reduced carbon dioxide and carbon monoxide emissions into the atmosphere.

- In the state of Gujarat for example, Holstein Friesian and Jersey crossbreds are widely being adopted in mixed farming systems. These crossbreds are having a considerable impact: 1.8 times more milk than Desi, Gir and Kankrej cows, acceptance by both tribal and non-tribal farmers in the farming systems, farm income, and employment opportunities (Patil and Udo, 1997). In recent years, dairying in integrated systems has significantly contributed to the development of peri-urban dairy operations through the supply of animals in several areas. Pregnant cows are often brought from the rural areas into peri-urban systems to promote milk production.

- Economic analysis of different farming systems (one hectare of irrigated land or 1.5 ha of un-irrigated land) indicated that under irrigated conditions, mixed farming with crossbred cows yielded the highest net profit, followed by mixed farming with buffalo, and arable farming. Mixed farming with Haryana cows made a loss (Singh et al., 1993).

- Comparative productivity and economies of dairy enterprises (mixed farming with three crossbred cows on one hectare of canal irrigated land versus mixed farming with three Murrah buffalo) indicated that mixed farming with crossbred cows under canal-irrigated conditions was more efficient for the utilization of land, capital, inputs and the labour resources of the farmer (Kumar et al., 1994).

- Baseline surveys in Gujarat, India, indicated that around 75 per cent of rural households kept cattle in the face of under-employment. More particularly, the farm surveys showed that cattle kept mainly for milk, contributed 32 per cent and 20 per cent for tribal and non-tribal ethnic groups respectively (Patil and Udo, 1997). By comparison to cows and buffaloes, lactating goats contributed between 54-68.9 per cent to total farm income through the sale of milk (Deoghare and Bhattacharyya, 1993 and 1994; Deoghare and Sood, 1994).

- Studies of Chinnusamy et al. (1994) between 1988-1993 on a 1 ha farm integrating crops (grain and fodder); silvipasture (trees and grass) and goats indicated that soil physical and chemical characteristics were all improved, along with the socio-economic conditions of the farmer.

Factors Influencing the Integration Process

The shaping of the integration process depends to a large extent on farm management. To arrive at more adaptable and sustainable development, all aspects of the farming system must be taken into account, as well as the context within which resource poor farmers operate (Fig. 3). In rainfed areas, resources are scarce and the agro-ecological situation is often vulnerable. Under these conditions farmers tend to follow a survival strategy: they are only prepared to adopt an innovation if they are reasonably convinced that they will not lose anything. Experience shows that any innovation is more acceptable to farmers, if it involves a step-by-step development of the existing farming system. The integration process is also promoted by further developing the role of man himself in animal production, especially that of women as well as by improving institutional services.

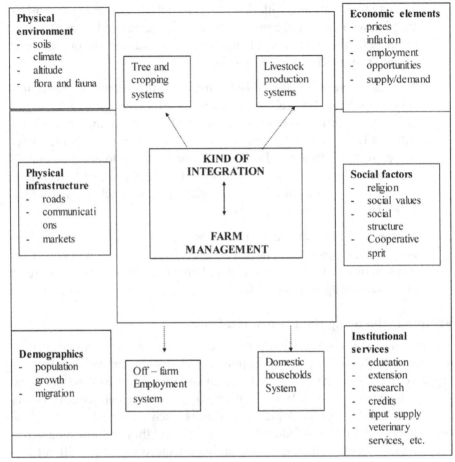

Fig. 3: Factors influencing the integration process

Constraints to Livestock Production

The productivity of the livestock in rainfed areas is lower than those of irrigated areas because of several factors. In most small farm situations, land is a limiting factor, but small farmers try to maximize production through diversification of the available resources and efficient use of low cost inputs. Within an integrated system where livestock is an important component, there are two major constraints to production, firstly the availability of improved genotypes and secondly, feeds and nutrition.

Concerning the availability of improved genotypes, cattle crossbreeding programs have lacked co-ordination and have been further constrained by problems of infertility, instability of the crossbreeds and inefficient artificial insemination services at the farm level. The level of exotic blood in the crossbreeds, is highly variable, and ranges from 25-75 per cent in small farms. The overriding issue is inadequacy of numbers and their instability, resulting in inability to intensify and expand commercially.

The availability and quality of feed resources and efficient nutritional management is the principal constraint related to feeds and feeding. The problem is also exacerbated by the higher nutrient demand of improved dairy animals, for example crossbreeds, for milk production. The feeds available include grasses and forages, crop residues, agro-industrial by-products and non-conventional feed resources.

Management Strategies for Improving the Livestock Productivity

There is need for an effective policy and planning that will optimize development resources and provide the necessary support and economic environment to livestock resources to express its full potential. An understanding of the production factors (livestock, capital, feed, land and labour) and processes (description, diagnosis, technology design, testing and extension) that affect animal production is a pre-requisite for livestock integration in land use diversification (Misra, 2006 and 2009). The livestock production tends to be more complex than crop production because livestock too often play a pivotal role in the overall farming system. Any constraint imposed on livestock may also restrict the system as a whole. Therefore, the integration of livestock production in rainfed farming must take all the factors of production in to account. In general, the aims of livestock production in integrated farming systems are, to:

- Conserve the natural resource base;

- Raise productivity through better utilization of available resources;

- Expand production where there is a sufficient demand and resources can be utilized at reasonable cost to the environment

- Optimize the allocation of development resources through rational management.

The selection and application of animal production technologies should confirm with the framework of sustainability criteria, they should be ecologically sound, economically viable, socially acceptable and technically appropriate. It is of course never easy to find a technology that will meet all these requirements. However, it is always most important to keep in mind that without true recognition and serious consideration of these prerequisites for future animal production, human survival will be at risk. Therefore any short-term gain must be regulated to avoid possible adverse effects on the long-term survival of the mankind.

Using these criteria, the production systems in existence do not pass this litmus test. The most sustainable system, if given a proper extension back-up and credit support in the long term is the crop-livestock mixed system at the small farm level. The major challenge of increased resource use by small farm system in rainfed areas concerns collective and holistic approaches, which can shift the current level of productivity from the traditional low-inputs system to an intermediate and market-oriented situation (Figure 4). The real challenge is the development of sustainable systems at intermediate level, and the shift to a market-oriented situation in which maximum productivity is consistent with agricultural growth, poverty reduction, food security, environmental sustainability and self reliance (Devendra, 2000 and 2011; Misra, 2006 and 2015)

Potentially important technologies that can make a significant increase in productivity both of crops and animals with in the system should be adopted, which consequently increase farmers income and total food will supply to meet the demand of raising human population (Bhat and Taneja, 1998). The following strategies of livestock production may be implemented in rainfed conditions for sustainable agricultural development.

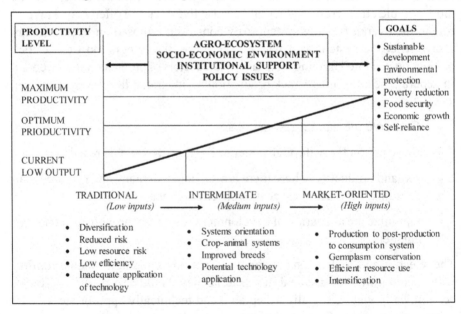

Fig. 4: Potential levels of productivity feasible from traditional market-oriented systems and the development of IFS for small farms

Strategies for increasing feed availability

The strategy to increase feed availability within small farm systems should have the final objective of developing sustainable all-year-round feeding system appropriate to the prevailing situations and the availability of feeds. In this quest, maximizing feed production is essential and the following approaches are feasible (Misra *et al.*, 2009 and 2015):

- Short duration fodder crops should be introduced in present cropping systems. Crop like sorghum, cowpea, guar are suitable as short duration crops for the rainfed areas.

- Quick growing leguminous crops like cowpea, sesbania should be grown in the inter spaces of widely spaced row crops of grain sorghum and maize.

- The degraded grazing lands should be converted into production systems like silvipasture. *Cenchrus* sp. amongst grassess, *Stylosanthis hamata* amongst range legumes and *Leucaena leucocephala* and *Sesbania sesban* as shrubs and trees have great potential to increase forage resources from the degraded lands.

- Effects should be made to identify and utilize non-conventional feed resources on a continuous basis to increases feed resource base.

- Timely harvesting and chaffing of fodder to reduce wastage, and judicious use at farm level.

- Appropriate fodder conservation measure during the period of abundance availability should be adopted for meeting the forage requirement during lean period.

The key elements in the use of these approaches are the quantification of the feeds produced from various sources, e.g. cropping systems throughout the year, and the efficiency with which they are utilized by the available animals.

Strategies for enhancing feed utilization

The scope for increasing the efficiency of feed utilization in effective production system is enormous. Innovative feeding practices are necessary that can sustain all-year-round feeding in more intensive systems of production. These could include the chopping, urea treatment, supplementation, use of multi-nutrient block, etc.

Chopping of fodder

Chopping of fodder should be popularized for judicious use at farm level. The advantages of feeding chaffed feed are that it avoids wastage and prevents selective consumption. Feeding of chopped roughage reduces the energy wasted while chewing, increases the feed intake and improves digestibility. The net biological value of the feed also improves.

Urea treatment of low-grade roughages

Low grade roughages such as paddy straw, wheat bhusa, sorghum kadabi, maize stover, dry grasses and other edible farm waste contain negligible amount of digestible crude protein and higher amount of non-digestible cell wall constituents. Through urea treatment, nutritional value of poor quality herbage could be improved and made more palatable. Four percent urea, 50 liters water per 100 kg straw/ herbage and 15 to 21 days incubation period are optimum for treatment. Urea treatment is quite flexible, as it can be adapted to local conditions. Urea treatment, apartfrom being a source of nitrogen for microbial synthesis, also provides additional energy due to the weakening/loosening of the lingo-cellulose bonds in the treated straw. Treatment improves dry matter intake by 7 to 10 units and digestibility by 4 to 15 units. For daily feeding required quantity of treated forage be removed from the heap and fed to the animals.

Strategic supplementation

A variety of supplements exist that can be used for feeding animals. These include oil meals and cakes as well as leguminous tree forages such as Leucaena and Gliricidia. Purchased concentrates (mainly energy and proteins) are expensive and their use can only be justified in relation to (i) scarcity or inadequacy of dietary nutrients for milk production (quantity and quality) (ii) restriction in energy uptake imposed by bulky roughages (iii) relatively low price of alternative mixed feeds, home grown or purchased concentrates (iv) increased milk yield where monetary value is greater than the cost of the concentrates required to produce it.

Urea molasses mineral block (UMMB) commercially available and can be used to supplement the low quality roughages to balance the deficient nutrients in the ration. UMMB contains soluble and fermentable nitrogen from urea, highly fermentable energy from molasses, and essential minerals. Natural protein sources such as groundnut or cottonseed extract have also been added to provide preformed peptides and amino acids. UMMB has been found to improve the dry matter intake of the basal roughage and the feed digestibility. The nutrients from the block are well utilized by the animals and UMMB supplementation improves reproductive performance of livestock due to enhanced availability and utilization of nutrients, particularly micronutrients.

Improved animal health care

Animal diseases seriously reduce animal productivity and also cause major economic losses. There are certain diseases, which are more common in rainfed areas need, more attention so as to prevent its outbreak. The most common diseases are Foot and Mouth disease, Hemorrhagic septicemia, Black Quarter,

Anthrax, Enterotoxaemia, Ring worm, Ascariasis, Tick infestation and mange, etc. To control and prevent these diseases, regular vaccination and deworming schedule should be followed.

The Way Forward

The integration of livestock with perennial crops in rainfed farming will increase both short-term benefits and long-term sustainability through optimum utilization of available resources. To sustain the production potential even by low producing animal's farmers should adopt the feeding management based on appropriate nutritional technologies. Integrated tree-crops-livestock systems are potentially very important, but are underestimated. The inclusion of animals provides the entry point for development, and has the twin advantages of increasing the supplies of animal proteins and also value addition to the horticulture crops. The benefits of integration are considerable and are mediated through positive crop-animal-soil interactions. A combination of clear policy, increased technology application, more intensive production systems, increased investments and private sector participation can significantly accelerate the wider adoption of the systems and demonstrable environmental sustainability. The systematic on-farm research is required in developing methodologies for the process of integrating ruminant species with tree crops, as well as studies on the nature (positive and negative), extent and impact of crop and animal interactions on environmental indicators. The looming effects of climate change which will be mediated through heat stress on animal performance and productivity remains largely unknown, and will need specific research effort.

References

Bhat, P.N. and Taneja, V.K. 1998. Sustainable animal production systems in India: *Issues and Appraoches.* 68:8: 701-712.

Chinnusamy, C., Sivasankaran, D. and Rangasawmy, A. 1994. Sustainable integrated farming systems for rainfed upland farm in Southern Peninsular India. In: Proceedings Third Asian Farming Systems Symposium, 7-10 November 1994. University of the Philppines, Los Banos. Phlippins, pp. 46. 1994.

Deoghare P.K. and Bhattacharyya N.K. 1993. Economic analysis of goat rearing in Mathura district of Uttar Pradesh. *Indian Journal of Animal Sciences* 63:439-444.

Deoghare P.K. and Bhattacharyya N.K. 1994. Economics of Jamunapari goat rearing in Etawah district of Uttar Pradesh. *Indian Journal of Animal Sciences* 64:1390-1393.

Deoghare P.K. and Sood S.B. 1994. Income and employment potential of goat rearing on farms in the rural households of Mathura district of Uttar Pradesh. *Indian Journal of Animal Sciences* 64:295-300.

Devendra, C. 1993. Sustainable animal production from small farm systems in South East Asia. FAO Anim. Prod. Health paper, No.106, Rome, Italy, IX.p.143.

Devendra, C. 1996. Overview of integrated animals-crops-fish production systems: achievements and future potential. Proc. Symp. Integrated Systems of Anim. Prod in the Asian Region, ,Chiba, Japan.pp.9-22

Devendra, C. 1997. Improvement of livestock production in crop-animal systems in the rainfed agro-ecological zones of South East Asia. ILRI (International Livestock Research Institute), Nairobi, Kenya. pp 107.

Devendra, C. 2000. Animal production and rainfed agriculture in Asia: potential opportunities for productivity enhancement. *Outlook on Agriculture*, 29:3: 161-175.

Devendra, C. and Thomas D. 2002. Smallholder farming systems in Asia. *Agril. Systems*, 71:1-2: 17-25.

Devendra, C. 2011. Integrated tree-crops-ruminants systems in south East Asia: Advances in productivity enhancement and environmental sustainability. Asian-Aust. *J. Anim. Scie.*, 24:5: 587-602.

Duckham,A.M. and O.D Masefiled. 1970. Farming systems of the World. Chatto and Windus, London, U.K.

Edwards, P., R.S.V. Pullin and J.A Gartener, 1988. Research and education for the development of crop-livestock-fish farming systems in the tropics. ICLARM Studies and Rev. No. 16. p.53.

Kumar H., Singh J.N., Kadian V.S., Singh K.P., Saxena K.K. and Kumar H. 1994. Comparative productivity and economics of dairy enterprises under mixed farming systems. Farming Systems 10:36-44.

Kurien V. 1987. True dimensions of dairy development. In: Gupta V. (ed), Dairy India. Third edition. New Delhi, India. pp. 21-24.

Misra and Mahipal, 2000. Strategies of livestock management for improving productivity under rainfed farming situations. In K.H.Vedini (eds). Management issues in rainfed Agriculture in India., MANAGE, Hyderabad. Pp.56-68.

Misra, A.K. 2006. Integration of livestock in land use diversification. In: Land use diversification for ustainable rainfed agriculture. K.D. Sharma and B. Soni (Eds). Atlantic Publishers & Distributors, New Delhi. 277-299.

Misra A.K., C.A. Rama Rao, and Y.S. Ramakrishna. 2009. Improving dairy production in India's rainfed agro ecosystem: constraints and strategies. *Outlook on Agriculture*, 38:3: 284-292.

Misra, A.K., B.K. Mathur, R.N. Kumawat, M. Patidar and M.M. Roy. 2015. Strategies for enhancing livelihood of small holders through livestock production in drylands of India. In: Livestock Production Practices for Small farms of Marginalized Groups and Communities in India. (Eds.: Girin Kalika and Prashta, Saikia).133-144

Patil B.R. and UdoH.M.J. 1997. The impact of crossbred cows in mixed farming systems in Gujarat, India: Milk production and feeding practices. Asian-Australasian *Journal of Animal Sciences* 10:253-259.

Ramaswamy, N. S. 1985. Draught animal power- socio-economic factor. In: Copland, J.W. (Ed.). Draught animal power for production. ACIAR Proceedings 10. ACIAR, Canberra, Australia, pp.20-25.

Ruthernberg, H. 1980. Farming System in the Tropics. 2nd Edn. Clarendon Press, Oxford, UK. P.424

Sere, C. and H.Steinfield, 1966. World livestock production systems: current status issues and trends, Anim. Prod. And Health Paper No. 127, FAO, Rome, Italy. P.82.

Singh, H.P., 1999. Management of rainfed areas. In G.G Singh and B.R. (eds), div. Od NRM, ICAR, Krishi Bhavan, New Delhi, pp. 339-576.

Singh K.P., Singh S.N., Kumar H., Kadian V.S. and Saxena K.K. 1993. Economic analysis of different farming systems followed on small and marginal land holdings in Haryana. *Haryana Journal of Agronomy* 9:122-125

17

Small Ruminants in Horti-based Farming System

B.K. Mathur

ICAR-Central Arid Zone Research Institute, Jodhpur, Rajasthan-342003

Indian hot arid zone which is about 12% of total geographical area of the landmass of 0.32 million km^2 has maximum covering in western Rajasthan i.e. 61% of the total area whereas the other areas in arid region are available in the states of Gujarat, Punjab, Haryana, Andhra Pradesh and Karnataka accounting for 20, 5, 4, 7 and 3% of hot arid area whereas the cold arid area of 8.4 million ha lies in the state of Jammu & Kashmir covering the Leh and Ladakh region.

The hot arid area is characterized by frequent droughts of 47% of frequency in the last century of moderate to severe nature. Due to higher occurrence frequency of droughts in this region of state, the livestock assumes great importance as a drought management measure as agriculture is at the mercy of rainfall pattern which is very uncertain in amount as well as distribution coupled with poor soil condition, higher evapotranspiration and higher wind velocity causing the soil erosion.

For arid region of Rajasthan the general climatic conditions, topographical features and biotic factors do not encourage agricultural operations in the absence of extractive industry the peasantry has to fall upon animal husbandry as their main occupation. Rearing some of the finest breed of cattle, camel, sheep and goats known for their endurance making much use of the meagre feed resources which are grasses ,herbs, shrubs, tree leaves and cultivated feed and fodder crops. Livestock farming has some in built superiority over crop farming as far as growth; stability and resource conservation are concerned. On an average, the region experiences 3 years of drought in every 10 years. The natural forces constituting the soil-climatic complex, which conspire to reduce the crop productivity and cause instability in agricultural production, have much less impact on livestock farming. This is due to differences in the nutritive value of natural vegetation, which mainly sustains livestock.

Livestock status in arid region of India

The hot arid zone of Rajasthan is comprised of 12 districts of the state lying in the western part and these are Barmer, Bikaner, Churu, Hanumangarh, Jaisalmer, Jalor, Jhunjhunu, Jodhpur, Nagaur, Pali, Sriganganagar and Sikar and this arid region of Rajasthan has livestock population of 27.55 million which is about 57% of the total population of the state. Of the total livestock species available in the state sizable number of crossbreed cattle (50.59), indigenous cattle (37.36%), buffaloes (30.66%), sheep (73.31%), goats (49.77) and camel (79.84%) are available in arid zone. Excepting the districts of Hanumangarh and Sriganganagar where the large ruminant production system of cattle and buffalo is dominant having about 54% of total livestock population but in all other districts of arid zone the small ruminants production system accounting from 55% to 77.5% of total livestock population is prevalent. The camels are essentially the animals of arid region having about 80% share of the total population of state (Livestock census, 2003).

According to the latest 2003 animal census, Rajasthan harbours 49.14 million heads of domestic animals while it was 54.65 million in 1997, thus a decrease of 10.08% between 1997 and 2003 mainly due to drought years (Table 1 and 2).

Table 1: Trend in Livestock Population (millions) of Arid and Non arid district of Rajasthan.

S.No	Livestock	Area of Raj.	1997 Census % of Raj	2003Census % of Raj	2007Census % of Raj	2012 Census % of Raj
1	Cattle	Arid districts	4.96	4.12	5.02	6.18
			(40.78)*	(37.97)	(40.46)	(46.38)
		Non Arid districts	7.20	6.73	7.39	7.14
		Total	12.16	10.85	12.41	13.32
2	Buffaloes	Arid districts	3.16	3.20	3.43	3.95
			(32.37)	(30.65)	(29.77)	(30.41)
		Non Arid districts	6.60	7.24	8.11	9.03
		Total	9.76	10.44	11.54	12.98
3	Sheep	Arid districts	10.46	7.35	7.94	6.88
			(73.30)	(73.5)	(70.37)	(75.75)
		Non Arid districts	3.82	2.65	3.34	2.20
		Total	14.31	10.00	11.28	9.08
4	Goat	Arid districts	9.53	8.36	11.77	12.79
			(56.25)	(49.76)	(53.80)	(59.03)
		Non Arid districts	7.41	8.44	10.11	8.88
		Total	16.94	16.80	21.88	21.67
5	Camel	Arid districts	0.53	0.40	0.35	0.287
			(70.66)	(80)	(82.14)	(85.19)
		Non Arid districts	0.21	0.10		0.048
		Total	0.75	0.5	0.43	0.325

Contd.

6	Total	Arid districts	28.57	. 7.5	29.08	30.18
	Livestock		(52.56)	(56.0)	(50.23)	(52.27)
	(including,	Non Arid districts	25.77	21.6	28.81	27.55
	horse,	Total	54.35	49.1	57.89	57.73
	pony, etc.)			(10.12)**	(10.94)**	(8.87)**

*Data in parenthesis is percent population of Rajasthan state
**Data in parenthesis is percent population of India

Table 2: Trend of Livestock Population 1997-2012 in Arid Zone of Rajasthan (Million)

Livestock species	Livestock Census Year				% change		
	1997	2003*	2007	2012	1997-12	2003-12	2007-12
Cattle	4.96	4.12	5.05	6.18	24.59	50.00	22.38
Buffaloes	3.16	3.2	3.45	3.95	25	23.44	14.49
Sheep	10.49	7.36	8.43	6.88	-34.41	-6.52	-18.39
Goat	9.53	8.37	11.7	12.79	34.21	52.81	9.32
Camel	0.53	0.39	0.35	0.28	-47.17	-28.21	-20.00
Others	0.22	0.18	0.13	0.11	100	-38.89	-15.38
Total	28.57	23.62	29.11	30.18	5.63	27.77	3.68

*Between 1997 -2003 drought years

Components of animal husbandry

Broadly scientific Animal Husbandry has three major components

1. Breeding
2. Feeding
3. Health

Livestock breeds in arid region of India

The superiority of livestock farming for development of arid region is highlighted by the fact that this region is endowed with some of the best breeds of livestock and drought-hardy perennial grasses. The local breeds of livestock have acquired certain characters to withstand the arid climate, and the characters have been transmitted through generations to make the present hardy breeds of animals. However, due to lack of proper nutrition, genetic potentiality of these animals has not been expressed to its maximum level. The major livestock breeds in the region are:

Cattle: Tharparkar, Rathi, Nagori and Kankrej.

Sheep: Marwari, Jaisalmeri, Chokla, Nali,Magra, Pugal and Sonadi.

Goat: Marwari, Kutchi and Parbatsari.

Camel: Bikaneri and Jaisalmeri.

Feeding materials for livestock in arid region of India

The Common available feeding material from natural rangelands (post monsoon) in arid region with crude protein percentage has been given in table 3.

Utilization of tree fodders resources

These resources include leaves of various fodder trees, shrubs and pods. A large number of trees and shrubs have been identified which are being lopped for fodder purpose. Important tree fodders and utilization of their nutritive value have been reviewed by many authors (Patnayak, *et al.*, 1995; Patnayak, 1992; Sharma *et al.*, 1992; Singh, 1981) (Table 4-7).

Although tree fodders are considered as emergency fodders, they serve as potential feed resources for small ruminants and provide about 50% of the total dry matter consumption. They are utilized as freshly loped leaves or by conserving as dry fodders. Dry matter in most trees varies from 12 to 20% on dry basis. Very high calcium and low phosphorus are a common feature in almost all tree fodders which when fed solely leads to a very wide Ca:P ratio. Dry matter intake from most of the tree leaves varies from 2.5 to 3.5 kg/100 kg B W. Although they are rich in protein, the digestibility of crude protein is very low (25 to 30%) in many cases. In spite of higher intake of nitrogen and calcium, retention of nutrients in general is poor and phosphorus balance is negative or marginally positive. Tree fodders, therefore, should be fed in mixtures with other feeds, specially containing high phosphorus, such as wheat bran, rice bran, grains etc.

Although tree fodders are consumed by all livestock, goats have special preference compared to others. The digestibility of nutrients in tree fodders in generally higher in goats with better growth rates than in sheep when fed dried tree leaves (Bohra, 1980; Singh and Bhatia, 1982). Low digestibility of protein is anti-nutritional factors such as mimosne, cyanogens, saponins, oxalates etc. The deleterious effects of various anti-nutritional factors and their possible amelioration have been reviewed by Kumar (1991).

Table 3: Common available feeding material from natural rangelands (post monsoon) in arid region with crude protein percentage

S.No.	Botanical name	Local name	C.P.%
	Grasses		
1.	Aristida funiculate	Lapra	4.38
2.	Brachiaria ramosa	Murat	11.37
3.	Cenchrus biflorus	Bhurat	15.31
4.	Cenchrus ciliaris	Dhaman	8-12
5.	Cenchrus setigerus	Moda Dhaman	8.77
6.	Cynodon dactylon	Dob	10.10
7.	Dactyloctenium aegypticum	Ganthia	10.10
8.	Eluesine comprass	Tantiya	13.99
9.	Eragrostis tremula	Chidi grass	6-10
10.	Lasiurus sindicus	Sewan	9.05
11.	Panicum antidotale	Gramna	7.1
12.	Tetrapogon tenellus	Kagiyo	13.36
13.	Urochloa panicoides	Kuri	12.02
	Weeds		
14.	Achryanthes aspera	Antipada	19.9
15.	Amaranthus viridis	Chandelio	4.9
16.	Celosia argentia	Makhamal	12.5
17.	Cyperus rotundus	Motha	9.1
18.	Digera muricata	Lolru	12.07
19.	Indigofera cordifolia	Bekario	18.0
20.	Trianthema portulacastrum	Sathi	14.31
21.	Tribulus terristris	Kanti	18.18
	Trees/shrubs		
22.	Acacia nilotica	Desibabul	13.9
23.	Acacia senegal	Kumat	17.48
24.	Acacia tortilis	Israili babool	14.6
25.	Azadirachta indica	Neem	15.8
26.	Calotropis procera	Aak	13.5
27.	Capparis decidua	Kair	2.05
28.	Cordia gharaf	Gundi	15.3
29.	Euphorbia caducifolia	Thor	5.61
30.	Prosopis cineraria	Khejri	14.63
31.	Tecomella undulata	Rohida	12.2
32.	Prosopis juliflora (Pods only)	Vilayati babul	13.8
33.	Salvadora persica	Metha Jal	10.5
34.	Ziziphus nummularia	Jharberi	14.1

Source: Mathur B. K. 2008

Proper methods of collection, conservation and processing are important aspects of utilization of tree fodders such as *Ailanthus excelsa, Z. nummularia* and *P. cineraria* were used in complete feeds for meat lambs (Bhatia and Patnayak 1982). The proximate analysis of some common top feeds particularly of desert region is given in table 4.

Table 4: Proximate components (% DM basis) of some desert top feeds

Scientific name	Local name	Crude protein	Ether extract	Crude fibre	Nitrogen free extract	Ash	Phosp-horus	Calcium	Magne-sium
Acacia nilotica	Babul	13.9	-	9.2	69.8	7.1	0.10	2.6	0.4
Acacia senegal	Kumut	10.3	-	9.5	65.7	16.4	0.05	6.9	0.6
Ailanthus excelsa	Ardu	19.6	3.7	13.5	47.7	15.5	0.20	2.3	-
Albizia lebbeck	Siris	16.8	4.0	31.5	36.2	11.5	0.15	2.57	-
Azadirachta indica	Neem	12.4	3.5	11.4	66.6	9.4	0.10	3.4	0.6
Dalbergia sissoo	Shisham	16.2	4.2	24.6	46.2	9.2	0.21	2.3	-
Ficus bengalensis	Bargad	9.6	2.6	26.8	51.6	9.2	-	-	-
Ficus religiosa	Pipal	9.7	2.7	26.9	45.8	9.3	-	-	-
Prosopis cineraria	Khejri	13.9	1.9	20.3	59.2	6.5	0.20	1.9	0.5
Prosopis juliflora	Vilayati babul	21.4	-	20.8	50.0	7.7	0.20	1.5	0.5
Salvadora oleides	Pieujal	9.6	-	9.3	40.2	40.8	0.10	11.9	0.7
Salvadora persica	Kharajal	14.2	-	9.4	44.9	31.4	0.20	8.8	0.7
Tecomella undulata	Rohida	12.2	-	15.8	63.0	8.9	0.20	3.4	0.9
Ziziphus nummularia	Bordi	11.7	3.0	16.0	65.0	7.2	0.20	1.6	0.3

Sources: Ganguli et al., (1964); Acharya and Patnayak (1977)

Nutritive value of some most common top feed: Leaves of *A. excelsa, Z. nummularia* and *P. cineraria*, the three most important top feeds, are also high in the TDN (total digestible nutrients) scale (64%, 50% and 41.0%, respectively), while the DCP (digestible crude protein) value of A. excelsa leaves in quite high (16%), whereas, that of the more palatable *Z. nummularia* or *P. cineraria* leaves, are considerably low (5.6% and 4.5%). Leaves of *A. excelsa, Z. nummularia* and *P. cineraria*, the three most important top feeds, are also high in the TDN (Total digestible nutrients) scale (64%, 50%, and 41.0% respectively), while the DCP (digestible crude protein) value of *A. excelsa* leaves is quite high (16%) whereas that of the more palatable *Z. nummularia* or *P. cineraria* leaves, are considerably low (5.6% and 4.5%). It is mainly attributed to presence of anti-quality factors like tannins and lignin in these feeds. Tannin content of some common desert feeds and forages, including tree leaves have been presented in Table 6.

Table 5: Palatability of the leaves of some desert trees

S. No.	Botanical Name	Local Name	Palatability rating (kg/100 kg live weight)	Test animal
1.	*Ailanthus excelsa*	Ardu	3.8	Sheep
2.	*Azadirachta indica*	Neem	4.0/6.0	Buffalo*
			0.33	Sheep*
3.	*Ficus religiosa*	Pipal	1.9	Bullock
			4.7	Goat
4.	*Prosopis cineraria*	Khejri	2.8	Goat
			1.9	Sheep
			1.4	Camel
5.	*Ziziphus nummularia*	Jharberi	3.3	Goat
			2.8	Sheep
			1.4	Camel

* Value are in kg/animal

Table 6: Tannin Content (% on dry matter basis) of Some Desert Feeds

S. No.	Plant Species	Tannin content
1.	*Acacia bevinosa*	2.4
2.	*A. nilotica*	8.0
3.	*A. salicina*	10.0
4.	*A. tortilis*	6.4
5.	*Bauhinia purpurea*	8.2
6.	*B. verigata*	5.4
7.	*Calligonum polygonoides*	22.1
8.	*Cassia auriculata*	19.5
9.	*C. marainata*	13.5
10.	*Colophospermum mopane*	10.0

Contd.

11.	*Corchorus depressus*	12.0
12.	*Dicanthium annulatum*	1.02
13.	*Fagonia cretica*, mature plant	12.8
14.	*Ficus bengalensis* (leaves)	5.4
15.	*Ficus bengalensis* (fruits)	2.2
16.	*Ficus drupacea var. pubescens*	18.9
17.	*Hardwickia binata*	9.1
18.	*Indigofera argentia*	7.1
19.	*Leucaena leucocephala*	4.7
20.	*Mytenus emarginatus*	11.2
21.	*Prosopis cineraria*	11.6, 14.8
22.	*Prosopis juliflora*	-
23.	*Ryncosia minima*	3.8*
24.	*Zizyphus mauritiana*	13.5
25.	*Z. nummularia*	11.7
26.	*Z. spinachristi*	9.6

*Plants growing on sand dunes contain tannins whereas cultivated plants devoid of tannins. 17% non-woody annuals, 14% of herbaceous perennials (P), 79% of deciduous woody P and 87% of evergreen woody P posses tannins. (Bohra and Ghosh,1977)

Un-Conventional Feed Resource of Desert Region as feed

Tumba (*Citrullus colocynthis*) Seed Cake - A Cheaper Feed Resource for Livestock Feeding in Arid Region

Tumba plant, an annual creeper, grows naturally in abidance in hot arid areas of the country with minimum possible water availability. It grows and multiply very fast during monsoon season and its fruits; Tumba - are available in the month of October-November. Mature fruits are golden yellow in colour, ball like, 10-12 cm in diameter. The taste of the fruit is very bitter. Tumba seed are rich in fat, having upto 16% oil. The Tumba (*Citrullus colocynthis*) seed cake (TSC) a by-product of the oil extraction industry is nutritionally rich as it contains 16 to 22% CP. Presently, TSC is available in abundance in arid regions and is being used as a fuel for furnances in factories, and its thus wasted.

Feeding trials were conducted in Central Arid Zone Research Institute, Jodhpur since 1986 on cattle to identify and evaluate TSC as a source of livestock feed. The studies clearly indicates that in cattle the conventional concentrates can safely be replaced by TSC to the extent of 25%, as a regular ingredient, which constituted guar (*Cymposis tetragonaloba*) korma and oil cakes (mostly til (*Sesamum indicum*) and cotton (*Gossipuim spps.*) seed cake) and also pelleted cattle feeds.

It is observed that the TSC replacement did not affect the palatability and intake of feeds and fodder. There was no significant (1>0.5) difference in milk yield pattern of control and treatment groups, in terms of quantity and quality.

TSC feeding to heifers upto calving and onwards did not show any ill effect on different reproductive parameters.

It can, therefore, be inferred that a simple practice of inclusion of tumba seed cake (TSC) in animal feed will definitely lower the cost of animal feeding by 18% to 20% without having any adverse effect on the production, general health and reproductive performance. In addition, it would result in the utilization of locally available non-conventional cheaper protein source, abridging the gape between demand and supply of the scarce protein, thus, benefiting the marginal farmers appreciably (Mathur *et al.*, 1989, Mathur, 1996).

Feeding Vilayati Babool (*Prosopis juliflora*) Pods Powder

Goats: Ten (10) goats in late lactation were divided into two group of 5 each i.e. group I control and group II treatment. Animals of each group were offered weighed quantity of roughage and concentrate on as such basis (consisting of 35% grinded bajara, 40% tumba seed cake and 25% groundnut cake). However, in treatment group 35% bajra in concentrate was replaced by Prosopis juliflora pod powder (PJPP) making ration near about isocaloric. Study showed that PJPP can be used up to 35% in the concentrate of goats (Mathur *et al* 2004). No significant effect was observed on blood parameters and milk yield of goats in late lactation, during extremes of summer in arid zone (Mathur *et al.*, 2003).

Sheep: A feeding trial was conducted to evaluate the acceptability and palatability of ration comprised of *Prosopis juliflora* pod husk (PJPH) in Marwari sheep. The results of the study indicated that *P. juliflora* pod husk can be used up to 50 percent level in the concentrate along with tumba (*Citrullus colocynthis*) seed cake as low cost ration of the sheep without any adverse effect on animal health (Mathur *et al.*, 2002).

Grazing and feeding of handmade Cheaper Concentrate
Mixture having Villayati Babool (*Prosopis juliflora*) pods powder

The acceptability and palatability of formulated concentrate mixture having *Prosopis juliflora* pods was high with no ill effect on health, increases milk and the cost of cattle production reduces significantly. (Mathur *et al.*, 2009).

Milk Yield of Lactating Marwari Goat Fed on *Prosopis cineraria* v/s *Colophospermum mopane* Leaves

Twelve (12) lactating Marwari goats were divided into 3 groups of 4 each namely T_1, T_2 and T_3. T1 and T_2 groups were stall fed for about five week while T_3 group was sent for grazing. Animals of T_1 and T_2 groups were offered balanced diet as per requirement comprising of concentrate feed and *C. ciliaris* as roughage. Additional T_1 group W.S offered with *P cineraria* leaves while T2 group replaced by *C. mopane* leaves. It was observed that T_1 group the palatability of P cineraria was 95% of the quantity offered while in T_2 group it was in reducing trend with 5% to 15% from first and fifth week of trial. The final body weights after V week of trial for T_1, T_2 and T_3 group was 31.95±1.099, 32.40 ±1.982; and 28.4 ±1.699 , kg respectively. Animals of T_1 and T_2 group maintained their body weight while the grazing group showed in average group weight. T_1 and T_2 group consumed the concentrate offered equally, however *Cenchrus ciliaris* grass kutter consumption was higher in T_1 group (1.65 to 1.94 kg/day/group) as compared to T_2 group (1.41 to 1.72 kg/day/group). Water intake in groups T_2 and T_3 ranged from 7.40 to 10.62 and 6.14 to 9.0 liters/day. The total milk yield on an average / animal for a period of 45 days in T_1, T_2 and T_3 was 28.07., 22.53 and 19.73 litres respectively. The milk yield was 19.74% higher in *P. cineraria* fed group (T_1) than *C. mopane* fed group (T_2) and 29.71% more than T_3 group maintained exclusively on grazing. Thus, it was observed that the palatability of dry *Colophospermum mopane* leaves was low and decreased with progress of feeding from 15 to 5% from 1st to 5th week and the traditional local *P. cineraria* leaves were better source of supplementation even in the dry form supporting the milk yield of goats (Mathur *et al.*, 2006). Although sole feeding of fresh *C. mopane* leaves is not possible in goats but it can very well be fed as a part of complete rations at 35-40% of whole ration (Patil *et al.*, 2007).

Thornless Cactus (Opuntia ficus indica)

A new fodder source *Opuntia ficus* indica- a thornless cactus introduced in Indian arid region was found to have the fodder value as maintenance feed and was observed to reduce the water requirement if fed along with the dry roughages in goats, sheep and growing cattle. In addition its high mineral content may reduce the mineral requirement, as arid animals are suffering from mineral imbalances (*Mathur et al., 2009*).

Thornless Cactus (*Opuntia ficus indica*) Feed a Mineral Rich Resource for Livestock

Cactus (*Opuntia ficus indica*) cladodes were having DM 08.00%, with total ash 26.55% composed of AIA 02.50 and ASA 24. Cacti is rich in minerals particularly Calcium. The mineral rich cactus (*Opuntia ficus indica*) cladodes have good acceptability, palatability and dry matter intake with higher body weight gain in cattle.

Feeding management of livestock during drought

- Preparation of ration utilizing top feeds
- Providing vitamin "A" doses
- Cheaper and balanced concentrate utilizing local non-conventional feed resources
- Addition of leguminous crop by products
- Chopping of straw
- Mineral mixture and common salt
- Deworming

Health management of livestock

Broadly common diseases are divided in to four groups:-

1. ***Viral diseases***: Rinder pest, (Cattle plague), Foot & Mouth disease, Rabies and Sheep pox.

2. ***Bacterial diseases***: Anthrax, Tuberculosis, Mastitis, Haemorrhagic Septicaemia, Black-quarter and Pneumonia.

3. ***Parasites and Parasitic diseases***: Ectoparasites (ticks and mites), Endoparasites (Protozoan, Nematodes, Trematodes and Cestods). Dew-

orming: Animal should be dewormed with broad spectrum anthelmintic(Albendazole/Fenbendazole/ Closantel) thrice yearly. In case of ectoparasites: Spray/dusting should be done at regular intervals with ectoparasiticide(,eg., deltamethrin). CLOSANTEL anthelmintic action is both endoparasiticide and ectoparasiticide. These simple treatment will increase the availability of nutrients to livestock.Miscellaneous diseases & Pathological conditions: Urinary calculi, Milk fever, Grass tetany, Cobalt deficiency, PICA, Vitamin A Deficiency and Paraplegia.

Future needs

1. Agricultural production in arid region is basically based on animal husbandry, hence agriculture policy should be readjusted.

2. Establishment of fodder banks at district/town levels and veterinary first aid facilities with pedigreed males can ensure tremendous increase in livestock productivity.

3. Establishment of perennial components in pasture lands and planning of fodder trees will be helpful.

4. Sustainable integrated farming system models should be developed for different situations to balance livestock, environment and human needs.

5. Greater emphasis is to be given on livestock farming rather than crop farming in arid region.

6. Low- producing males should be castrated and selection of genetically superior animals within local breeds may be ensured for future breeding programmes.

7. Conservation of local biodiversity is necessary through need based policies and programmes.

8. Rainfed farming should be banned in non- irrigated areas receiving upto 150mm precipitation annually. Instead, the farmers should be encouraged, educated and financed to grow endemic, nutritive and highly productive grasses.

9. Use locally available suitable alternate feed resources, e.g., agro industrial byproducts and value addition of top feeds, to bridge the gap between demand and supply of feed.

References

Administrative Bulletin .2008-09. Directorate of Animal Husbandry, Govt. of Rajasthan, Jaipur, Rajasthan.

Annual Report. 2008. Central Arid Zone Research Institute, Jodhpur

Anonymous. 2007. Feeding Lana (*Haloxylon salicornicum*) and Saji (*Haloxylon recurvum*) to Lactating Rathi Cattle. Annual Report, CAZRI 2007, pp.101-102.

Balak Ram. 2003. Impact of human activities on land use changes in arid Rajasthan: Retrospect and prospects. (in) Human Impact on Desert environment, pp 44-59. Pratap Narain, Kathju, S., Amal Kar, Singh, M.P. and Praveen-Kumar (Eds). Scientific Publishers, Jodhpur, India.

Bohra, H.C. and Ghosh, P.K. 1977. Effect of restricted water intake during summer on the digestibility of cell wall constituents, nitrogen retention and water excretion in Marwari sheep. *Journal of Agricultural Science* (Camb.), 89: 605-608.

Census of Livestock. 2012. Goverment of Rajasthan, Jaipur.

Chakrawarti Singh and Rollefson, I.K. 2005. Sheep pastoralism in Rajasthan-Still a viable livelihood option? Workshop report, LPPS, Sadri, Dist.-Pali, Rajasthan.

Dutta, T., Taneja, V.K. and Singh, A. 1995. Effect of climatic variables on average daily milk production in crossbred cattle. *Indian Journal of Animal Sciences*, 65(a): 1004-1007.

Gahlot, A.K. 2005. Animal health management in drought prone area. Seminar on "Sustainable livestock production in arid region", March 22-23, 2005.College of Veterinary and Animal Science, Bikaner (Raj.).

Gupta, A.K. 1985. Forage quality of *Cenchrus setigerus* Vahl. at different growth stages. *Annals of Arid Zone, 24*: 143-150.

Kaushish, S.K. 2006. Sheep breeds of India and Their production. In: Strategies for Sustainable Livestock Production (ed. Kaushish, S.K.., Bohra, H.C., Mathur, B.K, Patel, A.K. and A.C. Mathur). Kalyani Publishers Ludhiana. Pp. 323.

Kaushish, S.K., Bohra, H.C., Mathur, B.K. and Patel, A.K. 1998. In: Fifty Years of Arid Zone Research in India. (eds.). Faroda, A.S. and Manjeet Singh. pp.393-426.

Lahiri, A.N. 1978. A note on *Prosopis cineraria* (Linn) Macbride. Jodhpur: Central Arid Zone Research Institute.

Mathur B. K. 2008. Role of livestock in sustainability of farming systems in dry land and arid regions,pp16-24. In proceedings training programme on MTC on *"Farming Systems Approach in Dryland and Arid Areas"*16- 23 October,2008.79p. For District level Government officers from all over India.At Division of Agricultural Economics, Extension and Training, CAZRI, Jodhpur.

Mathur B.K. 2013.Juliflora Based Animal Feed. A Compendium of Agro Technologies. Agri-Tech Investors Meet ,18-19 July 2013,New Delhi.pp.122-123. National Agricultural Innovation Project, Krishi Anusandhan Bhawan –II. Pusa Campus,New Delhi -110 012. P. 127.

Mathur B.K. 2014. Nutritional Constraints and Health Management of Livestock. Assistance to States for Control of animal Diseases (ASCAD) Seminar 2014 on "New Horizons: - Effective and Productive Livestock Development" 19-20, February, pp.31-41.Animal Husbandry Department, Government of Rajasthan, Ajmer (Rajasthan). P.42.

Mathur, B. K. and Bohra, H.C. 1993. Nutritive value of *Prosopis juliflora* leaves and pods. Presented at the workshop on potential of *Prosopis* spp. for Arid and Semi Arid Regions of India, Jodhpur. Jodhpur: Central Arid Zone Research Institute, November 22-23, 1993.

Mathur, B.K, Bhati, T.K and Tewari, J.C. 2006b. Comparative palatability of *Prosopis cineraria* v/s *Colophospermum mopane* leaves in lactating Marwari goat in arid region. In: National Symposium on "Livelihood Security and Diversified Farming Systems in Arid Region".

January 14-16, 2006. Organized by Arid Zone Research Association of India, CAZRI, Jodhpur. No.3-2, pp.51-52.

Mathur, B.K, Patil, N.V., Mathur, A.C. and Patel, A.K. 2005. Impaction of Rumen of desert buck due to Polythene bags- a case study. *Livestock International*, 10(5): 17-20.

Mathur, B.K. 1996. Effect of replacement of cotton (*Gossypium* spp.) seed cake from Tumba (*Citrullus colocynthus*) seed cake and sun dried poultry droppings on the performance of sheep. Ph.D. thesis submitted to Rajasthan Agriculture University, Bikaner, Rajasthan.

Mathur, B.K. 2003. Livestock: human need for sustainability in arid environment. In : Human Impact on Desert Environment. (Eds. Pratap Narain, S. Kathju, Amal Kar, M. P. Singh and Praveen- Kumar). Arid Zone Research Association of India and Scientific Publishers (India), Jodhpur, India. pp. 506-514.

Mathur, B.K., Abichandani, R.K. and Patel, A.K. 1999. Higher incidences of twinning in Marwari goats by supplementary feeding in hot arid region. In "Proceedings IX Animal Nutrition Conference", 2-4th December 1999. Acharya N.G. Ranga Agricultural University, Rajendra Nagar, Hyderabad. Abstract No. 237., P. 144.

Mathur, B.K., Bohra, H.C., Patel, A.K. and Kaushish, S.K. 1998. In situ Rumen degradability of Khejri (*Prosopis cineraria*) leaves in Marwari sheep. Presented at the Golden Jubilee Seminar on. "Sheep, Goat and Rabbit Production and Utilization for Maximizing Production Efficiency, "Indian Society for Sheep and Goat Production and Utilization. Jaipur: CSWRI, April 1998. P. 27.

Mathur, B.K., Harsh, L.N., Mathur, A.C., Kushwaha, H.L., Bohra, H.C., Prajapat, Dinesh and Tiwari, J.C. 2009.Acceptability and Palatability of a Easily Processed, Balanced and Cheaper Concentrate Mixture Containing *Prosopis juliflora* Pods in Lactating Tharparkar Cattle. In: *"International Conference on Nurturing Arid zones for People and the Environment: Issues and Agenda for the 21ˢᵗ Century"*. Proceedings (Abstracts) of AZCONF 2009, 24-28 November, CAZRI, Jodhpur.pp.242.

Mathur, B.K., Mittal, J.P. and Prasad, S. 1989. Effect of tumba cake (*Citrullus colocynthis*) feeding on cattle production in arid region. *Indian Journal of Animal Sciences,* 59: 1464-1465.

Mathur, B.K., Patel A.K., Mathur, A.C., Abichandani, K.K. and Kaushish, S.K. 2000. Wheat straw (untreated treated) consumption by tharparkar heifers during summer under stall fed conditions in "Third Biennial Conference of Animal Nutrition Association" on "Livestock feeding strategies in the new millennium", Nov. 7-9, 2000. CCSHAU, Hissar. No. 282. P.206.

Mathur, B.K., Patel, A.K. and Kaushish, S.K. 2000. Utilization of tumba (*Citrullus colocynthis*) seed cake of desert for goat production, 70 (4): 431-433.

Mathur, B.K., Patil, N.V., Mathur, A.C., Bohra, H.C., Bohra, R.C. and Sharma, K.L. 2009. Comparative Mineral Status of Jodhpur District Villagers' Cattle v/s Institute Farm Managed Tharparkar Cattle in Hot Arid Zone. Proceedings of Animal Nutrition World Conference, 14-17 Feb., New Delhi, India. pp.61.

Mathur, B.K., Sirohi, A.S., Mishra, A.K., Patel, A.K., Mathur, A.C., and Bohra, R.C. 2013. Performance of goats fed on Ardu (*Alianthus excelsa*) and Neem (*Azadirachta indica*) leaves in arid region. *Veterinary Practitioner*, 14 (2): 348-350.

Mathur, B.K., Siyak S.R. and Bohra, H. C. 2003. Feeding of Vilayati babool (*Prosopis juliflora*) pods powder to Marwari Goats in Arid Region. *Current Agriculture,* 27(1-2): 57-59.

Mathur, B.K., Siyak, S.R. and Paharia, S. (2002). In corporation of *Prosopis juliflora* Pod husk as Ingredient of low cost ration for sheep in arid region. *Current Agriculture*, 26(1-2): 95-98.

Narain, P., Sharma, K.D., Rao, A.S., Singh, D.V., Mathur, B.K. and Ahuja, Usha Rani (2000). Strategy to combat Drought and Famine in the Indian Arid Zone. Central Arid Zone Research Institute, Jodhpur, P. 65.

Patidar, M., Mathur, B.K., Rajora, M.P. and Mathur, D. 2008. Effect of grass – legume strip cropping and fertility levels on yield and quality of fodder in silvipastoral system under hot arid condition. *Indian Journal of Agricultural Sciences,* 78: 394-398.

Patil, N.V, Mathur, B.K, Bohra, H.C. and Patel, A.K. 2005. Relative preference index for arid forages for goat and sheep. National Seminar on "Conservation, Processing and Utilization of Monsoon Herbage for Augmenting Animal Production", December 17-18, 2005. Organized by Indian Society for Sheep and Goat Production and Utilization, at RRS, CSWRI, Bikaner.pp.190.

Patil, N.V., Mathur, B.K., Patel, A.K. and Bohra, R.C. 2011. Nutritional evaluation of Colophospermum mopane as fodder. *The Indian Veterinary Journal,* 88 (1): 87-88

Venkatateswarlu, J., Aantharam, K., Purohit, M.L., Khan, M.S., Bohra, H.C., Singh, K.C. and Mathur, B.K. 1992. Forage 2000 AD The scenario for arid Rajasthan, CAZRI, Jodhpur, P. 32.

18

Conservation and Utilization of Plant Biodiversity in Hot Arid Ecosystem

J.P. Singh[1] and Venkatesan, K.[2]

[1]ICAR- Central Arid Zone Research Institute, Jodhpur, Rajasthan-342003
[2]ICAR- Central Island Agricultural Research Institute, Port Blair
A&N Island - 744 105

Biodiversity is a matter of concern to everyone as most of the economic sectors depends on it e.g. agriculture depends on it for wild genetic resources and biological control, human health depends on nature for many pharmaceuticals, industry requires raw materials which come from the nature, tourism increasingly base its attraction on natural amenities linked to biodiversity, water resources require intact vegetation to protect watershed and prevent siltation, disaster prevention needs natural vegetation to help people to respond to drought and the lists goes on and on (Mc Neely, 2004). Each of these sectors needs to be reminded of its reliance on biodiversity and provided with opportunity to invest in insuring the continued survival of the biological basis of their prosperity. Thus biodiversity has emerged as an integral aspect of human life and hence undoubtedly one of the keystone of sustainable development. Biodiversity in arid lands are under severe threats due to habitat destruction (Manuder and Clubbe, 2002), invasion of species (O'Kennen et al., 1999) dilution of customary conservation practices, climate change (Given, 1994; WCMC, 1992), increasing human population, unsustainable exploitation of natural resources, grazing pressure etc.

The hot arid region of India is also very important from biodiversity point of view. It is characterized by extreme temperature, low and erratic rainfall, high evapo-transpiration, intense solar radiation and strong wind. The region is endowed with variety of soils developed from alluvial and aeolian parent materials. Vegetation of the region is typified by sparse and scattered thorny trees/ shrubs and grasses and described as the Tropical Thorn Forest by Champion and Seth (1968). Six characteristic types of vegetation viz., mixed xeromorphic thorn forest, mixed xeromorphic woodlands, mixed xeromorphic riverine thorn forest, lithophytic scrub, psammophytic scrub and halophytic scrub are recognized on the basis of physiognomy in Indian arid zone (Satyanarayan, 1964). Each vegetation type comprises specific trees, shrubs, forbs, grasses and seasonals (Shankar and Kumar, 1988). Though, annuals dominate the vegetation composition, but the perennials (both herbaceous and woody) are important both from economic and ecological point of view. Among the perennials, shrubs occupy prominent place as they cover > 70 % of landscape from Aravalli to western boundary of Indian Thar Desert. By virtue of possessing peculiar morpho-physiological characteristics these shrubs are highly adaptable to harsh edapho-climatic conditions and high biotic pressure. Though shrubs are less in number of species, they are ecologically most successful biotypes (Kumar *et al.*, 2005) and have maximum share with respect to vegetation cover in the Thar Desert.

Plant (Floristic) Diversity

The region is not very rich in plant diversity as far as the number of species is concerned, but rich in the uniqueness of plant biodiversity found in the region, which is adapted to extreme agro-climatic conditions, in which agriculture is little developed, owing to the lack of availability of water and harsh climatic conditions. This phyto-geographical region is also unique being at meeting point of eastern and western flora. There is an overall dominance of the African element 37.1 %, as compared to the Oriental elements, which are only 20.6 %. Also there are the Iranian elements (5.5 %) and Saharo-Sindian elements (20%), presenting a wide spectrum of genetic diversity. In relation to the number of species, the region has around 682 species belonging to 352 genera and 87 families of flowering plants (Bhandari, 1990), of which around 25 are endemic. Of these, 9 families, 37 genera and 63 species are introduced. Poaceae and Leguminoceae are the largest families amongst monocotyledons and dicotyledons, respectively. Incidentally, Poaceae is the largest family of 57 genera and 111 species. The vegetation is dominated by stunted prickly shrubs and drought-resistant perennial herbs. The trees are few and their distribution is scattered.

Since ancient times, woody perennials have been the integral part of the traditional agro-forestry system (TAFS) in arid region. Various multipurpose trees and shrubs are integrated with crops according to prevailing edapho-climatic and socio-economic conditions, which intact represent extensive agro-forestry systems. These systems signify the diversity of indigenous shrubs/ trees e.g. *Acacia jacquemontii* Benth., *A. nilotica* (L.) Willd. ex Del. ssp. *cupressiformis* (J.L. Stewart) Ali & Faruqi, *A. nilotica* ssp. *indica* (Benth.) Brenan, *A. senegal* (L.) Willd., *Balanites aegyptiaca* (L.) Delile, *Calligonum polygonoides* L., *Capparis decidua* (Forsk.) Edgew., *Clerodendrum phlomidis* L.f., *Cordia gharaf* (Forsk.) Ehrenb. & Aschers., *Grewia tenax* (Forsk.) Fiori, *Leptadenia pyrotechnica* (Forsk.) Decne, *Lycium barbarum* L., *Maytenus emarginata* (Willd.) Ding Hou, *Mimosa hamata* Willd., *Prosopis cineraria* (L.) Druce, *Salvadora oleoides* Decne., *S. persica* L., *Tecomella undulata* (Sm.) Seem, *Ziziphus nummularia* (Burm.f.) Wt. etc. and the grasses e.g. *Lasiurus scindicus* Henr., *Cenchrus ciliaris* L., *C. setigerus* Vahl, *Cymbopogon jwaruncusa* (Jones) Schult., *Dichanthium annulatum* (Forsk.) Stapf, *Panicum antidotale* Retz., *P. turgidum* Forsk. etc. These drought-hardy species of the region represent population of variable and adapted types and have peculiar structural, functional and biological characteristics, which allow them to survive in harsh climatic conditions of hot arid zone (Singh *et al.*, 2005). Though a large number of trees and shrubs are grown in agricultural field, *Prosopis cineraria* and *Ziziphus nummularia* are two most important multipurpose woody component of traditional agro-forestry system of western Rajasthan (Mann and Saxena 1980; 1981). Some of shrubs like *A. jacquemontii*, *B. aegyptica*, *C. polygonoides*, *C. phlomidis*, *L. barbarum*, *M. emarginata*, *M. hamata* and *Opuntia elatior* Mill. are maintained on field boundaries as live fence. Thus woody perennials are integral part of the traditional farming system of arid region imparting stability and sustainability in production under harsh edapho-climatic condition. During recent past, introduction of tractors, extension of irrigation facilities and intensive cropping resulted into sharp reduction of population of these woody perennials in agricultural fields.

Economic Utilization of Plant Biodiversity

Food: A number of herbaceous to woody perennial species are used in arid region as a source of food during normal and scarcity periods and among them, some species are of great importance as emergency foods, while others as supplementary foods in famine. These are an important source of minerals, fibre and vitamins, which provide essential nutrients. Herbaceous species like *Amaranthus viridis* L., *Caralluma edulis* (Edgew.) Benth. & Hook., *Ceropegia bulbosa* Roxb., *Cucumis callosus* (Rottl.) Cogn., *Glossonema variens* (Stocks) Benth., *Portulaca oleracea* L., *Sesuvium sesuvioides* (Fenzl.) P. Beauv. etc are utilized by inhabitants and most of these are considered as "weeds of rangelands/agricultural fields". The seeds of grasses such as *Cenchrus ciliaris*, *C. setigerus*, *C. biflorus* Roxb., *Lasiurus scindicus*, *Panicum antidotale*, *P. turgidum*, *Brachiaria ramosa* (L.) Stapf., *Dactyloctenium sindicum* Boiss., *Ochthochloa compressa* (Forsk.) Hilu and wild legumes such as *Indigofera cordifolia* Heyne ex Roth, *I. hochstetteri* Baker are collected and mixed with pearlmillet grain to increase the bulk of the food during the famine period. Further, to improve the quality and quantity of pearlmillet flour the seeds of *Citrullus lanatus* (Thunb.) Matsumara & Nakai and *C. colocynthis* (L.) Schard. are also mixed. Woody perennial species like *Acacia senegal, Calligonium polygonoides, Capparis decidua, Cordia gharaf, Haloxylon salicornicum, Grewia tenax, G. villosa, Leptadenia pyrotechnica, Prosopis cineraria, Salvadora oleoides, S. persica, Ziziphus nummularia,* etc are the valuable source of sustenance and ecological significance in the region. Unripe fruits of *Capparis decidua* are used to prepare pickle and vegetables and important ingredient of famous *"Pachkuta"* mixture of dry vegetables of Rajasthan. Flower buds of *C. polygonoides* locally called as *"Phogla"* is used in preparation of local dish called *"Rayta"* which is supposed to have cooling effect and thus preventing ill effect of scorching heat. Tender pods of *Leptadenia pyrotechnica* are also used as vegetable in this region.

Ripe fruits of *Ziziphus nummularia, Grewia tenax* and *Salvadora oleoides* are very delicious and consumed by desert inhabitants. Ripe fruits of *G. tenax* are eaten fresh and are used as an iron supplement for anemic children because of high iron content (Gebauer *et al.*, 2007). The fruit of *Capparis decidua* contains fairly good amount of protein and mineral matter. The seeds of *H. salicornicum* are used for making bread by mixing with pearl millet during famine period (Singh *et al.*, 2009). Tree species like *Acacia senegal, Prosopis cineraria* are also important in hot arid region for their pods and seeds.

Since ancient times, cucurbits have played a vital role in subsistence farming and house hold nutritional security of inhabitants of Indian Thar Desert. Cucurbits like *Cucumis callosus* (Kachri), *Cucumis melo* L. var. *momordica* (Roxb.) Duthie (Kachra), *Citrullus fistulosus* Stocks (Tindsi), *Citrullus lanatus* (Mateera) were the integral component of rainfed cropping. Among these *C. lanatus* locally called "Mateera" is still an important constituent of traditional kharif cropping.

Fodder: Perennial fodder grasses are an important component of natural grazing lands. Sewan (*Lasiurus scindicus*), Murath (*Panicum turgidum*) and Bur (*Cymbopogon jwarancusa*) are suitable grass species for areas receiving < 200 mm annual rainfall, whereas, Dhaman (*Cenchrus ciliaris*) and Moda Dhaman (*Cenchrus setigerus*) are suitable grass species for areas receiving rainfall between 200-400mm. Leaf fodder of Khejri (*Prosopis cineraria*) locally known as *Loong* is one of the important tree fodder in the area. Rangeland shrubs constitute a major part of the diet of range grazing animals, particularly the small ruminants i.e. sheep and goat, which constitute largest share in total livestock population. Bordi (*Ziziphus nummularia*) is relished by all the livestock species; green shoots and young fruits of Phog (*Calligonum polygonoides*), Lana (*Haloxylon salicornicum*) and Kair (*Capparis decidua*) are eaten by camel and cattle. The shrubs in general have 10-12 % crude protein and 12-30

% crude fibre and thus their intake supplement the protein content of the total feed intake (Khan, 2005). Hence shrubs will compensate for deficiency of minerals, which is often came up in range grazing animals.

Fuel wood: Shrubs are the major source of fuel wood in arid region. *Calligonum polygonoides, Acacia jacquemontii, Ziziphus nummularia* from natural habitats and cropped lands are main source of fuel wood species. The calorific value of *C. polygonoides, Z. nummularia, C. decidua, Salvadora* and *Euphorbia* spp. are 7590, 7900, 7810, 6770 and 7790 BTU/ lbs, respectively (Muthana, 1984). Underground wood bowl of *C. polygonoides* is of high commercial value and has unique quality of burning instantaneously and provide intense heat, therefore extensively used by local gold and iron-smiths. It is also used in gypsum factories and brick-kilns in their furnaces. During recent past there is rapid decline in number of desirable fuel wood species due to their over exploitation and people of this region shifted to use of inferior quality fuel wood species like *Leptadenia pyrotechnica, Calotropis procera, Aerva tomentosa* etc.

Fibre: Many shrub species like *Leptadenia pyrotechnica, Calotropis procera* (Ait.) R.Br., *Crotalaria burhia* Buch.-Ham., *Abutilon indicum* (L.) Sweet of the region has fibrous nature and offer considerable scope for the extraction and use of fiber. *L. pyrotechnica* is an important fiber-yielding shrub and used for making cordage and ropes and supposed fairly resistant against rotting. Fiber of *C. procera* and *C. burhia* are traditionally used for making cordage and ropes. Floss from seeds of *C. procera* is soft and utilized in stiffing the pillows.

Medicine: The species like *Arnebia hispidissima* DC, *Aloe barbadensis, Asparagus racemosus* Willd., *Barleria prionitis* L., *Blepharis sindica* T. Anders., *Boerhavia diffusa* L., *Caralluma edulis, Citrulus colocynthis, Clerodendrum phlomidis, Capparis decidua, Cordia gharaf, Commiphora wightii* (Arnott.) Bhandari, *Convolvulus microphyllus* Choisy, *Cymbopogon jwaruncusa, Dipacdi erythraeum* Webb. & Berth., *Fagonia indica* Burm.f., *Glossonema variens* (Stocks) Benth, *Haloxylon salicornicum, Indigofera oblongifolia* Forsk., *Lycium barbarum* L., *Peganum harmala* L., *Tribulus terrestris* L., *Withania somnifera* etc. are among the important medicinal plants of the region. Recently, *L. barbaum* fruits have become more popular due to its acceptance as a "super-food" with highly advantageous nutritive and antioxidant properties (Amagase *et al.*, 2009). Resin of *C. wightii* is used for treatment of obesity and hyper-lipidemia. *Caralluma edulis* possesses high anti-oxidant properties (Ansari *et al.*, 2005), vitamin-C, calcium and iron content (Menon *et al.*, 1984, Rathore *et al.*, 2008) and can be suggested as a useful diet for obese persons.

Other economic Products

Oils and fats: The wild species like *Balanites aegyptaica* (L.) Del., *Citrullus colocynthis, Salvadora oleoides, S. persica* etc. are the source of oil/fat. *Salvadora*'s seed contains 35-47 % pale yellow, solid fat which is good for soap making.

Gum and resins: *Acacia jacquemontii* and *A. senegal* are important gum-yielding species of the Thar Desert. The local inhabitants collect the gum for their own consumption and also for sale in local market that fetches good price. *Commiphora wightii* locally called as 'Guggul' is another important resin producing species, which has high demand in pharmaceuticals sector.

Saji: Halophyte shrubs like *Haloxylon recurvum* (Moq.) Bunge ex Boiss., *Salsola baryosma* (Roem. & Schult.) Dandy and *Suaeda fruticosa* (L.) Forsk. absorb high amounts of sodium and chloride ions and store them in leaf vacuoles. When dried and burnt, these plants yield "saji" (a mixture of sodium and potassium bicarbonate), which is much used in the local *Papad* making industries (Singh *et al.*, 2005).

Dye and tanning materials: Henna (*Lawsonia inermis* L.) is an important dye-yielding shrub of the region. Lawson or 2-hydroxy-1-4-nephaquinone is its main dye component. It is used as herbal hair dye and staining body's skin. Bark of *A. jacquemontii* is also used in small-scale tanneries and imparts brown to black color in leather and has tan to non-tan ratio of 1:7. *Cassia auriculata* L. is also used for tanning leather.

Basket and mats: Indigenous shrubs are used to prepare baskets and mats in small-scale rural and cottage industries. *A. jacquemontii, Alhagi maurorum* Medic., *G. tenax,* etc. are widely used for this purpose.

Thatching materials / agricultural and household articles: The shrubs like *L. pyrotechnica, C. burhia, C. procera, Z. nummularia, C. phlomidis, C. decidua* are used as thatching materials. The wood of *C. decidua, Z. nummularia, A. jacquemontii* and *C. polygonoides* are used to prepare handle and beams of agricultural implements (plough, spade, sickles) and manufacturing of household articles viz. furniture, curd churners etc.

Conservation of Plant Biodiversity

Habitat destruction, invasion of species, climate change, over exploitation, high grazing pressure, dilution of customary conservation practices coupled with recurring drought resulted in sharp decline in the population of indigenous species of economic importance in their natural distribution cover. The intensity and method with which the parts/products of these valuable species are collected from their natural populations has seriously affected the process of regeneration of these desirable species. Moreover, the market value of various products e.g. guggal resin has increased enormously resulting in their over-exploitation and on the other hand has gradually resulted in depletion of these resources.

The plant diversity of the region can be conserved through establishing protected areas, habitat restoration via techniques such as reintroduction of lost species, introduction of livestock grazing system compatible with wild life, control of unwanted vegetation, *ex-situ* conservation; inclusion of these in various alternate land use systems (ALUS) and adopting sound harvesting practices. As pointed out by Mc Neely (2004) the major challenges in implementing the conservation of biological diversity (CBD) in arid land lies not so much in the biology of species concerned, but rather in the social, economic and political arena within which people operate. Therefore, local people participation is the key for conserving the plant biodiversity. Participation of people can be assured by educating the people about their economic and ecological significance of the desirable species and way of integrating them in their production systems in harmony with their socio-economic setup. The socio-religious movements like *Chipko* movement have been geared to meet conservation efforts. Efforts must be to initiate action which strengthens conservation programme capitalizing this ethos so as to ensure community participation. This way only conservation can be taken up in both grazing lands and agro-ecosystem (Kaul, 1993) in order to save precious diversity in the desert.

Multipurpose woody perennials need integration with crop-based production system for sustained economic gains. System based production systems having diverse type of plant species is a time tested strategy which provides stability in production along with catering diverse needs of farmers. Earlier experience shows that the selection of system-based production should in accordance with prevailing agro ecological and socioeconomic milieu (Singh *et al.* 2006). There fore these plants can be integrated in various production systems such as agri-horti, agri-silvi, boundary plantations in farming system perspective. Integration of these species in farming system not only conserve the diversity of these amazing plant species but also improve the socio-economic status of farmers of hot arid zone of Rajasthan.

Conclusion

Hot arid region of India endowed with rich plant diversity which occurs on a wide range of habitats. These highly specialized species are adapted to grow in harsh climatic conditions of this region with a great potential of economic and environmental importance. These highly adapted species have specific traits and are vital for mitigating anticipated adverse effects of climate change. The diverse gene pool of these species offers a good scope for selection for tolerance to a wide range of abiotic stresses such as moisture deficit, salinity and temperature in harsh arid environment. Further, plant diversity of the hot arid region is under severe and consistent stress due to occurrence of frequent drought and irrational exploitation. There is an urgent need to create greater awareness amongst the inhabitants/farmers for their sustainable utilization and conservation of arid plant biodiversity in farming system perspective.

References

Amagase, H., Sun, Buxiang and Carmia Borek. 2009. *Lycium barbarum* (goji) juice improves in vivo antioxidant biomarkers in serum of healthy adults. *Nutrition Research* 29: 19-25.

Ansari, N.M., Houlihan, L., Hussain, B. and Pieroni, A. 2005. Antioxidant activity of five vegetables traditionally consumed by South-Asian migrants in Bradford, Yorkshire, UK. *Phytotherapy Research* 19: 907-911.

Bhandari, M.M. 1990. *Flora of the Indian Desert*. MPS Repros, Jodhpur, Rajasthan, India. p 435.

Champion, H.G. and Seth, S.K. 1968. *A Revised Survey of the Forest Types of India* . Govt. of India, Delhi.

Gebauer, J., El-Siddig, K., El Tahir, B.A., Salih, A.A., Ebert, G. and Hammer, K. 2007. Exploiting the potential of indigenous fruit trees: *Grewia tenax* (Forssk.) Fiori in Sudan. *Genet. Resor. Crop Evol.* 54:1701-1708.

Given, D.R. 1994. *Principles and Practice of Plant Conservation*. Chapman & Hall, London.

Kaul, R.N. 1993. Strategies for conservation of biodiversity in the thar desert. In: *Biodiversity conservation: Forests, Wetlands and Deserts* (Eds. B. Frame, J. Victor and Y. Joshi), Tata Energy Research Institute, New Delhi, pp. 91-103.

Khan, M. S. 2005. Arid Shrubs for livestock. In: *Shrubs of Indiana Arid Zone*. (Eds. Pratap Narain, Manjit Singh, M.S. Khan and Suresh Kumar), Central Arid Zone Research Institute, Jodhpur, pp 87-90.

Kumar S, Parveen, F., Goyal, S. and Chauhan, A. 2005. Ethnobotany of shrubs of arid Rajasthan. In: *Shrubs of Indian Arid Zone* (Eds. P Narain, M Singh, M.S Khan and Suresh Kumar), Central Arid Zone Research Institute, Jodhpur. pp 15-26.

Mann, H.S. and Saxena, S.K. (eds.) 1980. *Khejari (Prosopis cineraria) in the Indian desert-Its Role in Agroforestry*. CAZRI Monograph 11 (CAZRI, Jodhpur) p. 73

Mann, H.S. and Saxena, S.K. (eds.) 1981. *Bordi (Ziziphus nummularia): A Shrub of the Indian arid Zone- Its Role in Silvipasture*. CAZRI Monograph 13 (CAZRI, Jodhpur) p. 93 .

Manuder, M. and Clubbe, C. 2002. Conserving tropical botanical diversity in the real world. In: *Plant Conservation in the Tropics: Perspective and Practice* (Eds. M. Manuder, C. Clubbe, C. Hankamer and Groves, M.), The Royal Botanic Gardens Kew, Cromwell Press Ltd., U.K, pp 29-48.

Mc Neely, J.A. 2004. Biodiversity and its conservation in arid regions. *Annals of Arid Zone* 43: 219-228.

Menon A.R., Bhatti, K.M. and Menon, A.N. 1984. Chemical analysis of *Caralluma edulis* Benth. and *Oxystelma esclantum* R.Br. *Plant J. Chem. Soc. Pak.* 6:71-72.

Muthana, K.D. 1984. Selection of species for fuel wood plantation in arid and semiarid areas. In: *Agroforestery in Arid and Semi arid Zones* (Ed. K. A. Shankarnarayan), CAZRI, Jodhpur, pp. 243-252.

O'Kennon, R.J. Barkley, T.M., Diggs Jr., G.M. and Lipscomb, B. 1999. *Lapsana communis* (Asteraceae): New for Texas and notes on invasive exotics. *Sida* 18 : 1277-1283.

Rathore M.S., Dagla H.R., Singh, M. and Shekhawat, N.S. 2008. Rational development *in-vitro* methods for conservation, propagation and characterization of *Caralluma edulis*. *World J Agric Sci* 4:121-124.

Satyanarayan, Y. 1964. Habitat and plant communities of Indian Desert. In *Proc. Symp. Probnlems of Indian Arid Zone,* Ministry of Education, Govt. of India, New Delhi. pp. 59-68.

Shankar, V. and Kumar, S. 1988. Vegetation ecology of the Indian Thar desert. *Int. J. Ecol. Environ. Sci.* 14: 131-156.

Singh, J.P., M.L. Soni and Rathore, V.S. 2005. Halophytic Chenopods Shrubs of Arid Zone. In: *Shrubs of Indiana Arid Zone.* (Eds. Pratap Narain, Manjit Singh, M.S. Khan and Suresh Kumar), pp 27-32. Central Arid Zone Research Institute, Jodhpur.

Singh, J.P., Mathur, B.K, and Rathore, V.S. 2009. Fodder potential of Lana (*Haloxylon salicornicum*) in hot arid region. *Journal of Range Management and Agroforestry* 30: 34-37.

Singh, J.P., Rathore, V.S. and Beniwal, R.K. 2006. Perennial medicinal plants for rainfed farming system in arid region. *Indian Journal of Arid Horticulture* 1, 8-14.

World Conservation Monitoring Centre (WCMC). 1992. *Global Biodiversity: Status of the Earth's Living Resources.* Chapman and Hall, London.

19

Climate Change and Its Impact on Fruit Crops

Dipak Kumar Gupta[1], Keerthika A.[1], Chandan Kumar Gupta[2]
Anil Kumar Shukla[1], M.B. Noor mohamed[1], B.L. Jangid[1] and
Chandan Kumar[3]

[1]*ICAR-Central Arid Zone Research Institute, Regional Research Station*
Pali, Rajasthan-306401
[2]*Agrometeorology and Environmental Science, Birsa Agricultural University*
Ranchi, Jharkhand
[3]*ICAR-Central Arid Zone Research Institute, KVK, Pali, Rajasthan-306401*

Global warming and climate change are becoming major challenges in the front of global community to deal with. Climate change is no more a myth but it is reality. The IPCC, 5th assessment report clearly stated that "evidence of climate-change impacts is strongest and most comprehensive for natural systems". Natural and human systems of almost all continents in the world have been affected by climate change in recent decades (IPCC, 2014). Agriculture is essential for sustainability of human life on earth but change in climate is threatening its sustainability. In recent decades all aspects of food security has been potentially affected by climate change, including food access, utilization, and price stability. It has been projected that in absence of adaptation measure, increases in local temperature by 2°C or more above late-20th-century levels will negatively impact production of the major crops in tropical and temperate regions.

Agriculture is one of the major economic activities responsible for food security and livelihood for millions of people. Success, production and productivity of agricultural crops are highly dependent on climate, soil and water. Thus climate change will have adverse effect on agricultural crops. High atmospheric temperature, unpredicted erratic rainfall, rapid soil degradation and evolution of new insect and pests due to climate change will offer new challenge to sustain agriculture. Many part of the world and India is already experiencing impact of climate change on agricultural crops (like early maturity, geographical shift and occurrence of new pests) and these impacts are predicted to increase rapidly in future. The increasing concern about adverse impact of climate change on food security has led to extensive research and documentation of impact of climate change on field crops like rice, wheat and maize. While the clear and holistic idea about impact of climate change on horticultural crops like vegetables and fruits are not well documented especially in developing countries like India.

Horticultural sector plays important role in nutritional security and economy of India. India found second position in production of fruit and vegetable in the world next to China. Due to diverse agro-climatic region in India, large numbers of region specific horticultural crops are grown that provide livelihood for millions of farmers. Some horticultural crops are highly specific to specific climatic condition (like cole crops, apple, plum, cherry, mango, orchids, black pepper etc.) thus predicted change in temperature and climate will adversely affect the horticultural production especially temperate crops. This chapter will provide information on climate change, its causes and impact on horticultural crops.

Global Warming and Climate Change

According to IPCC, climate change refers to statically significant change in mean state of climate and/or its variability extended over long period typically decade or more, while, UN framework convention on climate change (UNFCCC) define climate change as change in climate which is attributed directly and indirectly to human activities that alter global atmospheric composition and which are in addition to natural climate variability observed over comparable time period. Thus, UNFCCC makes a distinction between "climate change" attributable to human activities altering the atmospheric composition, and "climate variability" attributable to natural causes. Change in climate is noticed by long term shift in climatic parameter like atmospheric temperature, precipitation amount and pattern, relative humidity etc.

Therefore, climate change may be due to natural internal processes or external forcing of the earth system, or to persistent anthropogenic changes in the composition of the atmosphere or in land use. Natural causes include changes in output of solar energy, changes in orbit of earth around sun, volcanic eruption, greenhouse effect etc. Climate change is a natural process and earth has whiteness major changes in climate since its inception. Earlier shift in climate from ice age to warm climate was due to natural process. But present change in climate is being reported due to anthropogenic causes which are a result of global warming caused by rapid increase in atmospheric concentration of greenhouse gases.

Global warming is increase in the average temperature of earth surface, lower atmosphere and ocean due to presence of excess amount of greenhouse gases (GHGs) like CO_2, CH_4, N_2O, CFCs in the atmosphere by the phenomenon called enhanced greenhouse effect. GHG are radiative active gases that act as transparent medium for shortwave radiation coming from sun and partial opaque medium for thermal infrared radiation (IR) emitted by earth. They allow some of IR to escape in to space (atmospheric window from 8000 to 12000 nm) and absorb majority of IR radiation and again remit it in all direction towards earth surface. The natural concentration of GHGs like CO_2 is necessary for maintaining global average earth temperature at 15°C i.e. comfortable to live and thrive otherwise in absence of GHGs the average earth surface temperature would be -18 °C i.e. hostile for live and thrive. But due to several human activities (like industry, fossil fuel consumption, widespread deforestation and burning of biomass, as well as changes in land use and land management practices) in last 100 years the concentration of natural GHGs (CO_2, CH_4, N_2O) has increased as well as new synthetic GHGs (CFCs, HCFCs, SF6) has also entered in to the atmosphere. According to IPCC 5th assessment report 2014, concentrations of CO_2, CH_4, N_2O has increased to 391 ppm, 1803 ppb, and 324 ppb, and exceeded the pre-industrial levels by about 40%, 150%, and 20%, respectively and global average temperature has increased by 0.85°C over the period of 1880-2012. Melting of glaciers and ice cap, increasing sea level due to thermal expansion and ice melting, shorter winter and longer summer, hotter night, frequent extreme weather like flood, drought, frequent occurrence of natural calamities like cyclone are main evidence of present global warming and climate change.

Climate Change in India

According to Indian Meteorological Department (IMD), India is also witnessing rise in temperature and occurrence of extreme events like rainfall, cyclone and heat waves. Analysis of data suggests that annual mean temperature for the India as a whole has risen by 0.56°C over 1901-2009. This warming is primarily due to rise in maximum temperature as well as minimum temperature. Spatial pattern of trends in the mean annual temperature shows significant positive (increasing) trend over most parts of the country except over parts of Rajasthan, Gujarat and Bihar, where significant negative (decreasing) trends were observed. Rainfall is decreasing over most parts of the central India during the pre-monsoon season. However during the post-monsoon season, rainfall is increasing for almost all the sub-divisions except for the nine sub-divisions. The frequency of extreme rainfall (Rainfall > 124.4 mm) shows increasing trend over the Indian monsoon region during the southwest monsoon season from June to September and is significant at 98% level. A significant increase was noticed in the frequency, persistency and spatial coverage of both of these high frequency temperature extreme events (heat and cold wave) during the decade (1991-2000).

Limiting Factors and Law of Tolerance

Every organism requires unique set of environmental conditions or factors for its survival. Both abiotic and biotic factors influence the growth, reproduction, abundance, distribution and survival of organisms. Some factors exert more influence than the other. Temperature, light, soil, humidity, carbon dioxide and oxygen, pressures are important limiting factors. The mechanisms of limiting factors can be explained by two main laws.

The Law of Minimum: It was proposed by Liebig in 1840. According to this law, growth and reproduction of organisms are dependent on the factor which is present in minimum quantity in the environment.

The Law of Tolerance: It was proposed by Shelford in 1913. According to this law, organisms are exposed to a variety of environmental factors such as light, temperature, nutrients, etc. Each organism survives well only at a particular range of intensity of the factor. This is called tolerance. When an environmental factor shifts beyond the tolerance of an organism, it can become dormant until the return of favourable conditions, or migrate to a place with favourable conditions or it can acclimatise. When the activity of an organism in response to a range of environmental factor is plotted on a graph, a bell shaped curve is obtained (Fig.1). Depending on ability to tolerate, the curve is divided in to two zones: the zone of tolerance and zone of intolerance. In the zone of tolerance, organisms survive well. Zone of tolerance is further divided in three zones.

Optimum Zone: The zone where growth, reproduction and survival capacity is high. Maximum numbers of organisms are found in this zone.

Stress Zones: The stress zone is found on either side of the optimum range in which the activity slows down. These are the zones of physiological stress and only a few organisms are found in these zones.

Zone of intolerance or lethal zones: In this zone the intensity of the environmental factor is too low or high. No organism can survive in this zone. When the species has a narrow range of tolerance, the prefix 'steno' is added to the factor. For example, stenothermal means an organism that can tolerate a narrow range of temperature; when the species has a wide range of tolerance the prefix 'eury' is added to the factor. For example, eurythermal means an organism that can tolerate a wide range of temperature.

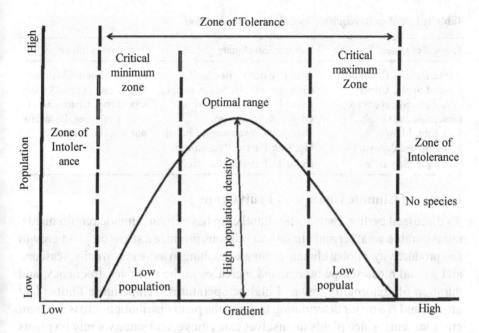

Fig. 1: Law of tolerance

Impact of climate on distribution of horticultural crops

Fruits have varying amounts of climatic adaptation. Most grow and yield good quality of fruits only in either the tropical, subtropical, or temperate zones (Table 1). For example, fruits such as apricot grows best in a temperate climate of uniformly cool winters and dry summers; citrus grow best in a subtropical climate without hard winter freezes and mango and pineapple grow best in a tropical climate without frosts. However, some fruits may be grown under more than one climate. For example, mango is grown under both tropical and subtropical climates. Grape and peach can be grown in both temperate and subtropical regions. In general, evergreen fruit trees (like Mango, Citrus, Guava, Sapota, Papaya etc.) grow under tropical and subtropical climate and they are not dependent on chilling requirement for floral bud differentiation and flowering. Whereas deciduous fruit trees (like Apple, Pear, Peach, Plum, Apricot, Cherry, Grapevine, Walnut, Almonds etc.) require dormancy during winter season and chilling units for floral initiation and have specialized modifications to protect sensitive growing tissue from unfavourable conditions. Therefore, change in the climatic condition of a region will affect growth and quantity and quality fruit.

Table 1: List of fruits requiring specific climatic condition

Tropical climate	Sub-tropical climate	Temperate climate
Aonla, Banana, Carambola, Custard apple, Citrus (mandarin, Sweet orange, Lime), Guava, Grape, Jack fruit, Mango, Mangosteen, Papaya, Pineapple, Sapota, etc.	Aonla, Avocado, Ber, Bael, Citrus (mandarin, Sweet orange, Lime, Lemon, Grape fruit), Grape, litchi, Mango, Date palm, Strawberry, Loquat, Fig, Kair Phalsa, Passion fruit Karonda, Pomegranate, etc.	Apple, Almond, Walnut, Pear, Peach, Plum, Apricot, strawberry, Cherry, Kiwi fruit, Pecan nut, Pistachio nut and Hazel nut etc.

Impact of Climate Change on Fruit Crops

As discussed earlier, each horticultural crop has certain climatic requirements, unfavourable weather and climatic conditions produce a stress on plant growth and productivity. Global climate change is resulting in increases in daily, seasonal, and annual mean temperatures, and increases in the intensity, frequency, and duration of abnormally low and high temperatures. Temperature limits plant growth and is a major determining factor in the plant distribution across different environments: since plants themselves can't move, and since we rely on plants to feed us, it would be very useful indeed if our plants were able to withstand and/or acclimate to the new environmental order.

Presently main reason behind changing climate is global warming due to increase in the concentration of Greenhouse gases (CO_2 CH_4, N_2O and chlorofluorocarbons etc.). CH_4, N_2O and chlorofluorocarbons (CFCs) have no known direct effects on plant physiological processes however CO_2 directly affect plant physiology. Air temperature directly influences metabolic activity of the plant and animals while atmospheric CO_2 concentration influence photosynthetic rate. Other factors affecting plants are changes in precipitation pattern; frequency of extreme weather events like heat waves, flood, drought and cyclone; changes in soil quality and emergences of new diseases and pastes.

Therefore, multidimensional impacts on growth and distribution of plants has been predicted and being observed due to climate change (Table 2). According to Intergovernmental Panel on Climate Change (IPCC), different parts of the world shows negative impacts of climate change on crop yields. The smaller number of studies is also showing positive impacts related mainly to high-latitude regions (where crop yield was less due to cold climate, but present warming has make possible to cultivate more crops and for longer duration).

Table 2: Impact of climate change on plants

Climate Change Impact				
Metabolism and phenology	Geographical distribution	Crop, insect, pest and disease interaction	Crop and soil interaction	Extreme weather events
• CO_2 fertilization effect	• Altitudinal shift	• Timing of flowering and pollination mismatch	• Low SOC	• Drought
• Early flowering	• Longitudinal shift		• Soil health	• Flood
• Early maturity	• Continental shift		• Soil erosion	• Hail storm
• High respiration			• Soil nutrients	• Heat waves
• Short life cycle		• New insect and pests		• Cyclone
		• New weed spectrum		
Impact on crop production (Quality and Quantity)				

Impact of rising CO_2 concentration on plants

The atmospheric concentration of CO_2 is continuously increasing. Before industrialization its level was about 280 ppm and presently it has reached 410 ppm in 2019. Carbon in the form of CO_2 is an essential plant 'nutrient', in addition to light, temperature, water and plant nutrients (Macro and micro nutrients). Therefore, higher concentrations of atmospheric CO_2 can have a positive influence on photosynthesis when other growth conditions (light, temperature, nutrient and moisture) are non-limiting. This effect of CO_2 on enhancing photosynthetic rate and crop yield is known as CO_2 fertilization effect. However, not all plant will be benefited equally from the rising CO_2 concentration. C_3 plants showed increased rate of photosynthesis under elevated CO_2 concentration while C_4 plants show no or less effect. It is mainly due to reduction or blockage of photorespiration pathway in C_3 plants under elevated CO_2 condition. In absence of sufficient amount of CO_2 in the atmosphere, C_3 plants suffer from photorespiration (Fig. 2). This process reduces photosynthetic activity of plant that reduces photosynthetic output by about 25%. The desired reaction in photosynthesis (Calvin cycle) is the addition of CO_2 to RuBP (carboxylation), however, in absence of proper CO_2 concentration, on the place of CO_2, oxygen is added to RuBP (oxygenation). This reaction account approximately 25% of reactions by RuBisCO which lead to synthesis of a product (2-phosphoglycolate) that cannot be used within the Calvin cycle (Figure 2). Most of the plants on which we rely for food, fodder, fibre and energy are C_3 plants for examples grain cereals (rice, wheat, soybeans, rye, barley); vegetables (cassava, potatoes, spinach, tomatoes, and yams); fruit tree (mango, apple, peach, and citrus etc.). Therefore, these plants may be benefitted by rising CO_2 concentration; however, net productivity will be controlled by rising temperature.

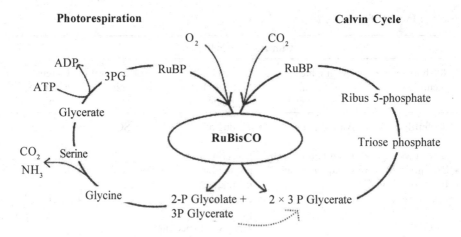

Fig. 2: Mechanism of photorespiration and calvin cycle

Impact of rising temperature

Temperature is one of the most important factor affecting physiology and metabolism of the plants. Most of the physiological processes work normally within temperature range of 0°C to 40°C. High temperatures can increase the rate of biochemical reactions catalysed by different enzymes. However, above a certain temperature threshold, many enzymes lose their function, potentially changing plant tissue tolerance to heat stresses. Therefore, temperature is of paramount importance in the establishment of a harvest index. The important impact are discussed below

Impact of temperature on biomass productivity and quality

Photosynthesis and respiration are proportional to temperature variations and increases with increase in the temperature up to certain temperature threshold. A general temperature effect in plants involves the ratio between photosynthesis and respiration. For a high yield, not only photosynthesis should be high but also the ratio of photosynthesis/respiration (P/R) should be much higher than one. The most favourable temperature for apparent photosynthesis in many plants is lower than that for respiration. It means temperature above favourable temperature for apparent photosynthesis reduces photosynthesis while enhances respiration. Therefore, exposure of plants to temperature higher than optimum temperature for photosynthesis reduces P/R ratio. At temperatures around 15°C, the above mentioned ratio is usually higher than ten, explaining why many plants tend to grow better in temperate regions than in tropical ones. Therefore, under climate change scenario effects of elevated CO_2 will be positive on the yield of C_3 crops however yield loss due to rising temperature may compensate these benefits in tropical regions. Further, production may also increase with temperature, but decrease as super-optimal temperatures are reached and in response to other stresses. Other influences of warming may affect fruit quality. For instance, in grapevines, air temperature plays a determinant role for grape maturation, including aroma and coloration. This in turn has an important effect on the wines. The effect is even more important during nights. Cool nights improve so much coloration that this has been used as a criterion to classify grape-growing regions worldwide.

Impact of temperature on plant phenology

Phenology is the timing of seasonal activities and life cycle events i.e. the sequence of the development stages (flowering, fruiting, ripening etc.). Plants require a specific amount of heat to develop from one phonological stage to another, such as from seeding to the four-leaf stage. This heat requirement is known as growing degree day or heat requirement for a particular phonological event like flowering. Thus, high temperature fulfil required amount of heat within short period of time leading to early flowering, maturity and over all shortening of life cycle of the species. Therefore, the higher the temperature during the growing season, the sooner the crop will mature or will attend particular phonological stage.

Temperature is one of the important factors influencing flowering at various stages of flower initiation to development of various parts. In most fruit crops higher temperature decreased the days interval required for flowering. Temperature not only influences the development of various parts of flowers but also determines the type of inflorescence. Cool temperature induces flowering in several tropical and subtropical fruit trees like citrus, litchi, macadamia, avocado, orange and Olive. In citrus, cooler temperature (20/15ÚC day/night) produces more leafless floral shoots while higher soil and air temperature enhanced production of leafy floral shoots. As hours of low temperature increase, the type of new growth arising from axillary buds changes from vegetative to mixed (flowers and leaves), to generative (leafless with at least one flower). In mango and litchi the number and ratio of hermaphrodite-to-male flowers changes in relation to temperature and other environmental conditions. In mango, development of six different shoot types [vegetative, flowering, mixed panicle, two transition stages (vegetative-to flower and flower-to-vegetative, and chimeral (flowers on one side and leaves on the other)] depends mainly on temperatures and its duration.

For temperate woody perennial crops such as apple, sweet cherry, wine grapes, etc., minimum winter temperature and heat requirements to produce a mature crop are the primary determinants of distribution. Other important factors include chilling requirements to break dormancy, risk of spring frost, and risk of high temperatures during fruit development. Trees must fulfil their chilling and heat requirements in order to break dormancy to produce leaves and flowers, and ultimately bear fruit in spring. Climate change is likely to affect chilling. The lack of chilling resulted in abnormal patterns of bud-break and development of trees with delayed flowering date and an extended flowering Period. Significant warming during the chilling requirement period may affect the completion of chilling requirement. In such cases, relatively long dormancy tended to delay flowering date despite the short length of the heat requirements. High temperature after budbreak and during the spring results in earlier flowering time and shortens the flowering period. In a warming environment, frost risk decreases as temperature less frequently fall below 0 °C. Climate warming sufficient to induce phenological advances should similarly shorten the frost season. If the last frost dates advance at a rate equal to or faster than the phenological shift, then leaf buds, flowers and resultant fruit yields will experience a decreased frost risk. If a drought or intense heat/cold damages trees and their buds, tree care, pruning and waiting for favourable weather next year is the only option.

Other effects of warm temperatures at the beginning of winter have been related with flower bud drop in peach. Warm pre-blossom temperatures have also been reported to decrease fruit set and fruit production in apricot due to abnormal flower development and non-viable flowers. This fact has been related to insufficient chilling accumulation.

Advancing trends in bloom dates of many trees indicate that dormancy breaking processes are indeed changing, most likely in response to climate change. Diurnal temperature ranges have decreased, leading to a lengthening of the freeze-free period in many regions. Studies in Europe and North America show a progressive increase in the timing of spring activities since the 1960s. Jammu & Kashmir, Himachal Pradesh, Arunachala Pradesh and Uttrakhand are major temperate fruit (Apple, plum, peach, Almond) growing states in the India. The annual mean temperature in Arunachal Pradesh and Himachal Pradesh is increasing significantly with @ 0.01 and 0.02°C/year and will have negative impact on temperate fruit yield. In a warming environment, frost risk decreases as temperature less frequently fall below 0°C. Climate warming sufficient to induce phenological advances should similarly shorten the frost season. If the last frost dates advance at a rate equal to or faster than the phenological shift, then leaf buds, flowers and resultant fruit yields will experience a decreased frost risk. In Germany phenological phases of the natural vegetation as well as of fruit trees and field crops have advanced clearly in the last decade of the 20th century. Some of evidence of phenological changes includes: Plant species are flowering 1.4 to 2.1 days per decade earlier in Europe over the last 30 to 48 years; early flowering of mango in North India is being observed; flower drops in Ber plants due to high temperature and heat waves in Rajasthan is becoming common problem.

Impact of on geographical distribution of plants

Climate change is shifting the habitat ranges of plants and animals, including agricultural crops. For example, as average global temperatures increase, plant and animal populations may move to new latitudes with more favourable climates. It is, therefore, possible that crops that used to be productive in one area may no longer be so or the other way around. For instance, an increase in temperature during coldest month has made mango cultivation possible in the valley areas of Himachal Pradesh and Uttarakhand. Climate changes are visible clearly in the shifting of apple cultivation from lower elevations to higher altitudes in Himachal Pradesh. Based on similar shifts in the past, it is reasonable to expect a 300 km shift in the temperate zone with a 3 degree increase. Species may also shift attitudinally as well as latitudinal. A 3 degree cooling of 500 meters in elevation is equal to 250 kilometres in latitude. As species move up on the mountains, they occupy smaller areas, have smaller populations and become vulnerable to genetic pressures. Species at the mountain tops may have nowhere to move to.

Impact on pest and disease dynamics

Community compositions are expected to shift with changes in climatic variables due to the changes in physiology, phenology, and distributions. As species distributions change, the composition of communities changes as well. There may be complex dynamics from a changing climate as a result of different responses from interacting species. Species that closely interact or compete may have different responses to climate change, influencing the outcome of their interactions. For instance, in California it is expected that the warming will expand the distribution of *Ceratitis capitata* and make this fruit fly more difficult to control (Gutierrez *et al.*, 2008). In the same way vine mealy bug will potentially move to northern locations and disrupt natural predators control over it (Gutierrez *et al.*, 2008). In many instances the control of pests will be more difficult and yield loses are expected because of this, until solutions are found. Increased photosynthetic rate in C_3 weeds has led to intense competition between C_3 weeds and C_4 crops.

Impact of extreme weather events

The frequency and intensity of extreme weather events like heavy precipitation, hailstorm, drought, heat waves, frost, and cyclone is increasing since last century. Occurrence of this extreme condition during flowering, fruiting and harvesting stage greatly affect quantity and quality of produce. Unpredictable rains during pre-flowering and flowering periods may cause poor fruit set and low pollinator activities in Mango and Litchi. In the changing climatic scenario, a major portion of the harvest may be wiped out by storms during later fruit development stage. Changes in rainfall patterns can adversely affect the quality and appearance of ripe mango fruits. Unseasonal rains encourage pests, which also lower fruit yield.

References

Basannagari, B. Kala, C.P. 2013 Climate Change and Apple Farming in Indian Himalayas: A Study of Local Perceptions and Responses. PLoS ONE 8(10): e77976. doi:10.1371/journal.pone. 0077976

C.L. Moretti, L.M. Mattos, A.G. Calbo, S.A. Sargent. 2010. Climate changes and potential impacts on postharvest quality of fruit and vegetable crops: *A Review Food Research International* 43: 1824–1832

Eike Luedeling. 2012 Climate change impacts on winter chill for temperate fruit and nut production: A review. *Scientia Horticulturae* 144: 218–229

IPCC 2014. 4th assessment synthesis report. Intergovernmental panel on climate change

L S Rathore, S D Attri and A K Jaswal. 2013 State level climate change trends in india 2013. India Meteorological Department Ministry of Earth Sciences Government of India 2013. Lodi Road, New Delhi- 3 (India)

NC Sharma, SD Sharma, Shalini Verma and CL Sharma. Impact of changing climate on apple production in Kotkhai area of Shimla district, Himachal Pradesh.

Qunying Luo. Temperature thresholds and crop production: Review *Climatic Change* (2011) 109:583–598.

W.B. Sherman T.G. Beckman. Climatic Adaptation in Fruit Crops. Ed. J. Janick. Proc. XXVI IHC – Genetics and Breeding of Tree Fruits and Nuts pp411-428.

20

Soil and Water Quality Issues in Arid Ecosystem and Their Management

Dipak Kumar Gupta[1], Chandan Kumar Gupta[2], Kamla K. Choudhary[1], Keerthika A[1]., M.B. Noor mohamed[1], Seeta Ram Meena[1], P.L. Regar[1], R.S. Mehta[1] and A.K. Shukla[1]

[1]ICAR-Central Arid Zone Research Institute, Regional Research Station Pali, Rajasthan-306401
[2]Department of Agricultural Physics & Meteorology, BAU, Kanke Ranchi, Jharkhand-834006

Dry areas specially arid and semiarid regions are characterized by harsh climatic condition with limited natural resources like fresh water and vegetation. Soil which is most important component for agricultural production is also problematic in this region due to high salt content. Further, these regions face dual water problems: very low rainfall and poor quality of available water. Most of the available groundwater in these areas has high salt content, which further enhance salt content in soil after irrigation with this water. Salinity of soil is one of the most severe abiotic factors limiting agricultural production in arid and semiarid regions of the world. Term 'salt-affected soil' refer to soils in which salts interfere with normal plant growth, and is broadly divided into saline, saline-sodic and sodic soil, depending on amounts of salt, type of salts, amount of sodium present and soil alkalinity. Excessive soluble salts in the saline soil limit the ability of plant roots to absorb soil water even under wet soil conditions, while high sodium content in sodic soil reduce the availability of Ca, Mg, other micronutrients and finally deteriorate physical condition of soil. Poor physical structure of soil results in difficult to till, poor seed germination due to soil crusting and restricted plant root growth. Due to the poor physical structure, sodic soils are also susceptible to wind and water erosion compared to saline soils. All these negative impact of salt accumulation on the soil finally affect growth and yield of the crops. As per estimates of ICAR-Central Soil Salinity Research Institute (CSSRI), Karnal, crop production loss to country due to sodicity is around 11.18 million tonnes while due to salinity is around 5.66 million tonnes. Furthermore, the area under this category is rising due to secondary salinization caused by faulty management of irrigation water, seepage loss, impeded drainage, rapid rise of water table causing water-logging. Certainly, these soils and saline water threaten the livelihood security of farming community. The area under cultivation is decreasing due to diversion of agricultural land in to other uses to meet needs of rising population and as such, agriculture has to make best use of poor quality land and water resources to meet the food demand. Therefore, management of salt affected soil become necessary for ameliorating negative impact and enhancing agricultural production in these regions. Each type of salt-affected soil has different characteristics and thus required different strategies for management.

Extent of salt affected soil and groundwater

According to Food and Agricultural Organization (FAO), globally the total area of saline soils was 397 million ha (3.1%) and that of sodic soils 434 million hectare (3.4%) during 1970-1980. Highest affected soils are present in in Asia & Pacific and Australia, accounting for 6.3 % and 8% of total area of this zone affected by soil salinity and sodicity respectively (Fig 1). According to Central Soil Salinity Research Institute, Karnal, in India, 6.74 million hectare area is salt affected under different agro ecological regions, out of which 3.79 million hectare is alkaline soil while 1.71 and 1.24 million hectare is saline and coastal saline area respectively. According to the available estimates, almost 15% of the total cultivated lands in India are influenced by salinity. Uttar Pradesh, Gujarat, Maharashtra, Tamil Nadu, Haryana and Punjab have about 80% sodic lands in the country. Gujarat (1.22 Mha) followed by Rajasthan (0.2 Mha) and Maharashtra (0.18 Mha) has highest area under saline soil, while, Uttar Pradesh (1.35 Mha) followed by Gujarat (0.54 Mha) and Maharashtra (0.42 Mha) has largest area under sodic soil (Fig 2). According to state agriculture department, about 0.42 and 0.45 million hectare of land are saline and sodic respectively in Rajasthan. Jaipur (74224 ha) followed by Sikar (59936 ha), Bhartpur (32613 ha), Bhilwara (27950 ha) and Pali (26374 ha) has highest saline soil while Alwar (97625 ha), Pali (57662 ha), Bharatpur (45217 ha), Dausa (38437 ha) and Sikar (30036 ha) districts has highest sodic soil in Rajasthan. Bharatpur, Pali and Sikar has largest area under saline-sodic soil.

The high salt content in the groundwater is major water quality issue in arid and semi-arid regions of India especially Rajasthan, Haryana, Punjab and Gujarat. According to central groundwater board (CGWB), about 2 lakh sq.km area in India has been estimated to be affected by saline water of electrical conductivity (EC) in excess of 3000 µS/cm and there are several places in Rajasthan and southern Haryana where values of electrical conductivity of ground water is greater than 10000 µS/cm at 25°C making it unsuitable for irrigation and drinking.

About 30 districts of Rajasthan, 21 districts of Gujarat, 16 districts of Andhra Pradesh and 15 districts of Haryana are severely affected by groundwater salinity with EC more than 3000 μS/cm (Fig 3).

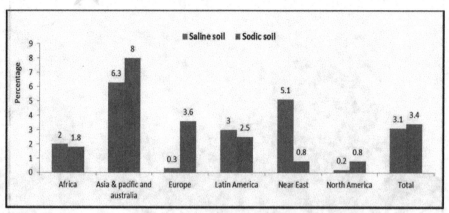

Fig. 1: Percentage area affected by soil salinity and sodicity in the different parts of the world (*Source*: FAO)

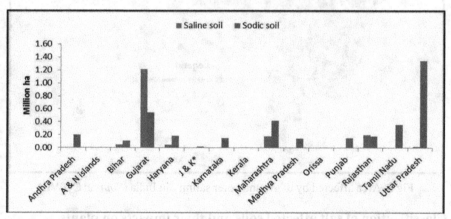

Fig. 2: Total area affected by soil salinity and sodicity in the different states of India (*Source*: CSSRI)

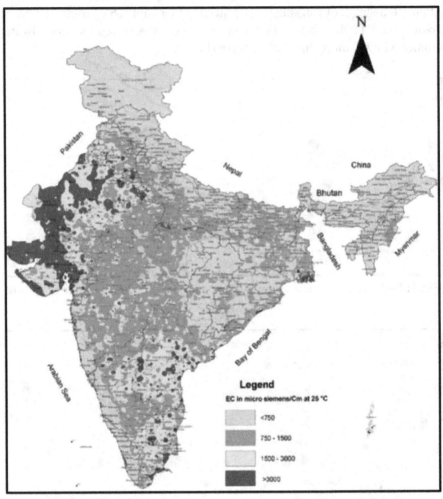

Fig 3: Area affected by of groundwater salinity in India (*Source*: CGWB)

Classification of salt affected soils and their impacts on plants

Salt affected soils can be visually identified, however, for proper management of these soils chemical analysis is required for the quantification of severity of problem and designing strategies for management. As earlier stated, on the basis of amount of salt, type of salts, amount of sodium present and soil alkalinity, salt-affected soil can be broadly divided into saline, saline-sodic and sodic soil (Table 1). The major cations and anions of concern in saline soils and waters are Na^+, Ca^{2+}, Mg^{2+}, and K^+ and the primary anions are Cl^-, SO_4^{2-}, HCO_3^-, CO_3^{2-}, and NO_3^-. When soluble salts accumulate, Na^+ often becomes the dominant cations on the soil exchanger phase, causing the soil to become sodic and dispersed. In sodic soil, predominance of Na^+ on the exchanger phase may occur due to precipitation of Ca^{2+} and Mg^{2+} as $CaSO_4$, $CaCO_3$ and $CaMg(CO_3)_2$. Sodium then replaces exchangeable Ca^{2+} and Mg^{2+} on the exchanger phase.

Saline soil is a term used to describe excessive levels of soluble salts in the soil water (soil solution), high enough to negatively affect plant growth, resulting in reduced crop yields and even plant death under severe conditions. Saline soil are those soils, for which saturated extract has electrical conductivity (EC) greater than 4 dSm^{-1} at 25 °C, exchangeable sodium percentage (ESP) less than 15 and pH lower than 8.5. The primary effect of excessive soluble salts on plants is to limit the ability of plant roots to absorb soil water even under wet soil conditions. A soil solution with high osmotic potential due to the higher concentration of soluble salts compared to the plant cells, does not allow plant roots to extract water from soil causing drought-like symptoms in the plants. Visually this type of soil can be identified by presence of white layer of salt deposition on soil (Fig 4). In contrast to saline soils, sodic soils have excessive levels of sodium (Na^+), high concentration of carbonate (CO_3^{2-}) and bicarbonate (HCO_3^-) and very low Ca^{2+} and Mg^{2+} adsorbed at the cation exchange sites leading to soil dispersion, poor soil structure, soil crusting and poor soil aeration and water infiltration. All these, negatively affect plant germination, root growth, crop development and yield. Sodic soils are chemically characterized with EC less than 4 dSm^{-1}, ESP and soil pH greater than 15 and 8.5 respectively. Visually this type of soil can be identified by soil crusting and presence of black color powder on soil due to dispersion of organic matter and a greasy (Fig 4). Saline-sodic soils are high in sodium and other salts. Such soil has EC greater than 4 dSm^{-1}, ESP greater than 15. Soil pH can be above or below 8.5.

Table 1: Chemical properties of salt affected soil

Soil chemical properties	Saline soils	Alkaline/ sodic soil	Saline–sodic soil
Electrical conductivity (EC)(dS m^{-1})	>4	<4	>4
pH	<8.5	>8.5	>8.5
Exchangeable sodium percentage (ESP)	>15	>15	>15

Fig. 4: a) Saline soil **b)** sodic soil

Factors responsible for development of salt affected soil

Most of the salt affected soil has been developed naturally due to climatic factors and parent's materials available. However, human intervention has also contributed towards increasing its extent due to irrigation; construction of water reservoir and canal areas, this process is known as secondary soil salinization. The major sources of soluble salts in soils are weathering of primary minerals and native rocks, residual fossil salts, atmospheric deposition, saline irrigation and drainage waters, saline groundwater, seawater intrusion, additions of inorganic and organic fertilizers, sludge and sewage effluents, brines from natural salt deposits, and brines from oil and gas fields and mining. The factors responsible for soil salt accumulation can be broadly divided into two: low rainfall and high evapotranspiration in arid and semiarid environment, poor drainage and irrigation with poor quality water (saline ground water, sewage waste water and industrial effluent).

In arid and semiarid climates, there is no enough rainfall to leach soluble salts which originates from parent material. Consequently, the soluble salts accumulate, resulting in salt-affected soils. An additional factor in causing salt-affected soils is the high potential evapotranspiration in these areas, which increases the concentration of salts in both soils and surface waters. It has been estimated that evaporation losses can range from 50 to 90% in arid regions, resulting in 2 to 20-fold increases in soluble salts in the soil. Poor drainage can also cause salinity and may be due to a high water table or to low soil permeability caused by sodicity (high sodium content) of water. The severity of soil salinity or sodicity occurs where there is shallow saline or sodic groundwater level. Groundwater within 6 feet of the surface can move through capillary action to or near the soil surface resulting in very low or no downward movement of excessive soil water and soluble salts. As close as groundwater level is to the surface, the greater level of salts or sodium will develop.

Salinization of soil also occur due to construction of large water reservoir and irrigation canal that results into seepage of water, rise in water table and water logged area in surroundings. Indira Gandhi Canal in Rajasthan has resulted into increase in density and area of vegetation cover due to afforestation, and the cultivated area has expanded due to irrigation. However, it has also resulted into rise in water table, waterlogging and irrigation induced soil alkalinity in command area and adjoining Ghaggar basin in Ganganagar District. Irrigation of non-saline soils with saline water can also cause salinity problems. These soils may be level and well drained. However, after they are irrigated with saline water drainage may become poor and the water table may rise resulting into accumulation of salt in root zone. Salinity problems are common in irrigated lands. One of the major problems in irrigated areas of arid region is that the irrigation water has high salinity, and when the soils are irrigated, the salts accumulate unless they are leached out due to low rainfall and high evapotranspiration.

Management of saline and sodic soil

The main strategies adopted for managing salt affected soil for better crop production is removal of excess soluble salts from crop root zone. This can be achieved by leaching out salt from root zone by good quality of irrigation water; improving soil drainage; soil tillage; application of soil amendment like gypsum, green manuring, organic manure etc. and cultivating salt tolerant plants.

a) Management of saline soil

Leaching and subsurface drainage

Saline soils can be reclaimed by scraping or flushing out the accumulated salt on the soil surface, however, these are less effective and costly affairs. Leaching with good quality (low EC) water has been reported most effective for removing salt from root zone. The water causes dissolution of the salts and their removal from the root zone. For successful reclamation, salinity should be reduced in the top 45 to 60 cm of the soil to below the threshold values for the particular crop being grown. The initial salt content of the soil, desired level of soil salinity after leaching, depth to which reclamation is desired and soil characteristics are major factors that determine the amount of water needed for reclamation. A thumb rule is that a unit depth of water will remove nearly 80 percent of salts from a unit soil depth. Thus, 30 cm water passing through the soil will remove approximately 80 percent of the salts present in the upper 30 cm of soil. However, in this method reclamation can be hampered by restricted drainage in area of high water table, low soil hydraulic conductivity due to restrictive soil layers, lack of good-quality water, and the high cost of good-quality water. Therefore, period of leaching should be selected during low evaporative demand, when water table is low and is better just preceding monsoon season to utilize rainfall effectively. Sub surface drainage is an effective tool for lowering the water table. In the absence of good soil drainage in place combined with a high groundwater levels, late-maturing, deep rooted and salt tolerant crops, like alfalfa, sugar beet and sunflower can also be excellent choices which withstand moderate salt levels and take water from deeper depths.

Irrigation management

Irrigation frequency is one of the major factors controlling negative effect of soil salinity on plants. Irrigation enhances soil moisture content and reduces salt concentration and osmotic pressure in root zone. Soil progressively dries out due to evapotranspiration resulting into increase in concentration of salts in the soil solution and, therefore, its osmotic pressure making the soil water increasingly difficult to be absorbed by the plants. Thus, infrequent irrigation aggravates salinity effects on growth. For these reasons crops grown in saline soils must be irrigated more frequently compared to crops grown under non-saline conditions so that the plants are not subjected to excessively high soil moisture stresses due to combined influence of excess salts and low soil water contents. Leaching of soluble salts is also accomplished more efficiently when the water application rates are lower than the infiltration capacity of the soil and such a condition cannot be achieved by flood irrigation methods. Therefore, sprinkler and drip irrigation has been found better in getting higher crop yield in saline soil due to frequent and continuous wetting of soil respectively.

Mulching and tillage

During periods of high evapotranspiration between the two irrigations and during periods of fallow there is a tendency for the leached salts to return to the soil surface. Soil salinization is particularly high when the water table is shallow and the salinity of groundwater is high. Any practices that reduce evaporation from the soil surface and/or encourage downward flux of soil water will help to control root zone salinity. Covering soil with straw mulch has been found effective in reducing this phenomenon. Straw mulch has reported to reduce soil salinity and enhances crop yield significantly as compared to bare soil surface. It has also found that periodic sprinkling of mulched soils results greater salt removal and therefore higher leaching efficiency than did flooding or sprinkling of bare soil. Deep ploughing of soil before monsoon season has found effective in leaching of soil below root zone after onset of monsoon as compared to unploughed land.

Selection of salt tolerant plants and varieties

Plants differ a great deal in their ability to survive and yield satisfactorily when grown in saline soils. To get good crop yield from saline soil, it is necessary to select crop and variety which are relatively salt tolerant. There are several crops, fodder, fruits and tree species has already been identified with different level of salt tolerance (Table 2). Also success has been achieved in developing salt tolerant verities of important crops like wheat and rice (Table 3). Crop like barley, sugerbeet and cotton etc. are classified as salt tolerant. While wheat, oats, sorghum, maize, sunflower, potato and soybean are classified as moderately salt tolerant and corn, alfalfa, pulses, many clovers, citrus, grape and most vegetables and fruit trees are sensitive to salt.

Table 2: Salt tolerant horticultural crops

Common name	Botanical name	Tolerance based on	Threshold (EC dS/m)	Rating
Vegetables				
Asparagus	*Asparagus officinalis*	Spear yield	4.1	T
Bean, lima	*P. lunatus*	Seed yield	-	M T
Cassava	*Manihot esculenta*	Tuber yield	-	M T
Beet, red	*Beta vulgaris* L.	Storage root	4.0	M T
Cowpea	*Vigna unguiculata*	Seed yield	4.9	M T
Squash	*C. pepo* L. var *melopepo.*	Fruit yield	4.9	M T
Winged bean	*Psophocarpus tetragonolobus*	Shoot DW	-	M T
Fruit crops				
Fig	*Ficus carica*	Plant DW	-	M T
Guava	*Psidium guajava*	Shoot & root growth	4.7	M T
Jamun	*Syzygium cumini*	Shoot growth	-	M T
Indian Ber	*Ziziphus mauritiana*	Fruit yield	-	M T
Natal plum	*Carissa grandiflora*	Shoot growth	-	T
Olive	*Olea europaea*	Seedling growth, Fruit yield	-	M T
Pineapple	*Ananas comosus*	Shoot DW	-	M T

T= tolerant; MT- medium tolerant

Table 3: Salt tolerant verities and root stocks of crops and fruits

Fruits	Rootstock/Variety
Citrus (*Citrus* spp.)	Rangpur lime, Cleopatra mandarin
	Rough lemon, tangelo, sour orange
	Sweet orange, citrange
Stone fruit (*Prunus* spp.)	Marianna
	Lovell, Shalil
	Yunnan
Avocado (*Persea americana* Mill.)	West Indian
	Mexican

b) Management of sodic soil

Strategies followed in reclaiming sodic soil is to first replace the excessive sodium from the cation exchange sites and its replacement by the more favorable calcium ions in the root zone. It can be achieved by either through Ca supplement or removal of sodium by plants and organic matters. Application of Ca supplements provides quick results, while amelioration through plant and organic manure but this is relatively slow process. Organic matter (i.e. straw, farm and green manures), decomposition and plant root action also help dissolve the calcium compounds found in most soils, thus promoting reclamation but this is relatively a slow process. The main management practices are

Application of Ca supplying amendments

This method includes replacement of sodium from exchange site with large amounts of calcium supplements followed by salt leaching. Like saline soils, sodic soils also require good soil drainage and low groundwater level. The only difference will be the application of calcium supplements and thoroughly mixing it into the soil. After replacement of sodium by calcium from the cation exchange sites, Na^+ get converts into Na_2SO_4 and leach out of the root zone. Sulfur and sulfuric acid can also be applied to correct a sodium problem in calcareous soils (soil in which $CaCO_3$ is present). Sulphur, indirectly through chemical or biological action, makes the relatively insoluble calcium carbonate available for replacement of sodium.

Several amendments like Gypsum ($CaSO_4$ $2H_2O$), $CaCl_2$, Sulphur, Pyrites, pressmud (a waste product from sugar factories which contains either lime or some gypsum) etc. has been found effective in managing sodic soil and enhancing crop yield. Among these, gypsum is widely used due to being cheapest and most abundantly available. Sulphuric acid has also been used extensively in some parts of the world, particularly in western United States and parts of USSR. The quantity of an amendment necessary to reclaim sodic soil depends on the total quantity of sodium that must be replaced. Replacement of each mole of adsorbed sodium per 100 g soil requires half a mole of soluble calcium. The quantity of pure gypsum required to supply half a cmole of calcium per kg soil for the upper 15 cm soil depth will be 1926 kg. This, in turn, depends on such factors as the soil texture and mineralogical makeup of the clay, extent of soil deterioration as measured by exchangeable sodium percentage (ESP) and the crops intended to be grown (Fig 5). Therefore determination of gypsum requirement is necessary for optimal use of gypsum and better crop productivity. Amendments like gypsum are normally applied broadcast and then incorporated with the soil by disking or ploughing. Gypsum mixed with the surface 15 cm is more effective than gypsum applied on the soil surface.

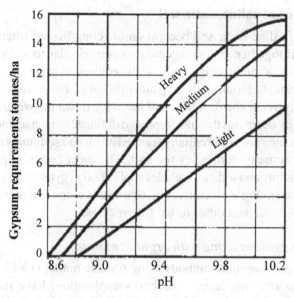

Fig. 5: Relationship between pH (1:2 soil-water) and the gypsum requirements of sodic soil

Other methods used for reclaiming sodic soil are application of organic manure, green manuring, long term reclamation with sodicity tolerant grasses and tree species. These methods are helpful in both reducing soil salinity and sodicity as well as. In a study conducted on highly sodic soils of kota in dryland situations, the application of gypsum @ gypsum requirement (GR) + FYM 10 t/ha proved very useful in lowering down the pH (9.38 to 8.15) and ESP (61.67 to 41.08) as well as seed and stover yield of Taramira (Table 4). Research at CAZRI, RRS, Pali showed that application of 8 ton per ha FYM and gypsum @ 50% GR was found to ameliorate the adverse influence of soil sodicity (pH 8.5-9.0) on pearl millet production. FYM @ 10 ton per ha or dhaincha as green manure crop have also minimized the effects of sodicity in pearl millet.

Table 4: Effect of amendments on the properties of sodic soils and seed yield of Taramira

Treatment	pH	ESP	Seed yield (qha⁻¹)	Stover yield (q ha⁻¹)
Control	9.38	61.67	2.5	08.1
FYM @ 10 t ha⁻¹	9.25	59.54	4.3	12.3
Gypsum 1/2 GR	8.80	52.72	4.8	14.9
Gypsum @ full GR	8.36	41.77	4.7	15.3
Gypsum @ 1/2 GR + FYM @ 10 t ha⁻¹	8.82	50.90	5.5	16.3
Gypsum @ Full GR + FYM @ 10 t ha⁻¹	8.15	41.08	7.6	23.1
CD 5%	0.12	2.28	0.32	1.91

GR: Gypsum requirement, FYM: Farm yard manure (Lodha *et al.,* 1997)

c) Management of saline-sodic soil

Management of saline-sodic soil becomes more complicated when both occur in the same soil together. Its management is very similar to sodic soils. Like sodic soils, remediation of saline-sodic soils requires an extra step of applying calcium supplements (commonly gypsum) followed by salinity remediation practices of improving soil drainage and lowering down the water-table level. As the soil salinity decreases through improved drainage or management, calcium and magnesium salts are preferentially leached away and sodium is more slowly leached and eventually dominates the soil. The most common practices for reducing both salinity and sodicity includes application of gypsum, organic manure and green manure along with cultivation of tolerant crops and other vegetation like grasses, fruit tree and other forest tree species.

Application of gypsum along with organic manure

The biological amelioration methods using organic matter (FYM, crop straw, green manuring and tree leaves and barks application) have two principal beneficial effects on the saline and alkaline soils reclamation: the improvement of the soil structure and permeability, thus enhancing salt leaching, reducing surface evaporation, and inhibiting salt accumulation in the surface layers; and the release of carbon dioxide during respiration and decomposition. Green manuring: Incorporation of green manuring into sodic soils continuously for four or five seasons improves the organic matter content of the soil, nitrogen fertility of the soil particularly if a legume is chosen as green manure crop and also improve permeability it helps to reclaim. *Sesbania aculeata* (daincha) has been found most effective green manure crop for sodic soil reclamation. Tree leave and bark: inclusion of bark and leaf of neem *(Azadirachta indica),* babul *(Acacia spp.)* and madar *(Calotrophis* spp.) powders are more effective in decreasing soil ESP, pH and CEC and exchangeable Ca in sodic soil. Straw mulching has been also found effective in amelioration of saline-sodic soil by reducing crust formation and evapotranspiration.

Selection of crop and cropping pattern

Select crop and varieties that can tolerate high to moderate soil sodicity. Crops are varying widely in their tolerance to soil exchangeable Na. In general, rice, dhaincha, sugarbeet, grasses like Karnal, Rhodes, Para and Bermuda grasses, forest tree species like *Prosopis juliflora, Tamarix articulata, Acacia nilotica, Jatropha curcas,* among fruit trees Ber *(Ziziphus maruritiana),* Jamun *(Syzygium cuminii),* Guava *(Psidium guajava),* Anola *(Emblica officinalis)* and Karonda *(Carissa carandus)* and Anar *(Punica granatum).*

Management of saline and sodic water

Application of saline and sodic water leads to secondary soil salinity and sodicity. Reclamation of this water is costly affair. The higher the amount of saline water use for irrigation, the higher is addition of salt in to the soil. The main strategy for using saline and sodic water should be to adopt efficient irrigation methods specifically drip for higher yield of crop. Therefore, their management strategies includes application of amendments in soil along with selection of suitable irrigation methods, amount of irrigation and time of application that leads to less salt buildup in soil. Research at CAZRI, RRS, Pali suggested surface application of gypsum @ 2 ton ha^{-1} in saline water (EC 2.7 to 12.0 dSm^{-1}) irrigated soils in premonsoon stage has increased 5-10 quintal higher yield of sorghum in gypsum treated soils as compared to untreated control (Table 5).

Table 5: Effect of gypsum application on the yield of sorghum on saline water irrigated soils

Irrigation water EC (dSm^{-1})	Sorghum yield (q ha^{-1})			
	1979 Rainfall 430 mm		1980- Rainfall 230mm	
	Control	Gypsum @ 2 t ha^{-1}	Control	Gypsum @ 2 t ha^{-1}
2.7	27.3	34.9	11.4	14.1
6	22.4	27.5	6.3	13.9
9	12.6	24.3	5.1	10.4
12	10.5	28.5	6.0	16.2

Some of the methods includes efficient irrigation system like drip irrigation, deficit irrigation and irrigation scheduling. Drip irrigation has been found very effective in giving higher crop yield when saline water was used for irrigation as compared to flood irrigation where good quality water was used as irrigation source.

In the drip irrigation method, water is supplied continuously at a point source and in the immediate vicinity of plant roots and useful when irrigating with water of high salinity. The method has the advantage that it keeps the soil moisture continuously high in the root zone, therefore maintaining a low salt level. The roots of the growing plants tend to cluster in the high soil moisture zone near the tricklers and therefore avoid the salts that accumulate at the wetting front. Although in the drip irrigation method appreciable salt accumulation is likely to occur between the rows depending on the inter and intra row space between the drip points. Irrigation scheduling and deficient irrigation practices helps in reducing amount of irrigation water applied to the crops and thus helps in reducing salt supply through irrigation water in to the soil. Some of the experimental result at CAZRI, RRS, Pali suggested that,

- Drip irrigation of pomegranate with saline water (EC 9.5 dSm^{-1}) resulted in better growth; canopy and fruit yield (2.6 kg per plant) of pomegranate over ring basin irrigation system.

- Saline waters of 4 to 6.0 dSm^{-1} EC can be safely used for establishment and irrigation of *ber* (Cv. *Seb &Gola*), Datepalm (Cv. *Khadrawi & Shamran*) and Pomegranate (Cv. *Khog & Jalore seedless*). Micro catchment for water harvesting in fruit crops help in rapid leaching of salts and amelioration of soil in root zone.

References

CGWB, 2010. Ground Water Quality in Shallow Aquifers of India. Central Ground Water Board Ministry of Water Resources Government of India, Faridabad.

Abrol P., Yadav, P.C. and Massoud, P.C. 1988. Salt-Affected Soils and their Management. FAO Soils Bulletin 39. Food and Agriculture Organization of the United Nations, Rome.

Ritzema H.P. , T.V. Satyanarayana, S. Raman, J. Boonstra. 2008. Subsurface drainage to combat waterlogging and salinity in irrigated lands in India: Lessons learned in farmers' fields. *Agricultural Water Management* 95:179–189

Rohilla, P.P., Rao, S.S. and Jangid, B.L. Research highlights, CAZRI, RRS, Pali (1959-2009). Central Arid Zone Research Institute, Jodhpur, Rajasthan, India

Sharma D.K and S.K. Chaudhari 2012. Soil and Crop Management Agronomic research in salt affected soils of India: An overview. Indian Journal of Agronomy 57 (3rd IAC Special Issue) : 175-185.

Sharma D.K., Anshuman Singh and Sharma, P.C. 2016. Role of ICAR-CSSRI in sustainable management of salt-affected soils-achievements, current trends and future perspectives. *4th International Agronomy Congress* (November 22–26): 91-103.

21

Depleting Ground Water- Scenario Consequences and Way Forward

B.L. Jangid, Khem Chand, P.L. Regar, Dipak Kumar Gupta
Keerthika, A., M.B. Noor mohamed and A.K. Shukla

ICAR-Central Arid Zone Research Institute, Regional research Station Pali, Rajasthan – 306401

Sustainable development and efficient management of water is an increasingly complex challenge in India. Increasing population, growing urbanization and rapid industrialization combined with the need for raising agricultural production generates competing claims for water. Ground water has an important role in meeting the water requirements of agriculture, industrial and domestic sectors in India. About 85 percent of India's rural domestic water requirements, 50 percent of its urban water requirements and more than 50 percent of its irrigation requirements are being met from ground water resources. Ground water is annually replenishable resource but its availability is non-uniform in space and time. Technically, the dynamic ground water refers to the quantity of ground water available in the zone of water level fluctuation, which is replenished annually. Hence, the sustainable development of ground water resources warrants precise quantitative assessment based on reasonably valid scientific principles. National Water Policy, 2012 has laid emphasis on periodic assessment of ground water resources on scientific basis. The trends in water availability due to various factors including climate change must also be assessed and accounted for during water resources planning. To meet the increasing demands of water, it advocates direct use of rainfall, desalination and avoidance of inadvertent evapo-transpiration for augmenting utilizable water resources. The National Water Policy, 2012 also states that safe water for drinking and sanitation should be considered as pre-emptive needs followed by high priority allocation for other domestic needs (including needs of animals), achieving food security, supporting sustenance agriculture and minimum eco-system needs. Available water, after meeting the above needs should be allocated in a manner to promote its conservation and efficient use.

Dynamic ground water resources of India (as on 31st March 2011)

1. Annual Replenishable Ground Water Resources - 433 bcm

2. Net Annual Ground Water Availability- 398 bcm

3. Annual Ground Water Draft for Irrigation, Domestic & Industrial uses - 245 bcm

4. Stage of Ground Water Development- 62%

5. Categorization of Blocks / Mandals/Firkka Talukas

- Total Assessed units-6607

- Safe -4530

- Semi-Critical- 697

- Critical -217

- Over-Exploited -1071

- Saline-92

Ground Water Level Scenario in the country

Ground water level is one of the basic data-element which reflects the condition of the ground water regime in an area. Ground water levels are being monitored by Central Ground Water Board and State Ground Water departments. CGWB monitors ground water level four times a year during January, April/ May, August and November. The periodicity of ground water level monitoring by the State Government varies from State to State. The primary objective of monitoring the ground water level is to record the response of ground regime to the natural and anthropogenic stresses of recharge and discharge parameters with reference to geology, climate, physiography, land use pattern and hydrologic characteristics. The natural conditions affecting the regime involve climatic parameters like rainfall, evapotranspiration etc., whereas anthropogenic influences include pumpage from the aquifer, recharge due to irrigation systems and other practices like waste disposal etc. CGWB monitors the ground water level and quality in a network of 15640 ground water monitoring wells located all over the country. This data along with State Government monitoring data is used for assessment of ground water resources. The ground water level scenario based on data from CGWB is presented in the following paragraphs.

Ground Water Level Scenario, 2010

The depth to water level map of India for Pre- Monsoon period (May, 2010) (Fig. 1) reveals that that in sub-Himalayan area, north of river Ganges and in the eastern part of the country in the Brahmaputra valley, generally the depth to water level varies from 2-10 meter below ground level (m bgl). In major parts of north-western states (Indus basin), depth to water level generally varies from 10-20 m bgl with pockets of deeper water level of more than 20 m bgl. In the western parts of the country covering the states of Rajasthan and Gujarat deeper water level is recorded in the range of 10-20 m bgl. In western Rajasthan and north Gujarat deeper water level in the range of 20-40 m bgl and > 40 m bgl have also been also recorded. In the west coast water level is generally less than 10 m and in western parts of Maharashtra State in isolated pockets water level in the range of 2-5 m has also been observed. In the east coast i.e. coastal Andhra Pradesh, Tamil Nadu and Orissa, water level in the range of 2-5 m bgl have been recorded. However South-eastern part of West Bengal recorded water level in the range of 5-10 m bgl. In central India water level generally varies between 5-10 m bgl, with patches where deeper water level more than 10 m bgl has been observed. The peninsular part of country generally recorded a water level in the range 5-10 m bgl. In some patches water level ranges from 10-20 m bgl. Isolated patches of water level of 10-20 m bgl and 20-40 m bgl have been observed. The Post-Monsoon (November 2010) depth to Water level map (Fig. 2) reveals that in Sub-Himalayan area, north of river Ganges and in the eastern part of the country in the Brahmaputra valley, generally the depth to water level varies from 2-5 meter below ground level (bgl). Isolated pockets of shallow water level less than 2 m bgl have also been observed. In major parts of north-western states depth to water level generally ranges from 10-20 m bgl. In the western parts of the country deeper water level is recorded in the depth range of 20-40 m bgl. In North Gujarat, part of Haryana and western Rajasthan water level more than 40 m bgl is recorded. In the west coast water level is generally less than 5 m and in western parts of Maharashtra State isolated pockets of water level less than 2 m has also been observed. In the east coast i.e. coastal Andhra Pradesh and Orissa, shallow water level of less than 2 m have been recorded. In eastern states, water level in general ranges from 2-5 m bgl. However South-eastern part of West Bengal recorded water level in the range of 5-10 m bgl. In central India water level generally varies between 2-5 m bgl, except in isolated pockets where water level more than 5 m bgl has been observed. Similarly pockets of shallow water level less than 2 m bgl is also observed along the west coast. The peninsular part of country generally recorded water level in the range 2-5 m bgl. In some patches water level ranges from 5-10 m bgl.

Stage of Ground Water Development

The overall stage of ground water development in the country is 62%. The status of ground water development is very high in the states of Delhi, Haryana, Punjab and Rajasthan, where the Stage of Ground Water Development is more than 100%, which implies that in the states the annual ground water consumption is more than annual ground water recharge. In the states of Himachal Pradesh, Tamil Nadu and Uttar Pradesh and UTs of Daman & Diu, and Puducherry, the stage of ground water development is 70% and above. In rest of the states / UTs the stage of ground water development is below 70%. The ground water development activities have increased generally in the areas where future scope for ground water development existed. This has resulted in increase in stage of ground water development from 61% (2009) to 62% (2011).

Categorization of Assessment Units

Out of 6607 numbers of assessed administrative units (Blocks/ Taluks/Mandals/ Districts), 1071 units are Over-exploited, 217 units are Critical, 697 units are Semi-critical, and 4530 units are Safe. Apart from these, there are 92 assessment units which are completely Saline (Table below). Number of Over-exploited and Critical administrative units are significantly higher (more than 15% of the total assessed units) in Delhi, Haryana, Himachal Pradesh, Karnataka, Punjab, Rajasthan and Tamil Nadu, Uttar Pradesh and also the UTs of Daman & Diu and Puducherry.

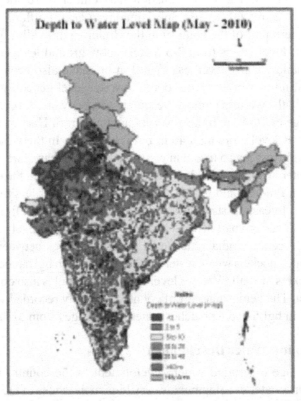

Fig. 1: Pre-monsoon depth to water level (2010)

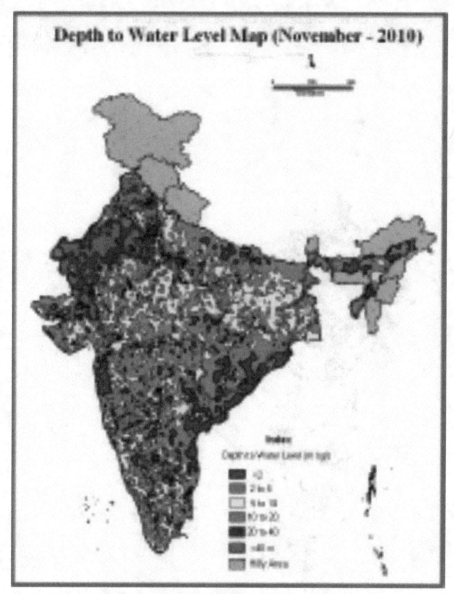

Fig. 2: Post-monsoon depth to water level (2010)

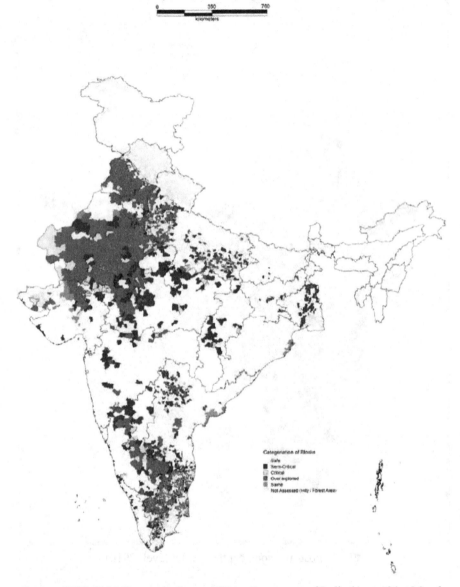

Source: GOI. 2014. Dynamic Ground Water Resources of India (As on 31st March 2011) Central Ground Water Board, Ministry of Water Resources, River Development & Ganga Rejuvenation, Government of India.

Categorization of Blocks/ Mandals/ Talukas in India (2010-2011)

Sl.No.	States / Union Territories	Total No. of Assessed Units	Safe		Semi-critical		Critical		Over-exploited		Saline	
			Nos.	%	Nos.	%	Nos.	%	Nos.	%	Nos.	%
	States											
1.	Andhra Pradesh	1110	877	79	97	9	15	1	83	7	38	3
2.	Arunachal Pradesh	11	11	100	0	0	0	0	0	0	0	0
3.	Assam	27	27	100	0	0	0	0	0	0	0	0
4.	Bihar	533	522	98	11	2	0	0	0	0	0	0
5.	Chattisqarh	146	125	86	18	12	2	1	1	1	0	0
6.	Delhi	27	2	7	5	19	2	7	18	67	0	0
7.	Goa	20	20	100	0	0	0	0	0	0	0	0
8.	Guiarat	223	171	77	13	6	5	2	24	11	10	4
9.	Haryana	116	23	20	7	6	15	13	71	61	0	0
10.	Himachal Pradesh	8	5	63	0	0	2	25	1	13	0	0
11.	Jammu & Kashmir	14	14	100	0	0	0	0	0	0	0	0
12.	Jharkhand	210	199	95	5	2	0	0	6	3	0	0
13.	Karnataka	270	152	56	34	13	21	8	63	23	0	0
14.	Kerala	152	126	83	23	15	2	1	1	1	0	0
15.	Madhya Pradesh	313	218	70	67	21	4	1	24	8	0	0
16.	Maharashtra	353	325	92	16	5	2	1	10	3	0	0
17.	Manipur	8	8	100	0	0	0	0	0	0	0	0
18.	Meghalaya	7	7	100	0	0	0	0	0	0	0	0
19.	Mizoram	22	22	100	0	0	0	0	0	0	0	0
20.	Naqaland	8	8	100	0	0	0	0	0	0	0	0
21.	Orissa	314	308	98	0	0	0	0	0	0	6	2
22.	Punjab	138	22	16	2	1	4	3	110	80	0	0
23.	Rajasthan	243	25	10	20	8	24	10	172	71	2	1

#	State / UT											
24.	Sikkim	4	4	100	0	0	0	0	0	0	0	0
25.	Tamil Nadu	1129	437	39	235	21	48	4	374	33	35	3
26.	Tripura	39	39	100	0	0	0	0	0	0	0	0
27.	Uttar Pradesh	820	559	68	82	10	68	8	111	14	0	0
28.	Uttaranchal	18	11	61	5	28	2	11	0	0	0	0
29.	West Bengal	271	217	80	53	20	1	0	0	0	0	0
	Total States	**6554**	**4484**	**68**	**693**	**11**	**217**	**3**	**1069**	**16**	**91**	**1**
	Union Territories											
1	Andaman & Nicobar	36	36	100	0	0	0	0	0	0	0	0
2	Chandigarh	1	1	100	0	0	0	0	0	0	0	0
3	Dadra & Naqar Haveli	1	1	100	0	0	0	0	0	0	0	0
4	Daman & Diu	2	0	0	1	50	0	0	1	50	0	0
5	Lakshdweep	9	6	67	3	33	0	0	0	0	0	0
6	Pondicherry	4	2	50	0	0	0	0	1	25	1	25
	Total Uts	53	46	87	4	8	0	0	2	4	1	2
	Grand Total	6607	4530	69	697	11	217	3	1071	16	92	1

Note

Blocks- Bihar, Chattisgarh, Haryana, Jharkhand, Kerala, M.P., Manipur, Mizoam, Orissa, Punjab, Rajasthan, Tripura, UP, UttaraKhand, WB

Taluks (Command/Non-Command) -Karnataka

Mandal - Andhra Pradesh

Taluks - Goa, Gujarat, Maharashtra

Districts (Valley) - Arunachal Pradesh, Assam, Himachal Pradesh, Jammu & Kashmir, Meghalaya, Mizoram, Nagaland, Sikkim, Tripura Islands - Lakshdweep, Andaman & Nicobar Islands Firka-Tamil Nadu Region - Puducherry

UT - Chandigarh, Dadar & Nagar Haveli, Daman & Diu Tehsil-NCT Delhi

Source: GOI. 2014. Dynamic Ground Water Resources of India (As on 31st March 2011) Central Ground Water Board, Ministry of Water Resources, River Development & Ganga Rejuvenation, Government of India.

Source: GOI. 2014. Dynamic Ground Water Resources of India (As on 31st March 2011) Central Ground Water Board, Ministry of Water Resources, River Development & Ganga Rejuvenation, Government of India.

Ground Water Resources Scenario of Rajasthan

The State of Rajasthan has diversified geology, ranging from Archean metamorphic to recent alluvial sediments. Based upon geological diversities, geomorphological setup and ground water potentialities, the state of Rajasthan can be divided into three broad hydrogeological units- (i) Unconsolidated formation (ii) Semi-consolidated formation (iii) Consolidated (Fissured formation). Large part of the State is underlain by Quaternary sediments (Thar Desert) consisting of clay, silt, sand and gravel of various grades. The fine sand and clay with or without kankar layers have formed multi layered aquifer system. The thickness may go up to 200-250 m below ground level, either resting directly over the basement rocks or over the Tertiary formation.

Drilling data of CGWB indicates presence of aquifer of poor quality down to a depth of 200 mbgl. Exploratory drilling data reveals that the yield vary from meagre to 10 m^3/day, transmissivity ranges between 80 and 300 m^2/day and storage co-efficient vary from 1.1x 10-5 to 3.9x10-6 in the state . In compact sedimentary rocks i.e. sandstone belonging to the Vindhyan formation is compact in nature and has low primary porosity. Ground Water occurs within the weathered residue and in the secondary porosity underneath. In general, the thickness varies from 5 to 10m.

Yield potential is limited due to compact nature of the formation. The limestone is also having low ground water potential. The yields of dug wells vary from 0.25 to 0.75 m^3/day with water levels ranging from 10 to 120 meters. The yield of the wells drilled in Vindhayan formation has been observed to be 15m^3/day, tapping fractures between 50-75 mbgl. In consolidated formation (Fissured) the thicknesses of the weathered zone vary from 5 to 50 m. Ground Water occurs under unconfined condition within the weathered zone. The results of the exploratory drilling carried out by CGWB in hard rock areas indicate presence of productive fractures down to a depth 40m and yield vary from 3 to15 m^3/day., whereas transmissivity vary from 3 to 30 m^2/day. The Ground water resources have been assessed block-wise.

The Annual Replenishable Ground Water Resource of the State has been estimated as 11.94 BCM and Net Annual Ground Water Availability is 10.83 BCM. The total Ground Water Draft due to all sources is 14.84 bcm. Projected for 2025 is 1.89 bcm while the net groundwater availability for future irrigation development is only 0.91 bcm and the stage of ground water development in the state is 137 %. Out of 243 blocks, 172 blocks has been categorized as 'Over Exploited', 24 as 'Critical', 20 as 'Semi-Critical' and 25 blocks as 'safe'. 2 blocks has been categorized as 'Saline'.

CASE OF PALI DISTRICT

Status of groundwater in Pali district

In the year 1998, five out of 10 blocks (panchayat samities) of the district, were in safe zone and during 2001 it came down to one. In 2004 none of the block was in safe zone. The district became semi-sensitive, sensitive and over exploited during the year 1998, 2001 and 2004, respectively. Since then the situation is worsening. Groundwater draft report for the year 2008 revealed that except two blocks viz. Pali and Raipur, which were in critical stage, all other eight blocks of district were over-exploited with regard groundwater development

Growth and instability in groundwater level in Pali district

Analysis of time series data for groundwater situation in district for pre and post monsoon periods for 26 years (1984 to 2009) revealed that Raipur, Jaitaran and Kharchi Block in Pali district experienced maximum groundwater table fall of 17.18 m & 17.48 m, 14.22 m & 14.51 m and 12.18 m & 13.10 m, respectively with district average of 08.81m and 08.54 m for pre and post-monsoon. The average per annum decline rate of around 0.3 m over the period, revealed a disgusting picture of groundwater availability and sustainability in these blocks.

However, the most alarming situation observed in Kharchi block, as it is also found most over-exploited (137.2% GW development) in terms of groundwater balance, followed by Rani (118.8 %) and Rohat (114.6%) blocks.

The highest annual compound growth rate (ACGR) in decline of ground water level was found for Raipur block, (2.62 % for pre-monsoon and 2.43% for post-monsoon); however, the highest instability index (23.16) was observed in Sumerpur block for pre-monsoon period and in Desuri blocks (28.77) for post monsoon period.

Change in cropping pattern in Pali District- as indicator of groundwater depletion

The coefficient of concordance (W) test was used to examine the change in cropping pattern for the period from 1982-83 to 2006-07 in Pali district. The entire time series data were split up into five Quinquennial periods and the coefficients of concordance (W) were estimated accordingly. The coefficient of concordance (W) varied from 0.84 in Period-III (1992-93 to 96-97) to 0.95 in Period-II (1987-88 to 91-92) for the district. For the overall period, coefficient was found with a modest value of 0.84. The coefficients were found highly significant at 1% level of significance. The highly significant value of 'W' indicates high degree of monocity in the cropping pattern. Hence, it could be concluded that there is no significant shift in cropping pattern within Quinquennial period, but over the entire period, a meagre shift in cropping pattern took place in the district.

The change in cropping pattern depends upon a number of factors other than the water availability. Ranks were allotted to crops based on the proportion of their area to gross cropped area. Pearl millet crop has dominated over all the Quinquennial in terms of acreage in the district, except Period-III and IV, in which Rape and Mustard replaced pearl millet crop. Pearl millet was followed by sorghum in the 1st and 3rd periods while it was followed by sesame in 2nd and 5th Quinquennial. During the last period, sorghum ranked 3rd. The water intensive crops such as cotton, wheat and chilli which possessed 3rd, 8th and 15th ranks during Periods-I, moved towards higher ranks in the subsequent periods. On the other hand, relatively less water intensive crops viz. Rapeseed & mustard and cumin shifted from higher to lower ranks during these two and half decades period. Rapeseed & mustard which was at the 5th rank in Period-I stood first during Period III and IV; however in last period it gone down drastically on 7th rank, similarly cumin was on rank 14th in period I and it came down to 11th in period V. The henna, a commercial crop of the district, has improved its rank from 11th in period-I to 6th in period-V. It indicates that the area under the crop in continuously increasing due to its ability to provide some income even under low rainfall and drought conditions. The sugarcane and paddy, two most water consumptive crops, though occupying meagre proportion in gross cropped area, their acreage declined continuously in district and they have almost gone out of cultivation.

The temporal change in area under different crops in absolute and percentage terms for the district was drastic. It was observed that area of all the irrigated crops declined over the time, clearly showing the impact of depleting groundwater. The area under barley declined continuously in the district during the period, similarly, for wheat, except Q3, the area declined in all the periods, even area under rape & mustard declined in Q4 and Q5, which is relatively low water requiring crop. It is again very important to find that except henna, no other crops has shown continuously increasing trend during this period, again underlining the importance of henna, as sustainable income generating agri-enterprise in the district. Similarly, cumin and Isabgol, another two very low water requiring high value spice and medicinal crops, shown continuous increasing trend, except in the last period recording significant fall in area (might be due to the fall in groundwater table), however, gained significantly over the period-I. Among the rainfed crop, the area under cluster bean (again a very important commercial crop), and green gram (very important pulse crop), showed continuously increasing trend, except period-III and II, respectively. Another crop, which gained momentum in the district is- Taramira, an oilseed crop grown mostly on conserve moisture with almost zero external input. Other crops have shown fluctuation trends in the district. If we look over the change in last Quinquennial to that of Q4 the area under green gram, cluster bean, henna, moth bean and sesame has increased in the district, whereas a declining trend in case of pearl millet, gram, sorghum, chilli, wheat, rape & mustard, cotton, cumin and Isabgol was noticed.

Demography of sampled farmers in Pali district for study of impact of depleting groundwater

- *Age and Education*: The average age of the sampled farmers was found to be 48.6 years and 3/4th (75%) of them were literate to educated up to above secondary class, however remaining 1/4th (25%) were illiterate .

- *Family type and size*: Majority (62.5%) of sampled farmers had joint family while remaining had nucleus. The average family size was found to be of 8.3 members with 4.4 male (2.3 adult and 2.1 child) and 4.0 female (2.2 adult and 1.8 child) members.

- *Occupation*: Of the sampled farmers all (100%) were having agriculture as their main occupation while 40.6% had income from some small business, 31.3% from service and 29.2% from labour.

- *Land holding and category*: The majority (63.6%) of sampled farmers were found to have small medium (2-4 ha) to large medium (2-4 ha) size land holding followed by those (16.7%) having large (more than 10 ha) land holdings. Around 20% were having marginal to small size of land holdings.

Groundwater resources with sampled farmers and their management

- *Number and type of wells*: The average number of wells with sampled farmers was 1.5. More than three-fourth (76%) majority of sampled farmers had open wells, while only 6.3% had only bore well and 17.7% had both open and bore wells.

- *Type and age of wells*: Of the sampled farmers' wells 81.3 % were open wells, 18.8 were bore wells and only 1% were open-cum-bore wells. Around 36% wells were very old (constructed more than 50 years back) 29.7% wells were dug/ constructed between 20-50 years and 34.5% were constructed/ dug in recent years (less than 20 years).

- *Management of wells*: The majority (65.5%) of farmers reported renovation of wells. Maximum farmers (37.9%) renovated their wells during last 6-10 years followed by those (34.7%) who renovated during 11-20 years back and more than 1/4th renovated their well during last 5 years.

- *Renovation type*: Of the total renovations of wells reported by the sampled farmers maximum 48.4% reported deepening of wells followed by those (21.1&20%) who gone for side boring, and deepening and side boring. Other renovation types were vertical bore, deepening and vertical bore and boring of tube well.

Groundwater available to farmers for irrigation

- *Average depth of open wells*: The average depth of open wells in the district varied from 30.2 meters (lowest) in Marwar Jn. Panchayat samiti to 47.5 meters (highest) in Raipur.

- *Average depth of tube wells*: The average depth of tube wells in the district varied from a mere 61.4 meters (lowest) in Pali Panchayat samiti to 204.4 meters (highest) in Raipur.

- *Water yield of wells and pump operation time*: The average water yield of wells in the district varied from 9412 liters per hours (lowest) in Pali Panchayat samiti to 45000 liters per hours (highest) in Marwar Junction. The pump operation ranged between mere 10-15 minutes to 20 hours.

- *Net irrigated area per well*: The net irrigated area per well ranged between at mere 0.01 ha to 9.6 ha in the district.

Utilization of available groundwater and consequences of depletion

- On the sampled farms irrigated area has decreased upto 78%. The groundwater available to farmers was mainly used to grow Rabi season crops viz. wheat, mustard, cumin etc. Where availability of groundwater was meagre it was used as drinking water only for people and animals and to grow some vegetables and fodder for livestock.

- Analysis of farm level data revealed drastic change in cropping pattern in the study area due to continuous fall in groundwater water and thus decreasing availability of same for irrigation. The Rabi season and other irrigated crops viz. cotton, chilli, wheat, vegetables etc. either vanished from the cropping pattern of area or area under these crop decreased severely. Farmers were mainly growing Kharif season rainfed crops viz. bajra, sorghum, cluster bean, moong bean etc. as a result; there was drastic decline in farmers' income.

Managing the groundwater

To manage the available ground water for irrigation the farmers were taking various measures to sustain farm income. The major ones were as follows-

- Deciding crops taking into consideration availability of water

- Quality of available water on most of the wells found acceptable

- Farmers were mostly using their own experience in deciding how much to irrigate, when to irrigate, and which crops to with limited water

- 53.0 percent were taking advice of agriculture supervisor and 30 percent from experienced farmers

- Farmers paid up to Rs. 40-50/- per hour for purchase of irrigation water

- In low rainfall years farmers grow less water requiring crops

- Farmers able to predict availability of GW for coming season according take decisions

- The majority of the farmers opined that there should be some control measures on extraction and utilization of groundwater viz. regulation of distance between wells, not growing crops of higher water requirement, adoption of sprinkler and drip irrigation system etc.

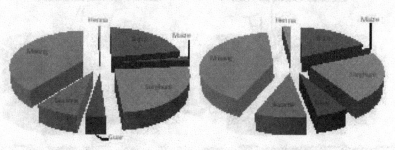

Change in *rain fed* cropping pattern at farm level in PS Raipur & Jaitaran

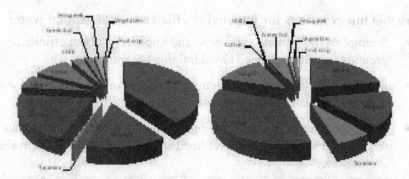

Change in *irrigated* cropping pattern at farm level in PS Raipur & Jaitaran

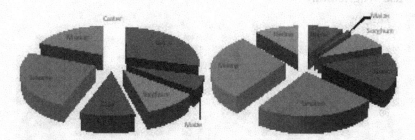

Change in *rain fed* cropping pattern at farm level in PS Pali & Marwar Jn.

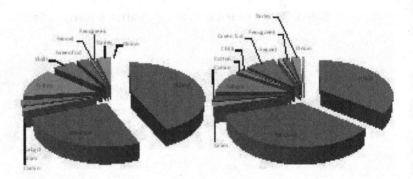

Change in irrigated cropping pattern at farm level in PS Pali & Manwar Jn.

Possible interventions for improving efficient use of ground water

- Strengthen the farmers awareness and knowledge about efficient use of groundwater through need based training programmes

- Promotion of efficient irrigation methods viz. micro-irrigation for efficient use of available groundwater

- Make the farmers aware of various alternate options of small scale enterprises for supporting the livelihood at farm level utilizing the available groundwater

- Socio-political intervention for rejuvenating the flow of various rivers/ rivulets so that groundwater is recharged

- Promoting rainwater harvesting for recharge of ground water and direct use in agriculture

22

Nursery Management for Fruit Crops

Akath Singh, P.R. Meghwal, Pradeep Kumar and
Pratap Singh Khapte

ICAR-Central arid Zone Research Institute, Jodhpur, Rajasthan-342003

There has been an increasing demand for horticultural crops more particularly fruit .With this, the demand for good quality planting materials has gone up and hence the nursery business has developed rapidly in the recent years in our country. Plant propagation aims at the reproduction of selected individuals or group of individuals, in this the basic principal of life impose certain requirements which must be met with in practices for successful propagation. Plants material of various fruit crops have not only to be raised efficiently and scientifically, for which skilled knowledge in raising and management of nurseries is needed, but also these have to be transported quickly from nurseries to place where these plants materials will be grown.

Availability of quality planting material is one of the most importanant requirement for increasing the production and productivity of any fruit crops. It becomes one of the great challenges for horticulturists to get the disease free quality planting materials of fruit crops from an efficient and reliable nursery.

Importance of Nursery

- Fruiting started after 4-5 years of transplantation in fruit trees and then only their actual quality may be known, hence to ensure the quality of particular species or variety, the role of reliable nursery for supplying true to type quality planting material is of paramount important.

- Regular supply of healthy, disease free and certified saplings is only possible from an ideal nursery.

- Since vegetative propagules viz. scion, buds, suckers, cuttings etc are taken from well managed productive, disease free true to type mother plants, If, the orchard established from such saplings their quality and productivity will be maintain for longer periods.

- Intensive care is possible in nursery therefore; saplings may be protected from insects –pests and diseases

- The seedlings require special attention during the first few weeks after germination. It is easier and economical to maintain the young and tender seedlings.

- Some of vegetatively propagated plants require more care, which is possible only in nursery under control condition.

- There is huge demand of skilled professionals for grafting, budding etc. nursery provides employment opportunities for technical and skilled labor. Therefore, nursery can itself be a very remunerative enterprise in the changing national scenario.

- Globalization has improved the chances of export of quality planting materials viz. asexually propagated, tissue culture plants to other countries.

Component of nursery

Nursery is a place where seedlings, samplings or any other planting materials are raised and sold out in gardens and orchards. The main component of an ideal nursery may be

Primary nurseries

Primary nursery is the place where seeds are sown and maintained for production of rootstocks. Flat/raised/sunken beds or pro trays or plastic crates etc are used for these purposes. The crops like papaya, guava, citrus having small seeds are sown in primary nursery and 15-20 days after germination it must be transplanted in secondary nursery. Primary nursery is also used for sprouting of stem cutting like pomegranate, fig etc.

Secondary nursery

Seedlings are replanted in secondary nursery or polythene bags (size 8x5") at 3-5 leaf stages. Secondary nursery is prepared by adding 2kg FYM, 8-10 g urea, 25-30g SSP and 15gMOP/m^2. Seedlings for root stock purposes are trained into single stem for budding or grafting and maintained till transplanting in field. Generally width of beds are 3' and length 15-20' but the length may vary depending upon conditions.

Mother plant

Plants used for taking bud, scion or part of vegetative part are termed as mother plant. This is the most important component in propagation of any fruit plant and these plants should be of guaranteed performance and have the known pedigree in respect to health, vigour, bearing and high yield with good fruit quality. Therefore, best plant of known should be secured for taking bud or scion shoots.

Tools, Implement and materials

- For use in propagation methods: budding/grafting knife, Secateurs, pruning knife, polythene strips, jute string/sutali, Moss, pro trays, Plastic crates etc.

- For use in seedbeds and nursery beds: Spade, fork, weeding hook/khurpi, bucket, watering can, etc.

- For use as potting/growing media: Rooting media soil, sand, vermiculate, water or even moisture saturated air. Soil free from organic matter should be used for rooting. Water is also used for rooting. It lacks aeration but aeration may be done artificially to get result. Running tab water is a good source for supplying oxygen, if need be.

- Sand, coco peat, compost, vermicompost, vermiculite, perlite, fertilizers, etc.

- Propagation structure: for hardening of seedling and production of transplant under protected condition like net house, mist house, low cost polyhouse, Hi-tech polyhouse, low tunnels etc.

Plant protection

i.   Common insecticides such as Sevin, Chlorpyriphos, Rogor, Furadan etc.

ii. Fungicides such as Bavistin, Dithane M-45, Blitox, Sulfex, COC etc.

iii. Micronutrients: Multiplex, Agromin (mixture of multi micronutrients)

iv. Growth regulators: Rooton (A, B, C), Rootex, Seradaux, IBA (root promoting hormones)

Basis of propagation

Plant propagation requires a complete knowledge of the scientific background of the life processes and of mechanical manipulation and technical skill needed to prove conditions essential for growth and regeneration. Its perfection needs a certain amount of practice and experience too. The manipulations include the art of breeding, grafting and the like operation.

The method used must be geared to the requirements of the particular kind of plant being propagated. The propagation of plants can be accomplished by a variety of ways depending open the kinds of plant involved and the propose of propagation. Propagation may be seeds, cuttings, bulbils, bulbs, corms, division, layering etc.

Seed may be classified as orthodox or recalcitrant depending on their ability to survive during storage. Orthodox seeds are capable of survival or reduction of their moisture by drying and their storability is increased. On the other hand seeds which lose their viability when dried to certain level of moisture contain consequent upon irreversible damage are known as recalcitrant seeds.

Disinfection of seed and soil

Propagation by seed, though common, considerable care and caution regarding freedom from disease, health, vigour, vitality and viability of the seed, well-drained soil free from fungi, nematodes, bacteria, and weeds for success at commercial scale. Disinfection of seed may be done by placing these in 0.6% acetic acid for 24 hrs at 21°C or in concentrated HCL for 30 minutes to free it from tomato mosaic virus.

Dry seed treatment: Dry seed treatment May be given to protect the seeds from infecting fungi and bacteria. Liquid suspension of organo-mercurial chemical like Agrosan-GN and cerasan and the non-mercurial compounds like Brassicol, Captan, Dithane M-45 and Thiram 0.3 % solution.

Hot water treatment: is very effective against fungi and bacteria infecting the seeds from inside. Dry seeds are immersed in hot water kept at 48-55°c for 20-30 minutes.

Formalin (40%) treatment: water 1:50 @ 2.5 1/30 cm² area to disinfect soil. The soil should be covered with polythene for 24 hrs after the application.

Chloropicrin (tear gas) may control all parasitic fungi but it is also highly phototoxic hence care must be taken to apply it through infection in the soil. No sowing should be done for about 2 weeks after the application.

Methyl bromide: Effective fumigant with deep penetration property. Effectiveness depends on the amount of soil moisture at the time application.

Vapam (sodium-N-methyl dithicarbamate,dehydrate): when sprayed as solution in water on the soil surface on the surface kills the nematodes ,fungi , and germinating weed seeds.

Use of growth regulators

Plant growth regulator may be employed to increase the percent success by hastening root initiation and increasing the number of developed roots per cutting. Rooting of plant species may be stimulated by IAA, NAA and IBA. The Indol-

Butyric Acid supposed to be better due to its non-toxic effect or wild range of concretion. All plant species do not respond to the same extent to different growth regulator. Infect the response depends on the nature of the cutting, its as physiological condition, among other things.

Dry application

The freshly cut and of the rooting material should be evently sprinkled with the commercial powder preparation to get good results

Soaking method

Solution of growth regulator of the concentration of 20-2000 ppm are used to promote rooting the lower portion of the cutting is dipped for about 24 hr to a length of 2-3 cm of the basal end just before planting in the rooting media.

Dip method

This method consists in increasing the concentration of growth regulating substation for a short period of say about 4-5 seconds. The concretion of the freshly made solution used in such cases is kept at 500-10000 ppm depending on the need of the plant. The cutting of transferred rooting media indiually after taking out from the solution for best results.

Parts Per Million (ppm) Conversions

ppm is a term used to denote a very, very low concentration of a solution. One gram in 1000 ml is 1000 ppm and one thousandth of a gram (0.001g) in 1000 ml is one ppm.

$1/1000^{th}$ a gram is 1 mg and 1000 ml is 1 liter, so that 1 ppm = 1 mg per liter = mg/Liter.

PPM is derived from the fact that the density of water is taken as 1kg/L = 1,000,000 mg/L,

Ppm (parts per million) to % (parts per hundred)

Example:

1 ppm = 1/1,000,000 = 0.000001 = 0.0001%

10 ppm = 10/1,000,000 = 0.00001 = 0.001%

100 ppm = 100/1,000,000 = 0.0001 = 0.01%

200 ppn = 200/1,000,000 = 0.0002 = 0.02%

5000 ppm = 5000/1,000,000 = 0.005 = 0.5%

10,000 ppm = 10000/1,000,000 = 0.01 = 1.0%

20,000 ppm = 20000/1,000,000 = 0.02 = 2.0%

(Parts per hundred) % to ppm

Example:

0.01% = 0.0001

0.0001 x 1,000,000 = 100 ppm

Production

Through seed

- The fruits like Papaya, lime and lemon, karonda, phalsa are usually propagated from seed. Similarly, the root stocks for different fruits are raised from seed. Well mature, uniform and healthy fruits should be selected for extraction of seed and it should be free from pulp-juice.

- Dormancy in some of the seeds may be broken by scarification, chemical treatment, soaking in water, stratification to improve germination.

- Seed are sown in primary nursery just after extraction from the fruits to a depth of 1.5-2.0 cm and 10 x 5 cm distance.

- Seed are covered with fine layer of wood ash or FYM mixture and mulched with dry leaves or paddy straw. Adequate watering with hazara is given at regular interval as per need.

- Mulch is removed after 35-40 days of sowing.

- Off type weak seeding are rouged out to maintain purity of cultivars.

- Repeat drenching with 1 % Bordeaux mixture or Bavistin 2.5 g / litre of water, if damping off disease is suspected.

- Regular spraying is given with Dimethote @ 2.5 ml/ litere, Dithan M-45 @ 2.5 g/liter or Blitox -50 @2.5 g/liter to avoid infestation of pests and diseases.

- Weeding and light hoeing is given as per need to avoid infestation of weed and to make soil porous.

- Two to three spraying of 300 ppm GA_3 +1 % urea is applied for better growth.

- Perforated polybags of 22.5x15 cm size with 100 gauge thickness can be used for sowing the seed. These bags are to be filled with well treated mixture of FYM, soil and sand in equal proportion. 2-3 seeds/bag are sown and after germination only one seedling is retained.

Through soft wood grafting

- Mango and mandarin can be propagated by soft wood grafting in July – August.
- Three to five months old rootstocks is used for softwood grafting.
- Previous season mature shoot is defoliated on mother plant 7-10 days prior to grafting.
- leaves are removed from the rootstocks leaving 2-3 leaves before grafting
- Defoliated scion inserted into prepared rootstocks by simple grafting and wrapped with polythene strip

Through cutting

- The propagation through cuttings is most rapid and commercial method of multiplication of Lemon, pomegranate, grapes, karonda, phalsa, fig.
- Cutting is taken 10-15 cm long and 2.5 cm thick from middle portion of one year old shoot. For best results, cuttings are collected during cooler part of the day, preferably in the morning.
- Top cut is made about 1 cm above the node and basal cut slightly below a node.
- Cutting is treated with IBA (2000- 5000mg/ liter water) solution for better results.
- Cutting is planted in June-July.
- The cuttings are set in moist coarse texture media provided 2 nodes below and 1 or 2 nodes above the media.

Air layering

- Litchi, sapota, guava, pomegranate, custard apple, and sweet lime can be propagated by air layering.
- Air layering is done either in spring or in monsoon season but the best time for layering is May to August.
- 1-2 year old shots are used for air layering.
- Leaves are removed from the basal of the selected shoots and stem

girdled by removing of bark about 2-3 cm wide at the place where leaves were removed.

- IBA @ 500 -1000 mg /liter water or a combination of IBA + NAA @ 500mg/liter water is applied on the girdled area.

- Girdled area is covered with handful of moist soil and sphagnum moss and wrapped with polythene strip.

- Rooted layer is cut just below the girdled portion and planted in pots or polythene bag or nursery beds in shady place.

Budding

- T-budding in ber, citrus and patch budding in bael, aonla, custard apple, are the commercially adopted method

- Budding is done either in Feb. - March or in July – August.

- The mother plant from which the bud-wood is obtained should be true to type, healthy, vigorous, free from pats and diseases.

- Rootstocks should be pencil thickness at the time of budding and 6 month to one year old depending upon crop.

- Cut for inserting the bud is made through budding knife in the bark at 15-20 cm above the ground level.

- Remove a patch of bark measuring 2.5 cm length containing a bud from the mother plant.

- The bud is inserted into the cut ("T" or patch) on the stock and tied with 2000-gauge polythene strip leaving the bud open.

- After the bud has sprouted, the top of the rootstock should be cut off above the union in the following dormant season to encourage the growth of scion.

- The polythene strip is also removed to avoid girdling.

Grafting

- Mango, guava, aonla etc. are commercially propagated by wedge grafting.

- One year old root stock raised in polybag having the pencil thickness is preferred for grafting.

- The scion shoot 15 to 18 cm long of pencil thickness (0.5-1.0 cm) with 3-4 healthy buds is used for grafting.

- Selected scion shoot should be defoliate on the mother plant, about one week prior to detachment.

- At the same time the apical growing portion of selected shoot should also beheaded, which helps in forcing the dormant buds to swell.

- Rootstock is headed back about 15 cm above the ground level and then, the beheaded rootstock is split to about 3.5-4.0 cm deep through the center of stem with grafting knife.

- A wedge shaped cut, slanting from both the sides (3-4.0 cm long) is made on the lower side of the scion shoot.

- The scion stick then inserted into the spilt of the stock and pressed properly.

- The union is then tied with the help of 150 gauge polythene strip.

- The best time for wedge grafting is during February – March or June-July.

Sucker (Banana)

Healthy sword suckers 3-4 months old, 1.5- 2.0 kg of banana should always be preferred for planting. Sword Suckers with a well – developed base and a pointed tip with narrow sword shaped leaf blades is considered as sword sucker. This is in early stages more vigorous and grows faster (than water suckers) and comes to bearing early. Whereas, water sucker is a small, undersized sucker of superficial origin, bearing broad leaves grown near clumps out of small, diseased or injured corms. It is normally not produced in a healthy banana clump. it arises if the planting is too old; overcrowded and shay not desuckered properly and not well looked after.

Propagation of pineapple

Suckers and slips are preferred for planting since they flower comparatively earlier than crown. Slips of 45-50 cm size with weighing from 350-450 g g gave an earlier uniform flowering and fruiting. Suckers of 55-60 cm, weighing 500-750 g. Before planting, the suckers and slips should be sun cured and dry leaf scales at the base should be removed. Planting material should be dipped in ceresan solution (4g in 1 lit. of water) or 0.2% Dithane M-45 before planting to protect the plants against bud rots.

23

Drip Irrigation in Horticultural Crops with Special Reference to Arid Zone

B. D. Sharma

ICAR-Central Institute for Arid Horticulture, Bikaner, Rajasthan

Micro irrigation normally called as drip irrigation was introduced in early seventies. Drip irrigation is one of the advanced methods of irrigation, by which water can be supplied directly into the root zone of the soil to achieve considerable saving of water through high water use efficiency, compared to surface irrigation method where water use efficiency is low due to losses in the water distribution system in the field. In micro irrigation method, water supplied at slow rate that is 1.0-2.0 kg/sq cm to the root zone of an individual plant through lets called emitters or drippers fitted with laterals pipes and filters at pump side. The success of micro irrigation is proven and efforts are being made in promoting the same extensively among the farmers in arid areas.

The hot arid regions of India are spread over about 31.7 million ha area mainly in the States of Rajasthan, Gujarat, Andhra Pradesh, Punjab and Haryana, which inhabit on an average 61 persons per square km making up a population of nearly 20 million peoples. The Indian arid zone is characterized by high temperature and low and variable precipitation which limit the scope for high horticultural productivity. However, these conditions greatly favour development of high quality production in number of fruits such as date palm, ber, pomegranate, citrus, aonla, bael, grapes and, guava. The optimized technologies and inputs could increase the existing low productivity. It is now realized that there is a limited scope for quantum jump in fruit and vegetable production in the traditional production areas. The amelioration of the extreme conditions is also considered vital for life support to the inhabitants of this area. The recent awareness regarding the potential of these ecologically fragile lands for production of quality horticultural produce has not only opened up scope for providing economic sustenance for the people of this region, but also for bringing new areas to increase production through horticulture.

Scope of expanding water supplies through development of water resources is very limited since suitable dam sites are fewer, development costs are prohibiting and environmental concerns are too strong. This has led to the realization that saving in the use of water is the option for expanding the water supply.

Advantages of Micro-Irrigation

The advantages of drip irrigation are as follows:

- Sophisticated technology
- Maximum production per mega litre of water
- Increased crop yields and profits
- Improved quality of production
- Less fertilizer and weed control costs
- Environmentally responsible, with reduced leaching and run-off
- Labour saving
- Application of small amounts of water more frequent

Disadvantages of Micro-Irrigation

The disadvantages of micro-irrigation are as follows:

- Expensive
- Need managerial skills
- Waste: The plastic tubing and "tapes" generally last 3-8 seasons before being replaced
- Clogging
- Plant performance: Studies indicate that many plants grow better when leaves are wetted as well

In India around 88% water is being used in agriculture sector, domestic use amounts to about 5% and industry use about 7% (Table-1). However the increased demand for water in industrial and domestic sectors which forces to reduce the percentage of area under irrigation. Expected annual water demand by various sectors in the year 2025 is 1093 Billion cubic metres by following advanced methods we can reduce that usage to 843 Billion cubic metres (Chand, 2002). As per the experts, water would be the single most critical limiting factor for India's economical development and water would be the greatest challenge for the 3rd millennium. The growing demand from the population calls for more efforts to enhance agricultural production activity. Water application systems are normally classified in three basic groups

Table 1: Usage statistics of water

Withdrawal	Word %	India %
Agriculture	69	88
Industry	23	7
Domestic	8	5
Total	100	100

Micro-Irrigation in Pomegranate Fruit Crop

The daily water requirement of fruit crops can be estimated by using different equations/formulae. However, the following formula is given in the FAO bulletin No. 24 which is commonly used for estimating daily water requirement of crop under drip irrigation system.

Volume of water= E_p x K_p x K_c x A x W (litres/day/plant)

Where

E_p : Pan evaporation (mm/day)

K_p : Pan coefficient factor, (0.7)

K_c : Crop factor as per the growth of crop (Canopy development)

A : Area per plant (Row-to-Row and Plant-to-Plant)

W : Wetted area coefficient

The crop factors (K_c) and wetted area suggested for determination of water requirement of fruit crops are presented in the table 2.

Total quantity of water = Quantity of water required/plant x Number of plants/ ha to be applied per hectare

$$\text{Time required to run the system} = \frac{\text{Total quantity of water required (Liters)}}{\text{No. of drippers per plant x No. of plants x dripper discharge}}$$

(OR)

$$\frac{\text{Quantity of water required/plant/day}}{\text{Dripper discharge x No. of drippers}}$$

The efforts have been made to estimate the daily water requirement of the pomegranate through drip on the basis of type of soil and canopy development and climate factors. On the experimental results, the crop coefficient and wetted are depending upon the crop age are given here (Table 2).

Table 2: Crop factors (Kc) and wetted area (per cent to the total crop area) for pomegranate fruit crops

Fruit crops	First year		Second year		Third year		Fourth year		Fifth year onwards	
	Kc	Wa	Kc	Wa	Kc	Wa	Kc	Wa	Kc	Wa
Pomegranate, (5x5m)	0.4	0.1	0.5	0.3	0.6	0.3	0.6	0.3	0.6	0.4

It has been reported that the results in respect of the quantity of water applied in pomegranate fruit crop on the basis of age of the plant and open pan evaporation (Table 3).

Table 3: Average water requirement of fruit crop by drip method

Name of fruit crops	Age (Y)	Water requirement (Litres/tree/day)											
		Jan	Feb	Mar	Apr	May	Jun	Jul	Aug	Sep	Oct	Nov	Dec
Pomegranate	1	-	-	-	-	-	-	13	10	10	12	9	8
	2	12	17	25	34	36	25	18	15	15	18	13	11
	3	16	23	34	46	50	34	18	15	15	18	13	11

Pomegranate (*Punica granatum*)

In pomegranate, irrigation at a particular place depends on the *bahar*, which has to be taken. In northern India, where *mrig bahar* crop is preferred, the irrigation period is May-June (after bahar treatment) and September to November (fruit development period). Similarly, in western and southern part of the country, where ambe and hasth bahar is common, irrigation through drip is essential just after the *bahar* treatment and during the fruit growth and development Depending upon the rainfall in monsoon season, it has been estimated that for optimum fruit production, 150-175 are the days where irrigation through drip system is required. It has beern reported that drip irrigation system may prove to be economic and water saving for pomegranate orchard. Pomegranate crop requires irrigation from January to July for fruit harvest and drip system with mulch saved about 78% irrigation water over check basin system. In pomegranate fruit crop drip irrigation method, applying water equivalent to 20 per cent area is superior method of water application over surface method and average annual irrigation requirement of pomegranate through drip method is 20cm. In cv. Ganesh, drip irrigation, in addition to water use efficiency, marketable produce increased as no cracking was observed. In arid conditions of western Rajasthan, Pomegranate crop, drip irrigation system saved 25% irrigation water and increases the yield by 87.5% over pipe irrigation system. It has also observed benefit cost ratio of pomegranate cultivation for drip method to be 5.17 compared to 2.80 in traditional irrigation system. In pomegranate about 45% water saving and 12% increase in fruit yield was achieved in drip irrigation system. Trickle irrigation on alternate day with 30% wetted area gave maximum fruit yield (57.3q/ha) of Ganesh cultivar of pomegranate. The data in table 4 revealed that although tree growth was not affected, the fruit yield and water use efficiency were considerably increased by drip irrigation in *cv*. Ganesh pomegranate.

Table 4: Effect of methods of irrigation on growth and yield of Ganesh pomegranate

Method	Water used Litres	Tree height(m)	Stem diameter(cm)	Fruit yield (q/ha)	WUEQ/ha cm
Check basin					
0.6 IW/CPE	4800	2.35	3.78	40.6	0.82
0.8 IW/CPE	6600	2.56	4.02	47.5	0.69
Trickle					
Daily					
20% WA	3580	2.16	3.79	52.5	1.25
30% WA	5322	2.44	4.02	55.9	0.72
Alternate day					
20% WA	3580	2.30	3.84	48.6	1.16
30% WA	5322	2.24	4.01	57.3	0.92

Conclusion

1. Water use efficiency and maximum production with minimum use of water

2. Economizing the cost of production

3. Waste and fallow lands can be brought under cultivation

4. Efficient use of poor quality and saline water and reduce environmental pollution

5. Improve soil health and scope for development of cheaper MI technologies achieve the food security.

24

Water Harvesting: A Sustainable Solution for Protected Cultivation for Economic and Nutritional Security of Farm Family in Arid Zone

Awani Kumar Singh

Indian Agricultural Research Institute, New Delhi-110012

In the present scenario of rapidly increasing population and decreasing cultivable land, there is urgent need to increase the agricultural productivity. According to Indian Council of Agricultural Research (ICAR), 280 million tones food grain would be required to match the country's demand by 2020. It will be a difficult task to meet this food requirement in near future from the present productivity levels. It is necessary to increase agricultural productivity, because about 60% of Indian agriculture is rainfed. Proper use of even a single drop of rain water is necessary to ensure the success of rainfed agriculture. In hills, agriculture is main source of livelihood of rural people. However, small and scattered land holding, difficult topography, women dependent agriculture, totally rainfed, typical socio-economic factors and migration of youth towards urban/plain areas are major constraints of hill horticulture.

In the high altitude areas, the river flow in deep valley at the toe of slopes rarely serve any purpose as far as domestic water supply and irrigation are concerned and hence, agriculture in these areas is mostly rainfed. However, these areas received high magnitude of rainfall but this rainwater is not possible to convert into soil moisture due to steep terrain of hills which provided minimum opportunity time for infiltration of rainwater. It has been estimated that only less than 15 per cent of the rainwater is able to percolate down through slopes and recharge the soil moisture. The rainfall pattern hill is erratic i.e. most of rain (approximately 70% of annual rain) received only during the two months of monsoon. Furthermore, the climate of high altitudes comprises several climatic anomalies such as wide variability in temperature (-3 °C to 35 °C), frequent snowfall, hail, sleet and erratic pattern of rainfall. Under such abrupt conditions crops grown in open conditions could not be sustained.

All these subtleties invoke there is urgent need to harvest rainwater and subsequently stored at proper place for long duration so that even a single drop of water can be utilized preciously. In present scenario, concept of poly-tank cum micro-irrigation system can be employed in protected cultivation. Such system is best way to provide irrigation, under low water availability areas with high water use efficiency. The micro-irrigation system saves water (40-50%), fertilizer (25-30%), labors expenditure (50%), soil-erosion cum fertility (90%), minimizes plant mortality (80%), enhances crop yield (25-75 %) and improves the quality of produce as compared to traditional pattern specially when employed under protected environment Keeping this in view, storage of rainwater in low cost adequate capacity poly-tank and use of this water for irrigation in crop grown in protected environment (Polyhouse or Tunnels) through gravity based drip irrigation system are being discussed in the following text. The success of such coupled system are strongly supported by the experiment conducted in high altitude areas i.e. at the research farm of Krishi Vigyan Kendra, Lohaghat, Champawat, a regional research and extension center of G. B. Pant University of Agriculture and Technology, Pantnagar, Uttarakhand.

1. Type of Water Conserving Structures

The following structures are commonly used for storage of water for domestic as well as irrigation purpose in hilly areas.

- Storage of water by construction of *kachha* tank.
- Storage of water by construction of cemented (RCC) tank.
- Storage of water by construction of poly-tank or LDPE (Low Density Polyethylene) tanks. Storage of water by installing syntax/ferro-cement tank.

2. Source of Water

Source of water of these water storage structures may be rainfall generated runoff, perennial streams, natural springs, Nalas and some time waste water from well and municipal hand pumps.

- Collecting water from the roof (due to rain) of residential areas using *"patnala"* (thin iron sheet bent in the shape of English alphabets "U" and "V") and storing the harvested water into tanks through PVC pipes and filters.

- Collecting surface runoff generated from rain by formation of ridges and gully on land surfaces and finally channelizing into the tanks.

- Diverting water from natural sources such as perennial streams, springs and *"Nalas"*.

- Collecting waste water of well/hand pump etc.

3. Why only Poly-tanks (LDPE) in Hills?

Based on the literature cited and experiment conducted at Krishi Vigyan Kendra-Lohaghat, a poly-tank has several advantages over other water conserving structures which are enlisted below.

- Majority of hill farmers are small and marginal with scattered land holding. Therefore, it is not practically feasible to construct cement tank at everywhere.

- Poly-tanks are less costly compared to cemented tanks and plastic tanks. The cost of construction of a poly-tank is 5-6 times less as compared to similar capacity cement tank.

- There are possibilities of cracking of cemented tanks due to landslides and earth quakes which is frequently occurring in hills. However, poly-tanks adjust their shape according to minor disturbance in land due to their flexible nature.

- There is significant seepage losses in *"kachha"* tank and cannot be recommended when water stored for long duration.

- Use of plastic tanks/ferro-cement tanks is restricted due to their limited capacity and handling problem.

- Construction of Poly-tank is so simple & theses can be constructed by simple farmers (Fig. 1).

4. Important points to remember during construction of Poly-tank

Following points are very important for construction of poly-tank:

A. Selection of site

Selection of site is one of the important steps in construction of a poly-tank. The following points should be kept in mind during selection of site:

- Site should be at a place where maximum amount of rain water can be collected. It is better to construct near any building so that water from roof of the building can be harvested and easily stored in the poly-tank.

- Site should be flat as possible and should not prone for erosion.

- Poly-tank should be constructed at least 2 m above the irrigated area so that sufficient head for drip irrigation system can be easily achieved through gravity.

- Site should be out of reach from animals and children.

B. Estimation of capacity of Poly-tank

Estimation of adequate capacity of poly-tank as per the requirement of the irrigation is also an important aspect. Construction of a poly-tank of lesser capacity than requirement may create water scarcity problem during lean season, whereas constructing a tank of larger capacity than the required is just wastage of money. Calculation for the capacity of tank required for giving 10 cm irrigation in 1 *nali* (200 m²) area is as follows:

Capacity of poly-tank (V) = Depth of irrigation (m) x Area (m²)

= 0.10 m x 200 m²

= 20 m³ or 20,000 liters

A 20,000 liter capacity poly-tank is adequate for providing 10 cm irrigation in one *nali* area.

C. Shape of the Poly-tank

Normally, tanks may be of three shapes i.e. square, rectangular and round. Geometrical studies revealed that round shape tank contain more volume of water than the other shape for same area. But the construction of round shape tanks is tough due to its complex shape, therefore square and rectangular shape poly-tanks are preferred. Once the required capacity of the tank is fixed, the length and width of the tank can be decided by the following formula for rectangular shape tank (Fig-1).

Fig. 1: Trench Opening

$$V = \frac{D}{2}\{LW + (L-2D)(W-2D)\} \times 1000$$

Where,

V = capacity or volume of tank (liter);

D = depth of tank (meter);

L = length of tank (meter);

W = width of tank (meter).

It is worth notable that a safety factor of 10% in required volume should be taken during the design of poly-tank. Accordingly the sides of 20,000 liter capacity tank (1:1 side slope), would be 5.1m (L) x 5m (W) at top, 2.1m (L) x 2m (W) at bottom with a depth of 1.5 m.

D. Slope of walls of Poly-tank

Adequate slope of the side wall of the tank is very important factor in construction of poly-tank. The main purpose of providing slope in the side wall is to divert the pressure of water through the side wall of the tank unless polythene sheet may be punctured or tear. Normally, the slope of side wall is kept 1:1 (Fig.1,2).

E. Depth of Poly-tank

In hilly areas, the depth of friable soil is less, so depth of tank should not be more than 1.5 m. Furthermore, digging below this depth becomes tougher and also costly job (Fig.1).

F. Type of Polythene Sheet

Life of a poly-tank largely depends upon the quality and thickness of the polythene sheet. Generally, polythene sheet is available in black, blue, white and other colours in market. Its thickness is measured in GSM, micron (μ) and sometime in gauge. A sheet having specification of 250 GSM means the weight of 1 square meter piece of this polythene sheet will be 250 gram or in other words 1 kg polythene sheet of 250 GSM specifications will cover 4 square meter area. Maximum size of polythene sheet is available in 50 m x 7 m dimensions in the market. A poly-tank of maximum 5 m width can be constructed by this sheet. However, the tanks larger than this width can be constructed by sealing two polythene sheets using heat sealer (joining two sheets at high temperature). Polythene sheet of desired width can be obtained directly from the manufacturing company on demand. Size of polythene (1:1 slope tank) can be measured by the following formulae:

Total length of polythene (m) = $2D\sqrt{2} + (L - 2D) + 1.5$

Total width of polythene (m) = $2D\sqrt{2} + (W - 2D) + 1.5$

5. Procedure for construction of poly-tank

Keeping in view the above points i.e. site selection, capacity estimation, shape and size of tank, selection of polythene of tank size etc., following procedure would be followed during the construction of a poly-tank (Fig.2):

Fig. 2: Construction of ploly tank

- The site for constructing poly-tank should be cleaned firstly so that the size of tank can be marked on the ground by lime. Once the rectangle/ square having side equal to side of top of poly-tank marked, another rectangle/square will be marked inside this. Notable that side of inside rectangle/square will be less than the outer rectangle/square by twice the depth of tank if side wall slope is 1:1.

- Now digging will be started straight downward through the sides of in- side rectangle/square up to the depth of the tank i.e. 1.5 m taken in this example. Now join the sides of outer rectangle with the bottom side of inner rectangle by cutting the soil in sloping manner. Finally we will get a tank having 1:1 side slope.

- All the stone, gravels, root and residue of vegetation should be removed from all surface of tank because these may puncture the polythene sheet. A spray of weedicides like Atrazine @ 0.4 gram/m^2 is recommended to avoid the growth of any kind of vegetation from the wall of the tank.

- Now, a mixture of clay, dung and straw is prepared and pasted on the side walls and bottom of the tank so that all these surfaces can be smoothened. Ultimate aim to stop all those chances due to which polythene sheet can be punctured.

- Now the polythene sheet of required size, calculated from the formulae discussed above, will be fitted in the excavated tank. The polythene sheet will be well covered by stone/brick and soil from all four upper corners.

- Pitching of boulders, tar-felt sheet, and some time bricks are recommended for long life of the polythene but it is not economical from framer point of view. However, now a day's good quality of polythene is available in the market which is resistant from sunlight, air, temperature variation etc.

- A small pit (1mx1mx1m) is recommended at the entry point of water so that soil and waste material with surface runoff should not enter into main tank.

Cost of construction of one 20,000 liter tank is estimated and presented in Table 1.

Table 1: Estimation of cost of construction of a 20,000 liter capacity poly-tank

Sl. No.	Particular	Quantity	Rate* (Rs/unit)	Amount (Rs)
1.	Digging of tank	22.3 m^3	50	1115.00
2.	Pasting of mixture of clay, dung and straw	1 job	One time	500.00
3.	Polythene sheet (250GSM)	54.4 m^2	75	4080.00
4.	Spray of Atrazine inside tank	1 job	One time	100.00
5.	Miscellaneous charges (@4%)	232.00		
	Total Cost (Rs)	6027.00		

Rate quoted in the table is based on the prevailing market price of Rudrapur, Uttarakhand and subject to vary place to place.

A poly-tank has usual life of 10 years while it is properly and carefully utilized. In this way, cost of construction of a poly-tank is around Rs. 603/year and Rs.50/month.

Case Study

Considering the socio-economic conditions of the farmers, land holding and natural resources in hills, an experiment was conducted at experimental field

(Fig.3) of Krishi Vigyan Kendra (KVK), Lohaghat, Champawat (Uttarakhand), a regional research and extension center of G. B. Pant University of Agriculture and Technology, Pantnagar, Uttarakhand during 2006-2010 to study vegetable production under polyhouse cum drip irrigation system. The experimental site was located at an altitude of 1750 meter above msl and receives an annual average rainfall of 1400 mm. A gable type polyhouse structure size 20m (L) x 5m (W) x 3m (CH) was selected for cultivation of hybrid capsicum (var. Tanvi) and tomato (var. Raksita) from March to October in each year. 360 plants were grown at 50 cm x 50 cm spacing in the polyhouse for the both crop. A Poly-tank tank (made by UV stabilized, blue colour Silpauline sheet of 250 GSM) of 20,000 liter capacity was constructed (tank size was 5.1m x 5m x 1.5m) in the upper side of polyhouse structure so that sufficient head due to gravity can be available for drip irrigation system. Rainwater harvested from the roof of the nearby building (effective roof area 120 square meter) was collected in the poly-tank by filter and plastic pipes. A drip irrigation system (dripper discharge 2.5 l/hr) was installed to this poly-tank for irrigation in the polyhouse. Study showed that 74 numbers of irrigations were required during the total crop period i.e. March to October. It is worth notable that drip irrigation system was operated ½ hour a day to maintain the uniformity during the study period. It was observed that maximum 15 number of irrigation were required in each month of May and June. Maximum water requirement of crops during May and June was recorded 18,750 liter in each month. Total quantity of irrigation water required was 92,500 liter to grow capsicum and tomato crop from the month of March to October and this quantity of water can be harvested from a single roof area of 120 m² in the similar rainfall area.

Fig. 3: Demonstration view of polyhouse vegetable production through rain water harvesting poly-tank water by drip system under rain fed condition of high hills at KVK, Lohaghat, Champawat, Uttarakhand

The total yield of capsicum and tomato from 100 m² polyhouse (Fig. 4) was recorded as 1050 kg and 1250 kg respectively, which give total return as Rs 45000 (capsicum @ Rs 25 kg⁻¹ and tomato @ Rs 15 kg⁻¹). However the total income obtained from the same area (100 sq. m) by the traditional method under open field condition were only Rs 2500. Study revealed that hill farmers can be obtained significantly high income i.e. as much as 18 times higher than the traditional method by adopting the techniques proposed here (fig.6-11). Study also indicates that a poly-tank of 20,000 liter capacity coupled with drip irrigation system was found adequate for irrigation in capsicum and tomato crop under 100 sq. m area prevailing the similar rainfall pattern. Integration of rain water harvesting+Poly-tank+drip-irrigation+polyhouse can be considerably enhanced the vegetable production in hilly area and probably the single best solution of the large adverse horticultural condition regularly occurred in the area. By this way horticulture can become beneficial in the hills which ultimately resulted in reducing migration of young hill youths towards urban/plain areas. Furthermore, such kind of techniques will be helpful in reducing the women drudgery in hill horticulture.

Impact of Technologies in hill farmers

After research and refinement, On & Off-farm demonstration's result and trainings about technology, the maximum farmers have adopted this technologies and getting net profit Rs.10-12thousand/year during off- season from 40-50m² small size polyhouse with the help of rainwater harvesting.

Fig. 4: Inside view of successive story of following farmer's polyhouses i.e. Shri Tara-Dutta Khargwal, Shri Ramesh Chaubey, Shri Mathura dutta Chaubey and Shri Ramesh Khargwal wherein they have produced off-season high value vegetable through rain water harvesting water under rainfed condition of high hill at Lohaghat, Champawat, Uttarakhand

25

Soil-less Cultivation Technology for Growing Vegetable Crops Nursery for Enhancing Economic and Livelihood of Farm Family in Arid Zone

Awani Kumar Singh and Balraj Singh

Indian Agricultural Research Institute, New Delhi-110012

Increasing susceptibility of vegetables to various biotic and abiotic stresses and very high cost of hybrid seeds has warranted the attention of the vegetable growers to improve the nursery raising technology of vegetables. Now-a-days protected nursery raising of vegetables has become a full flagged industry in several developed countries like Israel, Japan, Spain, Netherlands, USA etc. In countries like Israel no vegetable grower is growing his own nursery, but they are getting the required kind of seedlings from an established nursery owner on pre- order basis. Similar system of nursery raising is also prevalent in several other advance countries. In those situations where vegetables are being grown under protected conditions it becomes pre-requisite to raise the required vegetable seedlings only under protected environment to get virus free healthy seedlings.

Raising nursery from seeds provide an easy and convenient way to nourish tender and young seedlings in a well managed, small and compact area for better germination of small and costly seed. In general, vegetables crops are divided into three groups according to their relative ease for transplanting. Beet root, broccoli, brussels sprouts, cabbage, cauliflower, tomato and lettuce are efficient in water absorption and rapidly form new roots after transplanting. Vegetables crops that are moderately easy for transplanting are brinjal, onion, sweet pepper, chilli and celery which do not absorb water as efficiently as crops that are easy to transplant, but they form new roots relatively quickly. The vegetable crops which are difficult to transplant are: all cucurbits and sweet corn which require special care during nursery raising and transplanting. Seedlings not only reduce the amount of time of the crop but it also increases the uniformity of the crop and harvesting as compared to direct sowing of the crops. Transplanting of seedlings also eliminates the need for thinning and provides very good opportunities for virus free, vigorous and even off-season seedlings if grown under protected conditions in soil-less media. So it becomes necessary to know the protected structures which can be efficiently used for successful raising of vegetable seedlings, suitable types of containers and growing media, method of seeding in containers in soil-less media, growing conditions and water and fertilizer requirement of seedlings along with hardening and transplanting to the field. Protected nursery raising technology is highly suitable and can be established as a small scale industry in major vegetable growing areas of our country by progressive farmers especially in peri-urban areas. By this way the vegetable growers will get virus-free healthy and off season nursery as per their requirement and it generate extra employment in rural areas.

Types of plug-trays or pro-trays

Vegetable seedlings or nursery can be raised in a variety of containers ranging from molded trays to pots. Styrofoam and / or plastic trays are considered to be the standard one in many parts of the world. Plastic trays of the same size with same size of cells are fixed in styrofoam are mostly preferred because they encourage more uniform root zone temperature and moisture (Table 1). In USA only Styrofoam flats are often used as compared to plastic pro-trays.

Cell shapes vary with in the seedling flats and available shapes include inverted pyramid, round and hexagonal while it is to be investigated if the cell shape affect the field performance of the seedlings, but cell size does appear to have a clear cut effect, and recommended cell size to use for raising vegetable seedlings will vary with the crop. In general larger cell sizes, especially for longer cycle crops (tomato and watermelons), often results in larger yields in the fields. Generally, smaller cells are required for short cycle vegetable crops. The standard cell size in the industry is 1.0 inches by 1.0 inches or approximately 200 plants per plastic pro-tray. Economics and duration of stay in the nursery greenhouse also help to determine the cell size that is used for a particular crop. Container used for nursery raising of vegetables must have good drainage and be able to hold soil-less media and easy in handling.

Generally plastic trays or pro-trays having different sizes of cells are used for raising seedlings. Mainly two kinds of plastic pro-trays are used in raising the seedlings . One tray which may have cavities of 3.75 cm (1.5") in size, whereas the other tray may have cavities of 2.5 cm (1.0') in size. These trays helps in proper germination, provide independent area for each seed to germinate, reduce the mortality rate, maintain uniform and healthy growth of seedlings, are easy in handling and storing, reliable and economical in transportation . These plastic trays may be fixed in thermocol base trays having the same number and size of cavities before filling the media. In case, thermocol basis are not available, only trays placed on floor or firm base may be used.

Fig. 1: Pro-trays and plants grown on pro trays

Table 1: Recommended size of cell in plastic pro-trays fixed in Styrofoam for raising vegetable seedlings

S.N.	Vegetable crops	Cell size (inches)	Optimum size of cells for early production (inches)
1.	Tomato	1.5-4.0	3.0
2.	Brinjal	1.5-4.0	3.0
3.	Sweet pepper	1.0-4.0	2.0
4.	Chilli	1.0-3.0	2.0
5.	Watermelon	1.5-4.0	3.0
6.	Muskmelon	1.5-4.0	3.0
7.	Cucumber	1.5-4.0	2.0
8.	Summer squash	1.5-3.0	2.0
9.	Bitter gourd	1.5-3.0	2.0
10.	Bottle gourd	1.5-4.0	3.0
11.	Ash gourd	1.5-4.0	3.0
12.	Pumpkin	1.5-4.0	3.0
13.	Cabbage	1.0-2.5	1.5
14.	Cauliflower	1.0-3.0	2.0
15.	Broccoli	1.0-3.0	2.0
16.	lettuce	1.0-2.0	1.5
17.	Ridge gourd	1.5-4.0	3.0
18.	Sponge gourd	1.5-4.0	3.0
19.	Round melon	1.5-4.0	3.0
20.	Long melon	1.5-3.0	2.0
21.	Brussels sprout	1.0-3.0	2.0
22.	Knol khol	1.0-3.0	2.0
23.	Beet root	1.5-3.0	2.0
24.	Okra	1.0-3.0	1.5
25.	Celery	1.0-2.5	1.0
26.	Parsley	1.0-2.5	1.0

Mostly artificial soil-less media is used for raising healthy and vigorous seedlings of vegetable in plastic pro-trays. Mainly three ingredients viz., coco-peat, vermiculite and perlite are used as root medium for raising the nursery. These ingredients are mixed in 3:1:1 (V/V) ratio before filling in the required containers or plastic pro-trays.

Ingredients used as root medium

Mainly three ingredients viz. coco-peat, vermiculite and perlite are being used as a root medium for raising the nursery in the high-tech nursery greenhouse. These ingredients are mixed in the volume ratio of 3:1:1 before filling in the trays. The characteristics of these three ingredients are as under:

1. **Coco-peat:** It is prepared from the waste of coconut husk. This medium has good porosity, improved drainage and air movement activity. This medium is completely free from infestation of any pest or pathogen. Coco-peat is commonly being used as a medium under protected cultivation of ornamental crops like roses and gerberas and for raising the nurseries of vegetables and ornamental plants in the developed countries.

2. **Perlite:** It is a light rock material of volcanic origin. It is essentially heat-expanded aluminum silicate rock. The volcanic ore is heated to extreme temperatures of about 1800°F to cause the rock particles to expand and to produce the white granular product. It's role in a mix is to improve aeration and drainage. If this ingredient is used in a mix, the horticultural grade should be selected since it has larger particle size and is thus more effective. Perlite is neutral in reaction and provides almost no nutrients to the mix (except for small amounts of sodium and aluminum). A disadvantage of the use of perlite is its low weight, which makes it float when the medium is watered. During mixing, it produces dust, further, which can be eliminated by wetting the material before its use.

3. **Vermiculite:** It is heat-expanded mica. This mineral is heated at a temperature of about 1,400°F (760°C) to produce the folded structure associated with the material. It is very light in weight and has minerals (magnesium and potassium) for enriching the mix, as well as good water-holding capacity. Neutral in reaction (pH), it is available in grades according to sizes. Grade 1 includes the largest particles and grades 4 and 5 are fine in texture. The most commonly used grades are 2 and 4. Its fineness, incidentally, makes it prone to being compressed easily in the mix. To reduce its potential, a mix including vermiculite should not be pressed down hard.

Advantages of soil-less media

The major advantages of soil-less media include the following:

1. *Uniformity of mix:* The physical and chemical properties of a mix are uniform throughout the mix, a condition not found in field soil. The homogeneity of the medium makes it possible for plants to grow and develop uniformly, provided other growth factors are also consistent.

2. *Ease of handling:* Mixes are light in weight and easy to transport. Ingredients are easy to scoop and move around during manual preparation.

3. *Versatility:* Mixes can be made to order (custom mixed) for specific needs. They can be used to amend soils in the field by mixing into flower beds, lawns or garden soil. They are convenient for use in container plants.

4. *Sterility:* Mixes, initially at least, are free from diseases and pests. Seed germination is less prone to diseases such as damping off.

5. *Good drainage and moisture retention:* A mix can be custom-made to provide the appropriate degree of drainage and moisture retention.

6. *Convenience of use:* Mixes are ready to use when purchased.

Sowing of seeds and fertigation

Seeds selected for sowing in plastic pro-trays must be generally true to type and of good quality (high germination > 95 % and do not contain of other cultivars or weed seeds etc). Seeds should be purchased fresh for each growing season and should be typically sold in sealed packages that indicate the crop, variety, its germination percentage and status of seed treatment, if any. Seeds are usually planted/sown at a shallow depth after pressing the media with finger in a gentle way into the potting plugs or cells which are filled with artificial media. The actual depth of sowing depends on the crop and the size of the seeds. After sowing of seeds a thick layer of vermiculite is given to cover the seeds for better germination as this media is having water holding capacity. After seed sowing the pro-trays are kept in the germination room at the optimum required temperature for early and better germination. Usually one to two seed are sown per cell and after the seeds germinate during the seedling production process, the seedlings are thinned to one per cell the thinned can be gap filled in those cells where seed germination could not take place.

The size of the cavities depends upon the kind of crops to be sown in the nursery trays. For cucumber, muskmelon, tomato, brinjal, we use the trays having 187 cavities of 3.75 cm (1.5") size and for lettuce, cabbage, cauliflower and capsicum, we use the trays having 345 cavities of 2.5 cm (1.0") size are prepared. For raising good and healthy seedlings, the optimum electrical conductivity of irrigation water should be 1.6 to 1.7 with a pH of 6.6 to 6.7 and the optimum temperature should be 20°C during the winter. The optimum temperature for raising seedlings during summer season is 30°C. Water is applied after sowing of the seed in each tray through fine sprinkler boom uniformly in the trays. Hundred percent humidity is maintained and when level comes down irrigation is applied. Special treatment of growth regulators @ 140 ppm is applied just after germination (emergence of seedlings) of seeds in winter season and 70 ppm special treatment is applied for summer season nursery. Nutrients are applied in the form of N:P:K (1:1:1) @ 140 ppm once a week through the fine sprinkler to maintain the uniformity in application of nutrients. When temperature goes below 20°C in winter, application of special treatment is not recommended. For maintaining the optimum temperature we can use the germination room as per requirements of the crops.

Hardening of seedlings and days required for transplanting

Hardening of vegetable seedlings before transplanting in the main field is very essential for reducing transplanting shock and also to have better crop stand. Plants should be gradually hardened, or toughened by acclimating them to the anticipated growing conditions of fields, at least a week before planting them in main field. This is done by slowing down their rate of growth to prepare them to withstand such as chilling, drying winds, shortage of water, or high temperatures. Vegetables like cabbage, lettuce, onion and many others can be hardened to withstand frost; others such as tomatoes and sweet peppers cannot. With holding water or minimal water supply are the best ways to harden a plant. Generally seedlings of all vegetables become ready for transplanting in 28-30 days after sowing in plastic pro-trays. Under the greenhouse conditions, cucumber seedlings become ready for planting in main field or greenhouse in 25-30 days after sowing during peak winter whereas muskmelon in 30-35 days during peak winter season but most of the seedlings of cucurbits are ready for transplanting within 20 days in summer season. In case of tomato and brinjal, the seedlings are ready for transplanting in 30-32 days if the germination room is used for initial period for seed germination during winter season whereas the optimum temperature for seed germination (i.e. 24-25°C) is kept by using electric heaters.

Advantages

1. Seedlings can be raised under adverse climatic conditions where it is not possible under open field conditions.

2. Healthy seedlings can be raised in short period as compared to the time taken under open field nursery raising.

3. There is no chance of soil borne fungus or virus infection to the seedlings as the nursery is grown in soil-less sterilized media and insects can not enter under the protected conditions.

4. Drastic reduction in the mortality in transplanting of the seedlings as compared to the traditional system of nursery raising. No mortality, no transplanting shock and quick establishment of the seedlings due to perfect development of root system.

5. Early planting is accomplished by raising such nursery.

6. It is suitable for raising the nursery of sexually and asexually propagated vegetables and ornamental crops.

7. Management of insect/pests and diseases under greenhouse/protected conditions is quite easy particularly the infection of viruses.

8. Looking to the very high cost of hybrid seed of vegetables we can reduce the seed rate 30-40 per cent in comparison with the traditional nursery raising system, as individual seeds are sown in each cell which produces a very healthy seedling.

9. Easy for transportation after packing for long distances.

10. Farmers can get the nursery ready from such nursery greenhouse any time as per the requirement.

Low cost poly-house for raising off-season nursery of cucurbitaceous vegetables

Cucurbits form an important and large group of vegetable crops cultivated extensively in India during summer and rainy seasons. The plants of these crops are of climbing or trailing habit with the exception of summer squash, some varities of which are generally busy in growth. They are grown right from village house, thatches, and banks of rivers and ponds to the large scale commercial cultivation in nearby big cities. A low cost technology to grow off-season nursery of cucurbitaceous vegetables for raising extra early crop in spring-summer season in north Indian plains which is highly profitable to the growers is described below.

The low cost poly-house is a zero-energy chamber made of polythene sheet of 700 gauge supported on bamboos with sutli (strings) and nails. This size of poly-house depends on the purpose of its utilization and availability of space. It has only one opening kept ajar for 1-2 hours during the day. The structure depends on the sun for heating. The temperature inside poly-house is 6-10°C higher than outside. The cold waves during December-January do not enter in to the poly-house and inside environment becomes conducive for quick germination of seeds and growth of young seedlings. The polythene bags of 16.5x10 cm size are used for sowing the seeds and raising the cucurbit seedlings in the poly-house. The polythene bags are filled with the mixture of garden soil, sand and compost in the ratio of 1:1:1. Before filling these seedlings, 4 to 5 small holes are made at bottom and side of each bag to facilitate drainage keeping 2.3 cm vacant from the top. The filled bags are kept inside poly-house in groups in such a way that one person can easily sow the seeds and work. A poly-house 10x3.5 m size can easily accommodate 5000 polythene bags (16.5x10 cm) and an expenditure of 55 paise will be required for raising 1-2 seedlings per bag. One to two seeds are sown in each bag during last week of December or 1st week of January after treating them with captaf @ 2 g/kg of seed. After sowing, a thin layer of sand is put to fill the top of the poly-bags in order to facilitate proper germination. After sowing the seed, light irrigation is given by water can. After 25-30 days of sowing, the seedlings become 10-12 cm long and then they are kept outside the poly-house for 2-3 days for hardening. In the first week of February when danger of frost is over, the seedlings are transplanted on the northern slope of prepared channels in the field after removing the polythene bags with the help of blade without disturbing the earth ball. After transplanting, light irrigation is given for better establishment of plants.

26

Harnessing Solar Energy for Sustainable Farming System in Arid Regions

Priyabrata Santra, P.C. Pande, N.M. Nahar, A.K. Singh

ICAR-Central Arid Zone Research Institute, Jodhpur, Rajasthan – 342003

Energy and food are the two main requirements for human civilization; however the demands for these two resources are increasing in a fast rate. Development of any region is reflected in its quantum of energy consumption. In this context, the situation in arid region is difficult. People burn firewood, agricultural waste and cow dung cake for cooking food causing damages to the fragile eco-system of arid zone. There is inadequate electrical power supply and hence villagers of remote arid areas are unable to derive general benefits. The kerosene is used for lighting and fast depleting diesel oil is used for running agricultural machinery including pumps. In this context solar energy is considered to be one of the promising sources in India.

About 12% of electricity production capacity in India is met through renewable energy sources. In India, national solar mission was launched in November 2009 with a target of 4000 MW grid and 1000 MW off-grid electricity generation from solar energy by the end of phase II (2013-2017), whereas these targets are 20,000 MW and 2000 MW, respectively by 2022. The target has been revised in 2015 to a total grid connected solar power generation of 1,00,000 MW comprising of 40,000 MW roof top generation and 60,000 MW grid connected solar power plants (Resolution of MNRE, Govt of India, No. 30/80/2014-15/NSM dated 1st July 2015).

Table 1: National solar mission targets

Sl.No.	Application segment	Target for Phase I (2010-13)	Target for Phase II (2013-17)	Target for Phase III (2017-22)
1.	Grid connected solar power generation	1,100 MW	4,000 MW	1,00,000 MW[§]
2.	Off-grid solar applications	200 MW	1,000 MW	2,000 MW
3.	Solar thermal collectors	7 million sq. m.	15 million sq. m.	20 million sq. m.
4.	Solar lighting systems	5 million	10 million	20 million

[§]Revised target after July 2015
Source: Ministry of Renewable Energy Sources, Govt. of India

As on 31ˢᵗ March, 2015, grid connected solar power plant installation in the country is 3077 MW and off-grid installation including solar pumps, solar lighting system, decentralized system is 368.7 MW. India was ranked 4ᵗʰ in the world for new capacity addition in 2011 with a total installed capacity of 7.37 million sqm metres equivalent to 5144 MW. Achievement in capacity addition in the year 2012-13 was 1.4 million sqm equivalents to 980 MWth. The growth is largely supported by India's own ambitious mission to promote the use of solar energy under the Jawaharlal Nehru National Solar Mission (JNNSM). The target set under JNNSM by the year 2022 is 20 million sqm of solar thermal collector area.

The arid and semi-arid part of the country receive much more radiation as compared to the rest of the country with mean daily annual radiation received at and around Jodhpur, 6 kWh m⁻². Technologically solar energy can be harnessed either in the form of thermal energy by using flat plate collectors and concentrators, or by generating electricity using photovoltaic cells. Work on the utilization of immense, non-pollutant and inexhaustible solar energy has been carried out at CAZRI, Jodhpur for various domestic, industrial and agricultural applications in order to supplement the energy demand. This includes the development of solar still, solar dryer, solar cookers, solar water heaters, solar candle device, solar polish making machine etc. In addition, work on the development of solar PV based systems like solar insecticide sprayer, solar PV duster, PV pump operated drip irrigation system for growing orchards, PV winnower, PV dryer has been contemplated.

Basics of solar energy utilization

There are two primary groups of technologies that convert solar energy into useful forms for consumption.

Solar PV technologies consist of semiconductor or other molecular devices called photovoltaic or solar cells that convert sunlight into direct current (DC) electricity. PV modules consist of multiple cells assembled on a common platform, connected in series and sealed in an environmentally protective laminate. If the power provided by one PV module is not enough, then multiple modules are linked together to form an array to supply power ranging from a few watts to many megawatts. In addition to the modules, other components (for example, inverters, batteries, charge controllers, wiring and mounting structure) may be required to form a complete PV system.

Solar thermal technologies heat water or air, and other possible working fluids, for non-electricity uses of energy. Solar water heaters can displace conventional electrical water heaters in homes and in commercial establishments. Hot-air-based thermal collectors can displace fossil fuel use in cooking, agricultural drying and more generally in industrial heat processing. In India, solar water heaters have been commercialized and are an economically viable option for many regions. Other applications are still in nascent stages.

Apart from these two major solar technologies, concentrated solar thermal power (CSP) technologies can also be utilized which first convert solar energy into heat energy and then into electrical energy. Most CSP technology options, namely, the parabolic trough, Fresnel mirror system and central tower, are meant for utility-scale use. Further, with thermal storage, these three CSP technologies can provide electricity several hours after sunset. However, a less proven CSP option, the Stirling engine system, can be used in 10 to 25 kW decentralised applications, and can also be easily aggregated for utility-scale plants. Unlike other CSP technologies, this requires little water, but it also lacks inherent storage of thermal energy.

Solar PV technology

Solar PV pumping system

A solar PV system mainly comprises of i) PV panels (ii) mounting structure (iii) pump unit (AC/DC) and (iv) tracking system (Fig. 1).

Sizing of PV panel depends on the capacity of pump to draw water. If the suction head is about 4-5 m, which is applicable in case of a surface water reservoir, 1 hp capacity pump is sufficient which requires about 900 W_p panel in case of DC pump and 1400 W_p panel in case of AC surface pump. An example of installation of 1 hp solar pumping system with AC and DC pump is shown in Fig. 2. If the solar PV pump is to be used for drawing more deep water from wells or tube wells, panel size will be higher accordingly. The mounting structure for erecting the panels with an angle from horizontal surface, which is generally equal to the latitude of any place needs to be strong enough to withstand the wind forces. The pumps to be used in a solar pumping system may be either DC or AC type and surface or submersible type as per situation. As the PV panels generate DC current, additional DC-AC inverter system is required for

AC pumping system. To track the panel perpendicular to the sun, tracking system is required. Two types of tracking system are available i) one axis tracking which tracks the solar panel as per azimuthal rotation of sun from east to west, ii) in additional to azimuthal rotation PV panels can be tracked as per zenith angle of sun using a two axis tracking system. Both manual and auto tracking systems are available in the market. However, in case of auto tracking system there will be an additional cost of tracker. Cost of available solar PV pumping system in market with 3 hp capacity pump is about Rs 4.00 lakhs with additional cost of Rs 14,000/- for auto tracker and about Rs 8,000/- for providing lighting systems (Table 2).

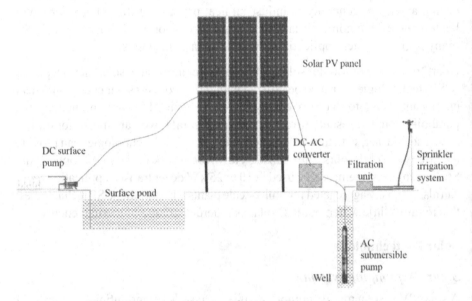

Fig. 1: Schematic diagram of a solar PV pumping system

For optimum use of harvested rain water in surface reservoir like farm ponds or tankas, small sized solar PV pumping systems with 1 hp AC and DC motor were experimentally tested at research farms of ICAR-Central Arid Zone Research Institute, Jodhpur (Fig. 2). Total suction head in both pumps was about 5 m. Among two installed pumps, the system with AC pump consisted of 1400 Wp (200 Wp × 7) whereas the DC pump consisted of PV array of 920 W_p (230Wp × 4). Each PV panel of AC solar pump was connected in series, which was further connected with an inverter to generate AC output of about 220-240 volt and 4-4.2 amp. In case of DC solar pump, panels are connected in parallel to generate DC output of 40-50 volt and 15-20 amp.

Table 2: Approximate base price of solar PV pumping system in Rajasthan state during 2015

Details	DC/AC	Mounting structure	Head (metre)	Base rate (₹)	
				3 HP	5HP
SPV surface pump	DC	Static	20 m	3,87,887	₹6,20,000
	AC	Static	20 m	4,10,000	₹5,70,000
SPV submersible pump	DC	Static	20 m	4,49,513	₹6,21,068
	AC	Static	20 m	4,14,578	₹5,80,000
Additional cost	50 m head over 20 m				₹6,500
	75 m head over 20 m				₹11,000
	Manual tracking system				₹4,050
	Auto tracking system				₹17,500
SPV domestic lighting system 37 W_p/40 Ah battery/9 W × 2 fixtures					₹7,999
Fencing around solar panels and structure					₹14,000

Fig. 2: Solar photovoltaic pumps at experimental fields of ICAR-Central Arid Zone Research Institute, Jodhpur, (a) solar pump with 1 hp AC motor and 7×200 W_p PV array, (b) solar pump with 1 hp DC motor and 4×230 W_p PV array.

Solar PV duster

The solar PV duster (Fig. 3) essentially comprises a photovoltaic panel carrier, storage battery and especially designed compatible dusting unit. The PV panel is carried over the head with the help of a light PV panel carrier, which provides shade to the worker and simultaneously charges the battery to run the duster. The battery is stacked in a bracket, which is fixed in situ to the panel carrier. Approximate cost of this device is about ₹ 9000/- and for dissemination of it, MoU has been signed between CAZRI and a private manufacturing unit. The unit has also the additional facility for lighting purpose during night time. The unit has been successfully tested for dusting sulphur dust and malathion powder.

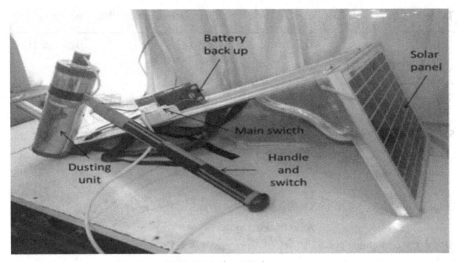

Fig. 3: Solar PV duster

Solar PV winnower cum dryer

The PV winnower cum dryer (Fig. 4) is a PV based device which may be used for winnowing threshed materials in the absence of erratic and unreliable natural winds and also for dehydrating fruit and vegetables more effectively and efficiently.

The system comprises of PV module and compatible especially designed winnower with dc motor - fan assembly. Thirty five to fifty kilogram grains could be separated using this device within 1 to 1.5 hours from threshed materials of pearl millet, mustard grain and cluster bean. The same fan of winnower is used in a dryer to use the system for dehydrating fruit and vegetables under forced circulation of air. Drying of mint, spinach, onion, mushroom, ber etc., can be accomplished with retention of colour and aroma.

Fig. 4: PV winnower cum dryer

PV winnower cum dryer with alternative materials such as slotted iron angle, fibre glass sheet etc., detachable pre air heating tunnel with fan and compatible PV module (75 W_p) with storage battery back-up (40 Ah) was also developed as a custom hiring model. With the use of the pre heating tunnel (3.7 m length) the inside thermal gradient of the drying chamber reduced from 5-6°C to 2-2.5°C leading to uniform drying. The battery back-up (40 Ah) helped continuous drying even during cloudy weather. Although, the thermal gradient changed as the drying progresses, the rate of drying was more uniform in different trays. The device has been tested for dehydrating fenugreek, mint leaves and khejri (*Prosopis cineraria*) pods. Retention of colour and aroma was observed in the dried product with more acceptability.

Solar PV operated protected agriculture system

Solar PV based protected agriculture system was developed with earth heat exchange thermal regulation facility for its use to grow crops during aberrant weather situations (Fig. 5A). Reduction in air temperature during summer season at the exit of the fan was in the range of 6-10°C and in winter rise in temperature was 2-4°C. These changes were close to predicted values developed through the mathematical model. Tomato plants were grown inside the PV clad structure to study the performance under load conditions (Fig. 5B). In order to regulate the temperature to less than 35°C, different blends of paraffin wax were prepared and heating and cooling profiles were studied to identify a suitable eutectic as a thermal storage material.

Energy pay back analysis of the PV clad structure was carried out by considering the energy used for operating the suction fan throughout the year and misting unit during summer and autumn months. The cost of PV clad enclosure was estimated to about ₹33,000. With the use of PV cells for operating the cooling fan in the PV clad structure, it can save about 200 kWh per year. Although PV module will last for more than 20 years, the system's life has been taken as 10

Fig. 5: PV clad structure for protected cultivation in arid region

years and assuming a modest benefit of ' 8,000 a year for nursery and growing vegetables and other crops, the benefit cost ratio has been worked out to be 1.5.

PV mobile unit

It is a portable moveable power system that can be custom hired for contemplating various domestic, small agricultural and other rural tasks in isolated cluster of houses (Dhanis). It comprises two PV modules (70 W$_p$ each) on a folding system with auto locking arrangement to fold PV panels inward while moving it and to open these at the optimum angle.

The mobile unit (Fig. 6) structure is provided with four wheels at the lower end, two on the rear side of the rectangular frame and the remaining two on a guiding trapezoidal frame attached to it in the front side with a flexible metallic

pulling sling, having handle to move it with ease. One 120 W DC motor with gear box is fixed on the platform with a pulley on the motor's shaft and another one fixed on to the rear wheel. The PV mobile unit was found to minimize exertion on the person pulling the unit with a speed of 1.4 km/hr. The system can be used to operate churner for butter extraction, cooler, blower, winnower etc.

Fig. 6: PV mobile unit operating aloe vera gel extractor

Solar thermal technology

Solar cookers

Solar cookers can be used for boiling rice, lentil, vegetables; roasting groundnut, potato etc., baking vegetables and cooking local food and feed for animals. With the use of such cookers one can save about 30-40% of fuel requirement. Various types of CAZRI solar cookers are: solar oven, which is better for cooking food for a family quickly but

Fig. 7: Animal feed solar cooker

requiring tracking of the device after every half an hour, double mirror box type solar cooker with sun tracking for extended time of three hours and a stationary cooker with optimised width to length ratio for using it without sun tracking.

The solar cooker for animal feed employs locally available materials such as clay and pearl millet husk (Fig. 7). The commercial material for its fabrication are plain glass, mild steel angle and sheet, wood and aluminium sheet cooking utensils. The cooker is capable of boiling 10 kg of animal feed, sufficient for five cattle per day. The efficiency of the cooker is 21.8%. The cooker saves 6750 MJ of energy per year.

Inclined solar dryer

Solar dryer is a convenient device to dehydrate vegetables and fruits faster and efficiently under control conditions while eliminating the problems of open courtyard drying like dust contamination, insect infestation and spoilage due to rains. A low cost tilted type solar dryer, costing about ₹3500 per m², has been developed at CAZRI and extensively tested for drying onion, okra, carrot, garlic, tomato, chillies, ber, date, spinach, coriander, salt coated aonla etc. The powdered products from some of these solar dried materials have been tested for instant use. Local entrepreneurs have adopted such inclined solar dryers (Fig. 8) of variable capacities (10–100 kg). About 290 to 300 kWh m⁻² equivalent energy may be saved by using the dryer by farmers and thus can accrue higher benefits from solar dried products.

Fig. 8: Inclined solar dryer

Solar system for purification of brackish water

Solar still is a convenient device for producing distilled water, which is required in the laboratories, battery maintenance and also for desalination of water making for making it potable (Fig. 9). Step basin tiled type solar stills of different sizes, (capacity 3 to 3.5 lit m^{-2} day^{-1}) have been adopted by railways, army units, schools and other commercial units. The production of distilled water is independent of season due to its special design. Rose water can also be prepared using such solar still for generating additional income.

Fig. 9: Multi basin tilted type solar still

A channel is provided at the bottom for collection of distilled water. The still can provide 8 to 10 litres of distilled water per day on clear sunny days. With the use of this still 20 lit day^{-1} of potable water (1500 ppm TDS) can be made available in a day from raw water containing 3000 ppm TDS.

Solar water purifier

A solar water purifier (Fig. 8) having a capacity to purify 30 litres of nadi water a day was designed, fabricated and tested. The double walled device has the outer box of galvanized sheet with dimensions 1295 mm × 585 mm × 200 mm and inner box made of aluminium sheet (22 SWG, 1200 mm × 500 mm × 100 mm) with glass wool in the annular space. Two clear window glass panes each of 4mm thickness have been fixed over it with an openable wooden frame. A 4 mm thick plane mirror is fixed as reflector with tilting arrangement that can be varied from horizontal surface up to 120 degree depending upon the season. The tilt is fixed once in a fortnight. The absorber area is 0.60 m^2 and 20 glass bottles each of 750 ml can be kept inside it for purification of water. The performance of the solar water purifier was carried out by measuring stagnation temperature and heating nadi water in the bottles, filled in 20 glass bottles and placed in the solar water purifier at 10 AM. The temperature of water in the bottles reached to 95°C by 2 PM. The average maximum stagnation temperature inside chamber of solar water purifier was 147.5°C when ambient temperature was 37.5°C. All harmful bacteria are destroyed at temperature more than 70°C (Fig. 10), making solar treated nadi water potable.

Solar water heaters

Substantial amount of fuel can be saved by using solar waters heaters for domestic purposes like bathing, washing of clothes and utensils and in industries like textiles

Fig. 10: Solar water purifier

and dairy. The natural circulation type solar water heaters with flat plate collector have been installed in hotels, hostels, guest-houses etc. Collector-cum-storage solar water heaters (Fig. 11) reduces the cost, almost half of the cost of conventional solar water heater, and provide 100-litre hot water at 50-60°C in the evenings and 40-45°C next day mornings (winter season) after covering the device with insulating cover.

Fig. 11: Collector cum storage type solar water heater

Integrated solar devices

With a view to using the same device throughout the year for one or other purposes, dual and multipurpose solar energy appliances have been developed at CAZRI, Jodhpur. A solar cooker cum dryer can cook food for 4-5 persons without sun tracking and in this dual purpose device fruit and vegetable can also be dehydrated. A solar water heater cum dryer can dehydrate 10-15 kg fruits and vegetables as a solar dryer and can provide 80 lit hot water of 55-60°C in winter afternoons. Integrated Solar Device (Fig. 12) is unique three in one solar device, which can be used to cook food round the year. The geometry of the device enables one to cook food for a family without sun tracking. Further the device can produce about 50 L hot water of 50-60°C utilising the low altitude position of sun during winter and thus having energy gain both from top and front windows. The device can also be used for drying fruit and vegetables for their use in off-season. The main feature of the device is that during dehydration the water also gets heated and the dehydration process continues even in night and simultaneously the temperature of the product is regulated in between 60-65 °C, optimum for dehydration of fruit and vegetables. This novel three in one solar device costing ₹6000/- finds utility throughout the year for one or other purpose and makes the system more practical and economical.

Fig. 12: Integrated three in one solar device

Solar candle device

The solar candle machine with 0.5 m² absorbing surface area is useful for farmers, rural ladies and unemployed youth. The device heat and melt paraffin wax for the production of candles to provide means for supplementary income. With this device (Fig. 13) 10-16 kg candles can be prepared conveniently during summer and 6-9 kg in winter and is a safe process having advantages of reducing labour and vaporization losses associated with conventional methods.

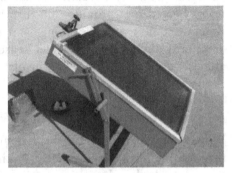

One can start the solar candle business with an initial investment of Rs. 12000/- generating an average income of × 2000/- a month. N.R.D.C. has licensed the system for commercialization through private entrepreneurs.

Fig. 13: Solar candle device

Passive cool chamber

The low cost passive cool chamber (Fig. 14) is useful to preserve different kinds of vegetables for short term period. It consists of a double walled chamber made of baked bricks and cement filled up with coarse sand. The water filled up in the annular side walls helps to maintain high humidity inside the inner chamber and reduces temperature.

Provisions were also made for water evaporation from the bottom side of the cool chamber by providing suitable drainage system which further enhances temperature reduction (up to 15°C) and maintains high humidity in the chamber (> 90%) and thus it enhances the shelf-life of the vegetables. The passive cool chamber has attracted wide attention of the farmers and vegetables growers. On the requests of the farmers/vegetables growers, 14 cool chambers have been commissioned in different districts.

Fig. 14: Improved passive cool chamber

Conclusions

Solar devices like PV duster, PV pump operated drip system, PV winnower, solar dryer, solar stills, water purifier, animal feed cooker, stationary solar cooker and integrated three in one solar device can be used in rural areas for supplementing the energy demand. Solar candle device and polish making system can be used for generating supplementary income while cool chamber can be used to enhance the storage of perishable fruit and vegetables. The solar devices are more relevant in connection with the development of desert where water is scarce, solar energy is plentiful and the places are not connected to electric grid. Therefore, integrated efforts are needed to harnessing solar energy, the plentiful natural resource available in arid region, for the sustainable development. Entrepreneurs should come forward for taking the manufacturing of these systems in the villages itself, providing employment to village skilled worker and contributing to a self -sustaining growth with ensured environmental protection.

27

Carbon Sequestration: Agroforestry is an Option to Mitigate Climate Change

M.B.Noor mohamed, Keerthika. A, Dipak Kumar Gupta,
A.K. Shukla, B.L. Jangid, P.L. Regar, Kamla K. Choudhary
S.R. Meena and R.S. Mehta

ICAR-Central Arid Zone Research Institute, Regioanl Research Station
Pali, Rajasthan-306401

Global climate changing has been a focus for a long time in many fields, the increasing CO_2 concentration in atmosphere was considered as one main driving force for global warming. The increases in greenhouse gasses in the atmosphere contributed to raise concerns over climate change enhanced the role of forests and plantations in the global carbon cycle are growing as a great issue. Carbon dioxide (CO_2) is one of the main greenhouse gases because a huge volume of CO_2 is added to the atmosphere, when compare to all other greenhouse gases and made itself as a primary agent of global warming. It contributed 72 per cent of the total anthropogenic greenhouse gases, causing between 9-26 per cent of the greenhouse effect (Kiehl and Trenberth, 1997). According to the Intergovernmental Panel on Climate Change (IPCC), there has been an unprecedented warming trend during the 20th century. Scientists now have estimated that the average global surface temperature is likely to rise by 1.4 to 5.8°C by the end of the 21st century (IPCC 2007).

The IPCC estimates that the level of carbon dioxide in today's atmosphere is 31 per cent higher than it was at the start of the Industrial Revolution about 250 years ago. An atmospheric level of CO_2 has risen from 280 ppm at the pre-industrial to the present level of 375 ppm. Most of the increase has occurred in the second half of the 20th century. The mitigation of climate change demands, determined commitment of scientists to develop strategies to effectively manage the issues of the changing climate through carbon sequestration. Carbon sequestration is ought to be a promising solution for reducing atmospheric carbon dioxide, an important greenhouse gas.

Sequestration of biomass carbon is considered as the most promising approach to mitigate the climate change (Kimble *et al.*, 2002). At global level, trees contribute 80-90% of plant biomass carbon and 30-40% soil carbon (Harvey, 2000). Therefore trees can play important role in carbon di oxide sequestration due to several reasons. The first is that the tree components fixes and stores carbon from atmosphere via photo synthesis. They can function as active carbon for the periods of many years and continue to store the carbon until they are harvested or die. The second reason is that trees can provide a good surface cover which minimizes the loss of nutrients from the surface soil, improve edaphic conditions, increase biomass production, decrease risk of soil degradation by erosion, leaching and nutrient depletion. Finally trees are one of the viable alternatives to increase forest cover which will widen the area of carbon sink.

Carbon management in forests is the global concern to mitigate the increased concentration of green house gases in the atmosphere. It is estimated that the world's forests store 283 Gt of carbon in their biomass alone. However the global forest cover is dwindling fast in view of great biotic pressure, industrialization, urbanization, land use changes for developmental activities and conversion of forests to agricultural land. Reviving forest cover and finding low cost methods to sequester carbon is emerging as a major international policy goal (Shively *et. al.* 2004). Agroforestry is widely considered as a potential way of improving environmental and socioeconomic sustainability. Agroforestry systems can play an important role in carbon mitigation programmes through carbon sequestration and can reduce the pressure on existing natural forests by providing fuel, fodder, timber and wood products directly to the farmers on the on hand and on the other it may provide many indirect environmental benefits such as soil and water conservation, biodiversity conservation, soil nutrients enrichment etc.

Carbon Sequestration

The term "carbon sequestration" is used to describe both natural and deliberate processes by which CO_2 is either removed from the atmosphere or diverted from emission sources and stored in the ocean, terrestrial environments (vegetation, soils, and sediments), and geologic formations.

Carbon sequestration can be defined as the capture and secure storage of carbon that would otherwise be emitted to or remain in the atmosphere. The idea is (1) to keep carbon emissions produced by human activities from reaching the atmosphere by capturing and diverting them to secure storage, or (2) to remove carbon from the atmosphere by various means and stores it. One set of options involves capturing carbon from fossil fuel use before it reaches the atmosphere. For example, CO_2 could be separated from power plant flue gases, from effluents of industrial processes (e.g., in oil refineries and iron, steel, and cement production plants), or during production of decarbonized fuels (such as hydrogen produced from hydrocarbons such as natural gas or coal). The captured CO_2 could be concentrated into a liquid or gas stream that could be transported and injected into the ocean or deep underground geological formations such as oil and gas reservoirs, deep saline reservoirs, and deep coal seams and beds. Biological and chemical processes may convert captured CO_2 directly into stable products. Atmospheric carbon can also be captured and sequestered by enhancing the ability of terrestrial or ocean ecosystems to absorb it naturally and store it in a stable form.

Carbon sink, carbon source, carbon flux; carbon neutral

Growth of trees is mainly due to absorption of carbon dioxide from air, and also nutrients and water from soil, and conversion of the product (photosynthate) into biomass. About 50% of the dry weight of a tree is carbon. A growing forest / plantation increase the total carbon content of land. Global carbon is held in a variety of forms such as swamps, wetlands, forests and forest soils, grasslands, agriculture crops etc. called stocks. A stock that stores carbon is called carbon sink; when it releases carbon, it becomes carbon source. If forests burn due to wildfires or deliberately burnt for agriculture purpose or exploited for timber, fuel etc, they become carbon source. As forest grows, biomass accumulates over decades and even centuries resulting in significant sequestration (storage) of carbon. This transfer of carbon from atmosphere to forest is called the carbon flux. Young fast growing forests and plantations have small stock. Other sources of carbon are fallen branch twigs and leaves on forest floor. Also, forest soils sequester or additional carbon out of decomposing plant litter. Mature forest balancing sequestration of carbon dioxide and release of oxygen is called carbon neutral.

Necessary Characteristics for Carbon Sequestration Systems

Any viable system for sequestering carbon must have the following characteristics.

a. **Capacity and price**. The technologies and practices to sequester carbon should be effective and cost competitive. This road map will focus on options that allow sequestration of a significant fraction of the goal.

b. **Environmentally benign fate**. The sheer scale and novelty of sequestration suggests a careful look at environmental side effects. For example, the long-term effects of sequestration on the soil or vegetation need to be understood. Until recently, dilution into the atmosphere was considered acceptable. Vast quantities of materials would be generated. The safety of the product and the storage scheme has to be addressed.

c. **Stability.** The carbon should reside in storage for a relatively long duration.

Types of Carbon Sequestration

- Oceanic Carbon Sequestration
- Geological Carbon Sequestration
- Terrestrial Carbon Sequestration

1. Oceanic Carbon Sequestration

Captured CO_2 could be deliberately injected into the ocean at great depth, where most of it would remain isolated from the atmosphere for centuries. CO_2 can be transported via pipeline or ship for release in the ocean or on the sea floor. The ocean represents the largest potential sink for anthropogenic CO_2. Discharging CO_2 directly to the ocean would accelerate the ongoing, but slow, natural processes by which over 90% of present-day emissions are currently entering the ocean indirectly and would reduce both peak atmospheric CO_2 concentrations and their rate of increase.

2. Geologic Carbon Sequestration

Geologic sequestration is putting CO_2 into long-term storage in geologic zones deep underground. Geological Carbon Sequestration is the practice of capturing CO_2 at anthropogenic sources before it is released to the atmosphere and then transporting the CO_2 gas to a site where it can be put into long-term storage.

Geologic sequestration involves three main processes: capturing carbon dioxide, transporting carbon dioxide, and placing the carbon dioxide in a geologic formation for permanent or semi-permanent storage. The carbon dioxide is placed into the geologic formation by means of a system of injection wells. An injection well is like an oil well or water well, except that instead of drawing material (oil or water) out of the ground, carbon dioxide is injected into the well. Injection wells are also used for the disposal of various types of wastes and to enhance oil recovery in some areas. The geologic formation should have a number of characteristics to make it suitable for placement and retention of the carbon dioxide. The formation should be deep enough and be covered with a formation that will prevent the carbon dioxide from migrating to the surface. The depth of injection would depend upon the depth of the formation being targeted for injection. Some underground formations that could be used for the geologic sequestration of carbon dioxide are:

- Depleted oil and gas reservoirs;
- Unmineable coal seams;
- Deep formations containing salty water; and
- Basalt formations.

Before geologic sequestration can be widely used, two issues need to be addressed:

1. Only a handful of specialized facilities like natural gas-processing plants, coal gasification plants, and ethanol plants currently have processes that separate CO_2 and make it available for geologic sequestration. Actions are under way now to develop economical methods of separating and capturing CO_2 at other large-scale systems like power plants that produce relatively large quantities of anthropogenic CO_2.

2. Although pure CO_2 has been stored as a gas in natural underground deposits for millions of years and oil field operators have safely pumped millions of tons of CO_2 underground into oil-producing formations to increase production (CO_2 flooding), we need validation demonstrations in geologic environments to ensure that we understand the best ways to site the systems as well as monitor the CO_2 in storage over the long term.

3. Terrestrial Carbon Sequestration

Terrestrial (or biologic) sequestration means using plants to capture CO_2 from the atmosphere and then storing it as carbon in the stems and roots of the plants as well as in the soil. In photosynthesis, plants take in CO_2 and give off the oxygen (O_2) to the atmosphere as a waste gas. The plants retain and use the carbon to live and grow. When the plant winters or dies, part of the carbon from the plant is preserved (stored) in the soil. Terrestrial sequestration is a set of land management practices that maximizes the amount of carbon that remains stored in the soil and plant material for the long term. No-till farming, wetland management, rangeland management, and reforestation are examples of terrestrial sequestration practices that are already in use.

Trees as source for carbon sinks

Photosynthesis is the biochemical process by which plants use sunlight to convert nutrients into sugars and carbohydrates. Carbon dioxide (CO_2) is one of the nutrients essential to building the organic chemicals that comprise leaves, roots, and stems. All parts of a tree which includes the stem, limbs and leaves, and roots contain carbon, but the proportion in each part varies enormously, depending on the tree species age and growth pattern. Nonetheless, as more photosynthesis occurs, more CO_2 is converted into biomass, reducing carbon in the atmosphere and sequestering (storing) it in plant tissue (vegetation) above and below ground. Several studies have established the fact that carbon sequestration by trees could provide relatively low cost net emission reductions (Parks and Hardie, 1995; Callaway and Mc Card, 1996 and Stavins, 1999). Trees stores about 80 per cent of all aboveground and 40 per cent of all below-ground terrestrial organic carbon (IPCC, 2001).

Carbon Sequestration in Agroforestry Systems

Carbon sequestration in different agroforestry systems occurs both below ground, in the form of enhancement of soil carbon plus root biomass and aboveground as carbon stored in standing biomass. Some of the earliest studies of potential carbon storage in agroforestry systems and alternative land use systems for India had estimated a sequestration potential of 68-228 MgC/ha, 25tC/ha over 96 Mha of land. But this value varies in different regions depending on the biomass production. Studies done by Jha *et al.*, 2003 showed that agroforestry could store nearly 83.6 t C/ha up to 30 cm soil depth, 26% more carbon compared to cultivation in Haryana plains. However, the magnitude of carbon sequestration from forestry activities would depend on the scale of operation and the final use of wood.

India has a long tradition of agroforestry, several indigenous agroforestry systems, based on people's needs and site-specific characteristics have been developed over the years. Agroforestry research was initiated in the country about three decades ago and several agroforestry technologies have been developed and tried on farmer's land. There is much concern that the increasing concentration of greenhouse gases (GHGs) and carbon dioxide in the atmosphere contributes to global warming by trapping long-wave radiation reflected from the earth's surface. The area under forests, including part of the area afforested is increasing and currently 67.83 mha of area is under forest cover. Assuming that the current trend continues, the area under forest cover is projected to reach 72 mha by 2030. Estimates of carbon stock in Indian forests in both soil and vegetation range from 8.58 to 9.57 GtC (Ravindranath *et al.*, 2008). The significance of agroforestry with regards to C sequestration has been widely recognized with an estimated global potential of between 12 and 228 Mgha[-1]. However, variability can be high within various agroforestry systems as biomass C stock depends on several factors including environmental conditions, soil type, magnitude of land degradation and the length of fallow period (Kaonga *et al.* 2009). Residue quality differences among agroforestry species further play a key role in regulating long term C build up, as the rate of soil organic matter decomposition is dependent on residue chemical quality which is mainly defined using various ratios of carbon, nitrogen, lignin and polyphenols.

Agroforestry systems can be better climate change mitigation option than ocean, and other terrestrial options, because of the secondary environmental benefits such as food security and secured land tenure, increasing farm income, restoring and maintaining above ground and below ground biodiversity, maintaining watershed hydrology and soil conservation. By including trees in agricultural production systems, agroforestry can increase the amount of carbon stored in lands devoted to agriculture, while still allowing for growing of food crops. The tree components in agroforestry systems can be significant sinks of atmospheric carbon due to their fast growth and high productivity. Thus, promoting agroforestry can be one of the options to deal with problems related to land use and global warming. The depend on the agroforestry system, the structure and function of agroforestry systems which to a great extend, are determined by environmental and socio-economic factors. Also tree species and system management can influence carbon storage in agroforestry systems. Agroforestry practices also have wide and promising potential to store carbon and remove atmospheric carbon dioxide through enhanced growth of trees and shrubs (Fig 1). Average sequestration potential in agroforestry has been estimated to be 25t C ha^{-1} over 96 million ha of land in India and 6-15 tC ha^{-1} over 75.9 Mha in China (Ravindranath et al., 2008). Watson et al., (2000) estimated carbon gain of 0.72 Mg C ha^{-1} yr^{-1} on 4000 million ha land under agroforestry, with potential for sequestering 26 Tg C yr^{-1} by 2010 and 45 Tg C yr^{-1} by 2040. Proper design and management of agroforestry practices can make them effective carbon sink. As in other land-use systems, the extent of C sequestered will depend on the amount of C in standing biomass, recalcitrant C remaining in the soil and C sequestered in wood products. Average carbon storage by agroforestry practices has been estimated as 9, 21, 50 and 63 Mg C ha^{-1} in semiarid, subhumid, humid and temperate regions (Montagnini and Nair, 2004).

Carbon sequestration potential by Indian forests

The total forest cover of India was approximately 76.87 Million hectare (Mha) (FSI, 2008.). The annual productivity of Indian forests was reported to have increased from 0.7 m^3 per hectare in 1985 to 1.37 m^3 per hectare in 1995 (FSI, 1997). An increase in annual productivity indicates an increase in the total standing biomass of Indian forests and its carbon uptake. As per FAO estimates, the total forest carbon stocks in India have increased over a period of 20 years (1986–2005) and amount to 10.01 GtC. The biomass carbon estimated by different studies has been given below (Fig.2 and Table 1&2).

Carbon sequestration by Indian forest soil

The total SOC pools in Indian forests have been estimated to be 4.13 Pg C (top 50 cm) to 6.81 Pg C (top 1 m soil depths) for the period between 1980-1982

(Chhabra *et al.*, 2002).. Based on different approaches, estimates of forest SOC stocks in India are in the range of 5.4-6.7 Pg (Dadhwal and Nayak, 1993). Ravindranath *et al.*, 1997 and Jha *et al.*, 2003 estimated that 9,815.15 M t of soil C is store in 62.43 M ha forest stands in India. The national average soil organic carbon per ha in forest soil was estimated as 182.94 t ha^{-1} (Jha *et al.*, 2003). They also observed maximum in central peninsular region (3,656.70 Mt) followed by north-east region (3,650.53 Mt) and least (174.33 Mt) in desert region (Fig.3).

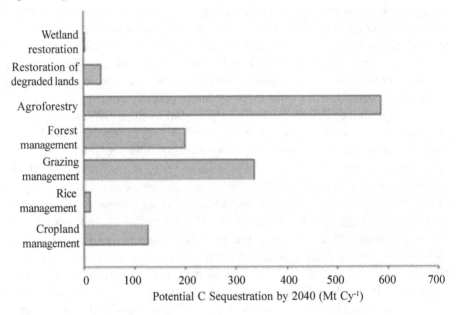

Fig. 1. Carbon content (%) in various tree species
Suryawanshi et al., 2014

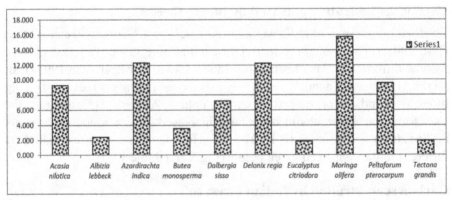

Fig. 2. Total organic carbon of trees in t/tree
Shamsudheen *et al.*, 2014

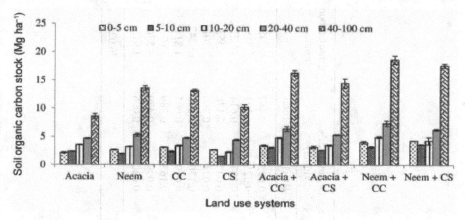

Fig. 3: SOC stock (Mg C ha^{-1}) in various land-use systems at different soil depths

Conclusion

Global climate change induced by increase in concentration of green house gases is likely to increase temperature, alter precipitation trends and cause more of extreme events. Under such circumstances, fragile ecosystems like arid zones are hit the most. Agro forestry is a climate smart production system and considered more resilient than monocroping in mitigating climate change. With food shortages and increasing trends of climate change, interest in agro forestry is gathering for its potential to address various on-farm needs, and full fill many roles in mitigation pathways. Agroforestry provides assets and income from carbon, wood energy, improved soil fertility and enhancement of local climate conditions. Performance of mitigation option in agroforestry will depend on the relative influence of tree species selection and management, soil characteristics, topography, rainfall, agricultural practices, priorities of food security, economic development options, etc. Therefore, Agroforestry in arid region are more important option for livelihood improvement, climate change mitigation and sustainable development. In arid region of India, the number of tree species is very few and moreover in hospitable environmental conditions don't allow much required germination and growth of tree species. In such situation adoption of improved agroforestry practices which can fit easily existing traditional agroforestry systems will play a crucial role as agroforestry itself provides many options for climate resilient agriculture.

Table 1: List of selected tree species and its average organic carbon potential

Sl.n	Trees	No. of Trees	Average GBH (cm)	Average Height (m)	Average Organic Carbon (t/individual)			Organic Carbon (ton/tree)
					AGB	BGB	Total	
1.	*Acacia nilotica*	32	19.1	12.84	0.367	0.095	0.462	9.248
2.	*Albizia lebbeck*	27	10.55	11	0.096	0.025	0.121	2.419
3.	*Azardiracta indica*	50	22.29	12.51	0.487	0.127	0.614	12.27
4.	*Butea monosperma*	15	12.73	11.11	0.141	0.037	0.178	3.553
5.	*Dalbergia sissoo*	22	19.1	10	0.286	0.074	0.360	7.207
6.	*Delonix regia*	13	22.29	12	0.486	0.126	0.612	12.247
7.	*Eucalyptus citriodora*	28	9.55	10.12	0.072	0.019	0.091	1.814
8.	*Moringa olifera*	14	25.47	12.31	0.626	0.163	0.789	15.77
9.	*Peltoforum pterocarpum*	17	19.1	13.27	0.38	0.099	0.479	9.576
10.	*Tectona grandis*	244	9.55	10.73	0.076	0.020	0.096	1.915
	Total	462						76.028

Suryawanshi *et al.*, 2014

Table 2: Biomass (Mg ha^{-1}) and carbon stock (Mg ha^{-1}) in selected land-use systems in Kachchh

Sl.No	Components	AGB	BGB	Total biomass	Carbon stock		
					Above ground	Below ground	Total
1.	Acacia	12.78	2.52	15.30	5.03	0.98	6.02
2.	Neem	7.79	1.85	9.64	2.92	0.71	3.64
3.	CC	6.26	4.70	10.96	2.44	1.82	4.26
4.	CS	2.78	1.75	4.53	1.04	0.71	1.74
5.	Acacia+CC	12.93	4.49	17.41	5.08	1.75	6.82
6.	Acacia+CS	12.55	3.14	15.69	4.91	1.24	6.15
7.	Neem+CC	9.60	3.79	13.39	3.53	1.39	4.91
8.	Neem+CS	9.35	3.12	12.48	3.65	1.22	4.81
	MS	38.53	3.71	50.17	6.13	0.54	7.88
	F7.23	21.85	16.73	16.80	21.65	14.75	16.88
	LSD 5%	2.30	0.82	2.99	0.92	0.33	1.18

AGB - Above ground biomass; BGB-Below ground biomass; CC- *Cenchrus ciliaris*; CS- *Cenchrus setegerus*
Shamsudheen *et al.,* 2014

References

Callaway, J.M. and McCard, B. 1996. The economic consequences of substituting carbon payments for crop subsidies in US agriculture. *Environmental and Resource Economics*, 7(1): 15-43.

Chhabra, A., Parila, S.and Dadhwal, V.K. 2002. Growing stock based forest biomass estimate of India. *Biomass and Bioenergy*, 22: 187-*194.*

Dadhwal, V.K. and Nayak, S.R. 1993. A preliminary estimate of biogeochemical cycle of carbon. Climate Change, 74: 191-221.

FSI. 2008. State of Forest Report, 2005. Forest Survey of India, Ministry of Environment and Forests Government of India, Dehra Dun, India.

Harvey, L. D. D. 2000. Global warming- the hard science. Singapore: Pearson Education: 336.

IPCC. 2007. Climate change 2007: Mitigation of climate change. Working Group III contribution to the Intergovernmental Panel on Climate Change, Fourth Assessment Report. Bangkok, Thailand. htt://www.ipc.ch/ipcreport/index.htm (accessed: May 2012).

IPCC. 2001. Climate Change 2001: Working Group I: The Scientific Basis. Cambridge University Press, New York.

IPCC. 2000. Land use, Land –use change and Forestry. A special Report of the IPCC. Cambridge University press Cambridge, UK. P.375.

Jha, M.N., Gupta. M.K., Alok. S. and Rajesh. K. 2003. Soil organic carbon store in different forests in India. *Indian Forester*, 129(6): 714–724.

Kiehl, J. T. and Trenberth, K. E. 1997. Earth's Annual Global Mean Energy Budget. *Bulletin of the American Meteorological Society*, 78(2): 197-208.

Kimble, J. M., Lal, R and Follet, R. F. 2002. Methods of assessing soil carbon pools. In Assessment Methods for Soil Carbon (eds Lal, R. *et al.*), Lewis Publishers, Boca Raton, FL :3–12.

Kaonga, M.L and Bayliss-Smith, T.P. 2009 Carbon pools in tree biomass and the soil in improved fallows in eastern Zambia. *Agroforest Syst.*, 76: 37-51.

Montagnini, F and Nair, P.K.R. 2004. Carbon sequestration: An underexploited environmental benefits of agroforestry systems. *Agroforestry systems*, 61: 281-295.

Parks, P.J. and Hardie, I.W. 1995. Least-cost forest carbon reserves: Cost effective subsidies to convert marginal agricultural land to forests. *Land Economics*, 71(1): 122-136.

Ravindranath, N.H., Somashekhar, B.S. and Gadgil, M. 1997. Carbon flows in Indian forests. *Climatic Change*, 35B: 297-320.

Ravindranath, N.H., Chaturvedi, R.K. and Murthy, I.K. 2008. Forest Conservation, afforestation and reforestation in India: Implications for Forest Carbon Stocks. *Current Science*, 95(2): 216-222.

Shively, G.E., Zelek, C.A., Midmore, D.J. and Nissen, T.M. 2004. Carbon sequestration in a tropical landscape: an economic model to measure its incremental cost. *Agroforestry Systems*, 60: 189-197.

Stavins. 1999. The costs of carbon sequestration: A revealed – preference approach. *American Economic Review*, 89(4): 994-1009.

Watson, R.T., Noble, I.R., Bolin, B., Ravindranath, N.H., Vetardo, D.J. and Dokken, D.J. (Eds.). 2000. Land use, Land use change, and Forestry. A special report of the Intergovernmental Panel on climate change. Cambridge University Press, NY.

Suryawanshi. M.N, Patel, A.R., Kale, T.S. and Patil, P.R. 2014. Carbon sequestration potential of tree species in the environment of North Maharashtra University campus, Jalgaon (MS) India. *Biosci. Disc.*, 5(2): 175-179.

28

Nutrient Management in Arid Fruit Crops

M.L. Soni, N.D.Yadava, Birbal and Subbulakshmi, V.

ICAR-Central Arid Zone Research Institute, Regional Research Station Bikaner, Rajasthan

The arid region of India covers about 12 % of countries geographical area and occupies over 3.17 lakh Km^2 of hot desert located in the parts of Rajasthan (61%), Gujarat (20%), Punjab and Haryana (9.0%), Andhra Pradesh (7%), Maharashtra (0.4%) and Karnataka (3.0%). In addition to this an area of about 78 thousand Km^2 of cold desert in Jammu & Kashmir and Himachal Pradesh presents an entirely different set of agro climatic conditions as compared to hot arid zone. The hot arid region is characterized by low and erratic distribution of rainfall (< 100 to 400 mm), high temperature, low relative humidity, high potential evapo-transpiration, high sunshine and high wind speed during summer. Soils of this region are dry for most part of the year and suffer from water stress. Besides growing annual crops, a great emphasis is being laid on the development of agroforestry systems in this region which also includes cultivation of arid fruit crops in cropping systems. Horticulture based production systems are considered to be the most ideal strategy to provide food, nutrition and income security to the people (Chundawat, 1993). Integration of annual crops with fruit trees yields multiple outputs that ensure production and income generation.

India has been bestowed with wide range of climate and physio-geographical conditions and as such is most suitable for growing various kinds of horticultural. Its horticulture production has increased by 30 per cent in the last five years. This has placed India among the foremost countries in horticulture production, just behind China. During 2012-13, its contribution in the world production of fruits & vegetables was 12.6 % & 14% respectively. Horticulture sector in India is the fastest growing sector within agriculture. "For every nutritional problem there is a horticultural solution". Hidden in this is the understanding of the dynamics of nutrients in horticulture. A due importance to nutrients is essential as they affect the productivity, quality and profitability. After achieving food security it is being increasingly felt that India needed to achieve nutritional security for betterment of its population. In achieving this cultivation of horticultural crops especially fruit crops plays a vital role. To achieve nutritional security our country needs 92 million metric tons of fruits. At present, India accounts for nearly 10 per cent of world production of fruit crops with annual production of about 81.2 million tonnes from an area of 6.5 m ha. Whereas, the second advance estimates put the production at 84.4 million tonnes for 2013-14 (GOI, 2014). Challenges to fulfill the needs of the ever increasing population and consequent to the recognition of the need for attaining nutrition security would therefore require a tremendous effort to increase the production of fruit crops in the country through a rational and balanced use of production inputs, the mineral fertilizers in particular. All fruit crops have strict requirement for a balanced fertilization management, without which growth and development of the crop is poor and the harvested part (roots, leaves, tubers, flowers or fruits) lacks required qualities. Many horticultural crops are heavy removers of nutrients and high yields can only be sustained through the application of optimal doses in balanced proportion. Among the major nutrients, potassium not only improves yields, but also benefits various aspects of quality. Ber and Amla are the two tree crops, which are becoming increasingly popular in dryland areas. Pomegranate is yet another important crop for dryland areas. Citrus (which includes orange, lime, lemon etc.) is also an important horticulture crop in the country.

Essential nutrients for fruit crops

The Nutrients are chemical elements which are absorbed by the plants in more or less quantity to transform light energy into chemical energy and to keep up plant metabolism for the synthesis of organic materials. These materials constitute among other things, foods for humans and animals and a range of raw materials for various industrial uses. Feeding of plants with nutrients is termed as nutrition. Successful growth and production of the plants in general requires a proper supply of the at least 16 elements. These elements are regarded as essential to life in higher plants. The criteria laid for categorising nutrients essentiality to plants are: (1) Complete or partial lack of the element in question must make normal plant growth impossible (2) Deficiency symptoms must be reversibly by the addition of the elements in question (3) The element must play specific role in the plant metabolic symptom. They are:

1. **Basic elements** (03)　: Carbon (C),Hydrogen(H) and Oxygen(O)

2. **Macro elements**(06)　: Nitrogen (N), Phosphorus(P) Potash(K), Calcium(Ca) Magnesium (Mg) and Sulphur(S)

3. **Micro elements** (07) : Manganese (Mn), Molybdenum (Mo), Chlo
rine (Cl), Zinc (Zn), Boron (B), Copper (Cu)
and Iron (Fe)

Macro elements are required in relatively large quantity and micro elements are required in relatively less quantity.

Table 1: Nutrients harvested by fruit crops

Fruits	Nutrients uptake (kg/tones of fruit)		
	N	P	K
Citrus	1.1	0.6	6.3
Pomegranate	1.6	0.5	5.0
Ber	1.4	0.4	4.8

Different fruit crops have different nutrient requirements (Iyengar, 1993). In general, the fruit plants need higher amount of nitrogen and potassium and smaller quantity of phosphorus. To ensure high economic productivity to sustain the available soil nutrients status at the desired level, correct dose of fertilizers and manures must be applied based on suitable diagnostic tools to avoid shortage and excess of nutrients.

Soil analysis

Planting an orchard is a large investment. For this reason, it is important to thoroughly assess soil properties before the orchard is established. The costs for an in-depth assessment of soil properties are much lower than the costs accruing from planting an orchard at an inappropriate site. Soil analysis can be used to assess the nutritional status for tree crops provided that a particular site selected for sampling does not fall outside the range of fertility and the sample size is adequate for the crop and soil. Soil tests have a role for correcting or avoiding problems such as acidity, salinity and nutrient interactions and toxicities, which are not directly related to plant composition. The analysis should preferably be taken with a leaf test. Soil samples should preferably be taken at the same time as leaves collected for tissue analysis. Soil sampling before planting serves three major purposes: (1) Determine soil properties such as pH, texture, nutrient availability, or salinity. (2) Identify unsuitable areas due to physical barriers to root growth or drainage. (3) Assess the variability in soil properties within the field to develop nutrient management plans.

While taking soil samples before planting an orchard, divide each field into blocks based on soil survey data, slope, or cropping history. Soil survey data provide an overview of soil properties. A convenient way to gain access to soil survey data is to use the interactive application. Even when a field appears to be uniform, it is worth dividing it into several blocks which are sampled and analysed separately. Ideally, a field is divided into blocks of 2-5 acres and a composite sample of five cores from each block is taken. When larger blocks are samples, 15 to 30 cores should be taken from each block for a composite sample. When soil survey data suggest possible physical limitations, a series of backhoe pits should be dug to identify textural changes and layers that restrict root growth and drainage. This information helps to determine the best method of soil modification and whether the site is suitable at all for orchards. As an alternative to backhoe pits, undisturbed soil cores may be taken with a soil probe. Soil sample cores are taken from the entire field or management area in a W-shaped sampling pattern or by walking a zigzag course around or through the area. Mix the cores thoroughly; remove large stones, pieces or roots and other foreign material. Sample by foot increments to a depth of 2 to 4 feet or deeper if restrictive layers may be encountered in the subsoil. Map the field based on soil survey data and the results from the soil analyses of the different blocks. This map will help to determine the blocks for soil and tissue samples in the following years.

Soil sampling in established orchards

Soil samples to determine the availability of potassium, phosphorus, micronutrients and salt content can be taken any time of the year. However, fall sampling is usually preferred as it allows for enough time to adjust the fertilization program for the following year. Taking soil samples every 3-5 years is usually adequate. In recently planted orchards, annual sampling may be done until the soil fertility program is established. To monitor available nutrients over the years, samples should always be taken during the same season, but preferably in fall. Samples need to be taken before fertilizer is applied.

Sampling procedure

Divide each field into blocks based on soil survey data, slope, cropping history, variety, rootstock, age, growth pattern, or irrigation system. Plant residue from the sample spot is removed. Samples are best taken with a soil probe or auger. The sample is taken halfway between the trunk and the drip line and within the wetting zone of the sprinkler/emitter. Cores are taken from the entire area of the field or management area in a W-shaped sampling pattern or by walking a zigzag course around or through the area. Mix the cores thoroughly; remove large stones, pieces or roots and other foreign material. Sample by foot increments to a depth of 2 feet. When diagnosing a problem, deeper cores may be recommended. To obtain an accurate estimate of the nutrient availability, between 15 and 20 cores should be taken from each block for a composite sample. Collect the samples in a clean plastic bucket. Galvanized or rubber buckets may contaminate samples with zinc. When all the cores for an area are taken, mix them thoroughly. Very wet samples should be air-dried before packaging. Do not dry the samples in an oven or at abnormally high temperature. Put about one quart of soil in a clean bag and label it clearly. Follow the instructions of the laboratory that will do the analysis. To receive accurate fertilizer recommendations, the sample information sheet needs to be filled out carefully. Include the information sheet within the package submitted to the test lab.

Leaf analysis

Leaf analysis is the only technique according to which sensible fertilisation can be applied. In order to obtain the best results from this analysis, care must be taken to ensure that climatic, soil type and irrigation conditions as well as cropping practices are normal. While taking sample, following points should be kept in mind: (a) The best time for taking samples is 1-2 weeks prior to setting fruits. Do not take samples from obviously good or weak trees, (b) Sample must be taken from the leaves as per recommendations so that findings may always be comparable, (c) To be true representative, samples should be taken from some 20 trees, well-spaced an area of out over 3 ha. The trees should be homogenous in appearance and be average bearers, (d) The results should be supported by a record of leaf colour, tree vigour and yield so that fertilizer management can be adjusted for the next crop, (e) The leaf sample should be accompanied by a soil sample.

Deficiency Symptoms

When nutrient is not present in sufficient quantity, plant growth is affected. Plants may not show visual symptoms up to a certain level of nutrient content, but growth is affected and this situation is known as hidden hunger. When a nutrient level still falls, plants show characteristic symptoms of deficiency. These symptoms, through vary with crop, have a general pattern. These are generally masked by diseases and other stresses and so need careful and patient observation on more number of plants for typical symptoms. The deficiency symptoms appear clearly in crops with larger leaves. The deficiency symptoms can be distinguished based on the (1) region of occurrence, (2) presence or absence of dead spots, and (3) chlorosis of entire leaf or Interveinal chlorosis. The region of appearance of deficiency symptoms depends on mobility of nutrient in plants. The nutrient deficiency symptoms of N, P, K, Mg and Mo appear in lower leaves because of their mobility inside the plants. These nutrients move from lower leaves to growing leaves thus causing deficiency symptoms in lower leaves. Zinc is moderately mobile in plants and deficiency symptoms, therefore,

Table 2: Leaf sampling technique and nutrient norms for some of the fruit crops

Crop	Plant part, age, stage and position	Optimum leaf nutrient norms (%)		
		N	P	K
Ber	5-leaf from tip of secondary and tertiary shoot in June	1.50-2.20	0.14-0.45	1.60-2.00
Pomegranate	8-leaf pair from tip in April flush or February crop and from August flush for June crop	1.20-1.40	0.10-0.20	1.00-1.40
Acid Lime	Basal leaf at 5-month-age from current season's growth	1.96-2.30	0.12-0.29	1.60-1.90
Sweet Orange	Basal leaf 6-month-old leaf from current season's growth emerged in March or September	2.00-2.20	0.10-0.11	0.40-1.20
Lemon	Basal 6-month-old leaf from current season's growth	2.20-2.70	0.15-0.30	1.00-2.00
Phalsa	4-leaf from growing tip one month after pruning.	1.50-1.60	0.15-0.20	1.60-2.00
Fig	Basal leaf from mid-summer growth	2.00-2.50	0.09-0.10	0.70-0.90
Guava	Third pair of leaf from age in August-December	1.60-2.40	0.15-0.30	1.30-1.70
Papaya	Petiole from sixth leaf from top, 6 months after planting	1.01-2.50	0.22-0.40	3.30-5.50

Source: Bhargava, 1999

appear in middle leaves. The deficiency symptoms of less mobile elements (S, Fe, Mn and Cu) appear on new leaves. Since Ca and B are immobile in plants, deficiency symptoms appear on terminal buds. Chlorine deficiency is less common in crops.

Nutritional disorders

Nutritional disorders are basically physiological disorders in the plants that affect productivity as well as the quality of fruits. Disturbance in the plant metabolic activities resulting from an excess or deficit of environmental variables like temperature, light, aeration and nutritional imbalances result in disorders. In fruit crops, the deficiency of micronutrients causes many more disorders than that of macronutrients. Nutritional disorders have become widespread with diminishing use of organic manures, adoption of high density planting, use of root stocks for dwarfing, disease and salt tolerance, unbalanced NPK fertiliser application and extension of horticulture to marginal lands. To get high quality fruit and yields, micronutrient deficiencies have to be detected before visual symptoms are expressed. The deficiencies of Zn, Mn and B are common in sweet orange, acid lime, guava and papaya in India. To correct both visual and hidden micronutrient deficiencies, appropriate foliar and soil applications are necessary. Some crops are more sensitive than others to the deficiency of a micronutrient and it can be inferred that the critical concentration of a nutrient is not same for all the crops. The susceptibility or tolerance rating of crops to nutrient deficiencies shows considerable variation due to wider hereditary variability within a crop species. When a nutrient element insufficiency (deficiency and/or toxicity) occurs, visual symptoms may or may not appear, although normal plant development will be slowed. When visual symptoms do occur, such symptoms can frequently be used to identify the source of the insufficiency.

Symptoms of deficiency may take various forms, such as:

- Stunted or reduced growth of the entire plant with the plant itself either remaining green or lacking an over-all green colour with either the older or younger leaves being light green to yellow in colour.

- Chlorosis of leaves, either interveinal or of the whole leaf itself, with symptoms either on the younger and/or older leaves, or both (chlorosis is due to the loss or lack of chlorophyll production).

- Necrosis or death of a portion (margins or interveinal areas) of a leaf, or the whole leaf, usually occurring on the older leaves.

- Slow or stunted growth of terminals (rosetting), the lack of terminal growth, or death of the terminal portions of the plant.

- Reddish purpling of leaves, frequently more intense on the underside of older leaves due to the accumulation of anthocyanin.

Visual symptoms of toxicity may not always be the direct effect of the element in excess on the plant, but the effect of the excess element on one or more other elements. For example, an excessive level of potassium (K) in the plant can result in either magnesium (Mg) and/or calcium (Ca) deficiency, excess phosphorus (P) can result in a zinc (Zn) deficiency and excess Zn in an iron (Fe) deficiency. These effects would compare to elements, such as boron (B), chlorine (Cl), copper (Cu), and manganese (Mn), which create visual symptoms that are the direct effect of an excess of that element present in the plant. Some elements, such as aluminum (Al) and copper (Cu) can affect plant growth and development due to their toxic effect on root development and function.

Hidden hunger

In some instances, a nutrient element insufficiency may be such that no symptoms of stress will visually appear with the plant seeming to be developing normally. This condition has been named hidden hunger, a condition that can be identified by means of either a plant analysis and/or tissue test. A hidden hunger occurrence frequently affects the final yield and the quality of the product produced. For fruit crops, abnormalities, such as blossomed rot and internal abnormalities may occur, and the post-harvest characteristics of fruits and flowers will result in poor quality and reduced longevity. A generalized visual leaf and plant nutrient deficiency symptoms in horticultural crops are given in Table 3.

Climatic and other causes: The occurrence of the symptoms may not necessarily be the direct effect of a nutrient element insufficiency. For example, stunted and slowed plant growth and the purpling of leaves can be the result of climatic stress, cool air and / or root temperatures, lack of adequate moisture, etc. Damage due to wind, insects, disease and applied foliar chemicals can produce visual symptoms typical of a nutrient element insufficiency.

Table 3: Deficiency Symptoms of nutrient in horticultural crops

Element/status	Visual deficiency symptoms
Nitrogen (N)	Light green leaf and plant colour with the older leaves turning yellow, leaves that will eventually turn brown and die. Plant growth is slow; plants will be stunted, and will mature early.
Phosphorus (P)	Plant growth will be slow and stunted, and the older leaves will have a purple coloration, particularly on the underside.
Potassium (K)	On the older leaves, the edges will look burned, a symptom known as scorch. Plants will easily lodge and be sensitive to disease infestation. Fruit and seed production will be impaired and of poor quality.

Contd.

Calcium (Ca)	The growing tips of roots and leaves will turn brown and die. The edges of the leaves will look ragged as the edges of emerging leaves stick together. Fruit quality will be affected with the occurrence of blossom-end rot on fruits.
Magnesium (Mg)	Older leaves will be yellow in colour with interveinal chlorosis (yellowing between the veins) symptoms. Plant growth will be slow and some plants may be easily infested by disease.
Sulfur (S)	A general overall light green colour of the entire plant with the older leaves being light green to yellow in colour as the deficiency intensifies.
Boron (B)	Abnormal development of the growing points (meristematic tissue) with the apical growing points eventually becoming stunted and dying. Flowers and fruits will abort. For some grain and fruit crops, yield and quality is significantly reduced.
Chlorine (Cl)	Younger leaves will be chlorotic and plants will easily wilt. For wheat, a plant disease will infest the plant when Cl is deficient.
Copper (Cu)	Plant growth will be slow and plants stunted with distortion of the young leaves and death of the growing point.
Iron (Fe)	Interveinal chlorosis will occur on the emerging and young leaves with eventual bleaching of the new growth. When severe, the entire plant may be light green in colour.
Manganese (Mn)	Interveinal chlorosis of young leaves while the leaves and plants remain generally green in colour. When severe, the plants will be stunted.
Molybdenum (Mo)	Symptoms will frequently appear similar to N deficiency. Older and middle leaves become chlorotic first, and in some instances, leaf margins are rolled and growth and flower formation are restricted.
Zinc (Zn)	Upper leaves will show Interveinal chlorosis with an eventual whiting of the affected leaves. Leaves may be small and distorted with a rosette form.

Manure and fertilizers recommendations

Soil chosen for planting any fruit crops must be brought at an optimum fertility levels before planting the fruit saplings after its chemical analysis (Chopra and Kanwar, 1986). To optimize the soil fertility status and to keep the plant nutrient status above the critical levels, a number of recommendations for fruit crops are available region wise (Tondon, 1987). The recommendations based on All India Coordinated Research Project on Fruit crops are given in Table 4. Besides these applications, regular applications of micronutrients are also necessary in the soils or foliar applications, where such deficiencies were observed. For effiecient utilisation of nutrients, it is essential to know the active root zone, so that the nutrients are applied around this zone, to make available to the plants (Chundawat, 1997).

Table 4: Manure and fertilizers recommendations for different fruit trees

Fruit species	Manure / Fertilizers (per Plant)				During first year and increase the doses in same proportion every year upto 5 years of age
	FYM(kg)	N(g)	P_2O_5(g)	K_2O(g)	
Ber	10	100	50	50	For first year, whole dose of FYM during the month of May after pruning and fertilizer in two split doses: 2/3 during July and rest 1/3 in Sept-Oct. Increase the doses in same proportion every year upto 5 years of age.
Aonla	10	150	75	100	Full dose of Phosphate and potash and half dose of nitrogen in the month of February and rest half dose of nitrogen in August during first year and increase the doses in same proportion every year upto 10 years
Pomegranate	10	125	50	50	During first year, full dose of Phosphate and potash and half dose of nitrogen along with first irrigation in spring and rest half dose of nitrogen 3 weeks after first irrigation are applied. Organic manure 23-30days before first irrigation. Increase the doses in same proportion every year upto 5 years of age upto 5 years
Citrus	10	120	60	80	During first year, full dose of Phosphate and potash and half dose of nitrogen in February and rest half dose of nitrogen is applied in April-May after the fruit set. Organic manure is applied in December-January. Increase the doses in same proportion every year upto 5 years of age.
Datepalm	20	100	50	50	During first year, full dose of Phosphate and potash and half dose of nitrogen in February and rest half dose of nitrogen is applied in April-May after the fruit set. Organic manure is applied in December-January. Increase the doses in same proportion every year upto 10 years of age.
Papaya	30	250	250	250	Papaya is a heavy feeder of fertilizers. Full dose of Phosphate and organic manue is applied at the time of planting. 1/6 dose of nitrogen and potash is applied at bi-monthly interval after planting. Increase the doses in same proportion every year upto 5 years of age.
Guava	10	75	60	75	During first year, full dose of Phosphate and potash and half dose of nitrogen is applied in June, rest half dose of nitrogen is applied in September. Increase the doses in same proportion every year upto 5 years of age.
Bael	10	75	50	50	FYM plus half dose of nitrogen and Potassium and full dose of Phosphorus should be applied just after rain. Remaining dose should be applied in the last week of August.
Fig	15	75	50	50	FYM plus half dose of nitrogen and Potassium and full dose of Phosphorus should be applied just after rain. Remaining dose should be applied in the last week of August.

References

Bhargava, B.S. 1999. Leaf analysis for diagnosing nutrient needs in fruit crops. *Indian Horticulture*, 43(4): 6-8.

Chopra, S.L. and Kanwar, J.S. 1986. Analytical Agricultural Chemistry. Kalyani Publishers, Ludhiana.

Chundawat, B.S. 1993. Intercropping in orchards. In: Advances in Horticulture Vol.2 Fruit crops: part 2. Chadda, K.L. and Pareek, O.P. (Eds). Malhotra Publishing House, New Delhi. pp. 763-767.

Chundawat, B.S. 1997. *Nutritional Management in Fruit Crops*. Agrotech Publishing Academy, Udaipur.

Government of India. 2014. *Handbook on Horticulture Statistics*. Ministry of Agriculture Department of Agriculture and Cooperation, New Delhi, India.

Iyenger, B.R.V. 1993. *Fertilizer management*. In: Advances in Horticulture Vol 2 . Fruit crops: part 2. Chadda, K.L. and Pareek, O.P. (Eds). Malhotra Publishing House, New Delhi. pp. 1055-1072.

Tondan, H.L.S. 1987. *Fertilizer Recommendations for Horticultural Crops in India*. FDCO, New Deli. pp. 16.

29

Management and Utilization of Degraded Lands and Poor Quality Water for Production of Spices, Medicinal and Aromatic Plants

O.P. Aishwath

ICAR-National Research Centre on Seed Spices, Tabiji, Ajmer, Rajasthan

According to FAO estimates, about 50% total geographical area in India is under various degraded hazards and about 2.1 million ha of land is getting degraded and deforested annually (Sehgal *et al.* 1998). Out of that underground water in arid and semi-arid areas is saline or sodic in nature. While spices, medicinal and aromatic plants may be a boon for these areas, which not only withstand under adverse conditions but also could be productive and helps in mitigating these problems. However, other crops are remain less or unproductive in these areas.

Spices are low volume and high value crops and have tremendous export potential, which is witnessed by several invasions faced by this Indian peninsula during ancient time. Medicinal and aromatic plants have an equally market potential with the world demand of herbal products growing at the rate of 7 per cent per annum (Aishwath and Tarafdar, 2007). Seed spices and aromatic plants are the natural source of flavour, perfumes and fragrance widely exploited by essential oil industries in the world. Owing to considerable diversity of edaphoclimatic conditions ranging from temperate to tropical and arid to sub-humid that prevails in India and an array of medicinal and aromatic flora occurs in wild state. Many exotic essential oil bearing plants, including mints, citronella, geranium, patchouli, lavender and chamomile have been naturalized in India. As the country has emerged as one of the main producers of seed spices and some of the essential oils, theses plants are significant foreign exchange earners. Some of the aromatic plants especially grasses can sustain both biotic as well as abiotc stresses, even in the areas unsuitable for conventional crops (Aishwath *et al.*, 2008). Most of the seed spices crops are seasonal crops and comes up well with limited moisture and nutrients. Therefore, under utilized natural resources can be utilized profitably by crop diversification.

1. Saline alkali soils resources optima for spices and aromatic plants

About 7.2 million ha soils are saline and alkaline out of which 3 million ha is alkali in India. Due to poor physical properties, excessive exchangeable sodium and high pH, most of these lands support a very poor vegetation cover. Most of these areas are under arid or semi-arid regions. This waste area is not utilized properly by the traditional farming. There is lot of scope for utilization of our valuable soil resource by seed spices and aromatic plants for the cultivation of these high-value crops with marginal inputs.

i)   **Saline soil:** Dill can be grown up to EC 3.3 dSm^{-1} (Sharma *et al.*, 2001). In another study on dill with salinity shows that it comes up well with EC 4-5 dSm^{-1} (Rao *et al.*, 2001 and Singh *et al.*, 1995). The beneficial effect of salinity on lemongrass (*Cymbopogon citrates*) and Java citronella (*Cymbopogon winterianus*) was observed at EC 5 dSm^{-1} as compared to lower or its higher levels of salinity (Patra & Singh 1995; Pal *et al* 1989). Lemon grass is a very hardy crop which grows successfully on cultivable waste land and the salt affected soil having EC up to 11dSm^{-1}. The essential oil of German chamomile (*Matricaria chamimilla*) popularly known as blue oil can withstand up to salinity 12.0 dSm^{-1} without leaving any adverse effects on yield. Palmarosa comes up well in various textural types of soils. The salinity upto 5.5 dSm^{-1} found beneficial with respect to its productivity. Based on the pot culture study, there was no adverse effect on yield and quality of palmarosa observed upto EC 11.5 and beyond that soil salinity may adversely affect the yield and quality of palmarosa. Fig. 1 enunciates that vetiver, jamarosa and citronella could withstand up to salinity 12, 12.5 and 5.5 EC dSm^{-1}, respectively (Singh and Anwar, 1985, Patra and Singh, 1995). In aromatic trees, various species of eucalyptus have wide range of salinity tolerance ie 3.0 to 20.0 EC dSm^{-1} (Fig. 2) as reported by Gupta and Gupta (1997). Some of the aromatic plants are salt includer type salinity tolerant has potential for hyper accumulation of salt resultant helps in reclaiming the saline soil such as palmarosa is an hyper salt accumulator (Aishwath and Pal, 2004). Palmarosa grown on the soil having EC 4.8 dSm^{-1}, after two years of cropping, the EC of the soil reduced down to 0.64 dSm^{-1} (Prasad and Singh, 1998).

ii)   **Alkaline soil:** Fennel performs well on heavy textured soils and can also be cultivated on salt affected black soil with salinity of 8-10 dS m^{-1}

(Singh, 1999). Fennel and coriander grown on partially reclaimed sodic soil having Exchangeable Sodium Per cent (ESP) 20 with various N and P levels yielded well and large accumulation of Na in fennel stover indicating its tolerance to sodicity. Reduction in soil pH and ESP and increase in exchangeable calcium indicates its capability for reclaiming sodic soils (Garg *et al.*, 2000). Dill (*Anethem graveolens*) can tolerate the pH up to 8.6 (Farooqi, and Sreeramu, 2001). It has been reported that aromatic grasses such as, palmarosa (*Cymbopogon martinii* Roxb. Wats.) and lemon grass (*C. flexuosus* Steud. Wats.) could successfully be grown on moderately alkali soils having pH up to 9.0 and 9.5, respectively, while vetiver (*Vetiveria zizanioides* L. Nash) which withstands both high pH and stagnation of water, could successfully be grown without significant yield reduction on highly alkali soils (Fig. 1&3). In some cases the yield and quality of vetiver was better on sodic soil (Patra & Singh, 1995, Gupta & Gulati, 1965 and Farooqi & Sreeramu, 2001). These grasses not only produce essential oils used for industrial purpose but also ameliorate the soil (Dagar *et al.* 2004). Palmarosa, lemon grass and vetiver grown for two years on sodic soils having pH 10.6, 9.8 and 10.5, respectively, the reduction in pH was noticed as 9.4, 8.95 and 9.50 in each soil, respectively (Prasad and Singh, 1998; Patra *et al.*, 2002 and Anwar *et al.*, 1996). Japanese mint (*Mentha arvensis*) also comes up well in slightly alkaline soil than acid sedentary soil (Prasad & Chattopadhyay 1999). Among the aromatic grasses like palmarosa, lemongrass, vetiver, citronella and jamrosa, the jamrosa is a high pH tolerant (pH=10) and citronella is least (8.5). However, under well supplied organic matter conditions (Fig. 3), even pH up to 9.7 has no detrimental effect on growth, yield and oil content in citronella plant as reported by Gupta and Gulati (1965). In case of amrette (*Abelmoschus moschata*), chamomile (*Matricaria chamomilla*), and Lavender (*Lavandula species*) can tolerate upper pH level as 8.6, 9.0 and 8.4, respectively (Farooqi, and Sreeramu, 2001). An another study it was found that dill could be grown up to soil salinity pH 8.8 with slightly yield reduction (Sharma *et al.*, 2001). German chamomile can produce 2.5-3.0 t ha[-1] of flower yield in alkaline soil having pH 9.5 and also have the high sodium uptake efficiency could help in improvement of sodic soil (Patra and Singh, 1995). Seed spices like cumin, anise and nigella are fairly tolerable to higher pH (Fig. 4).

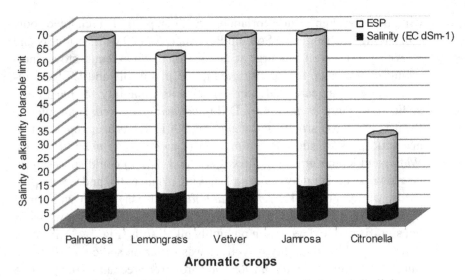

Fig. 1: Adaptability of aromatic crops with salinity and alkalinity

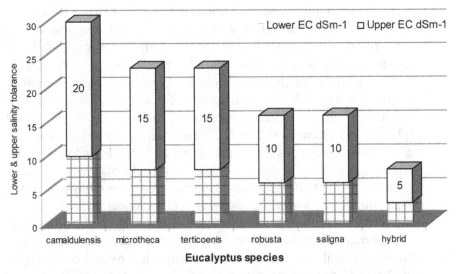

Fig. 2: Salinity tolerant limit of various species of *Eucalyptus sps.*

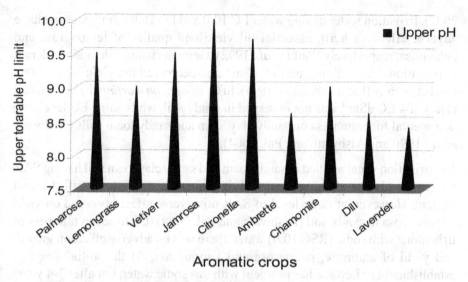

Fig. 3: Upper permissible limit for adaptability of aromatic crops with pH

2. Poor quality irrigation water and productivity of spices, medicinal and aromatic plants

Underground water always contains some soluble salts. Continuous use of such water may leave some harmful affect the crop growth and soil properties depending upon the quality and quantity of salts present. Accumulation of salt in soil profile depends upon the texture, structure and the depth of the soil (Kelly, 1963 and Aishwath and Pal, 2004), which directly affects the yield and composition of the crops (Aishwath and Pal, 2000). The major constraints of underground water uses for irrigation are (i) Salinity and (ii) Sodicity.

Fennel can be cultivated profitably in the salt affected black soils by using saline water of 4 dS m⁻¹. It is also revealed that the water of sub-marginal quality with salinity of 4 dS m⁻¹ can be used without any significant yield reduction in dill. It is found to tolerate a pH of 8.6 (Rao *et al.*, 2001) Usage of saline irrigation water (ECiw 8-10 dS m⁻¹) on highly calcareous saline soils, *Eucalyptus tereticornis* and aromatic grass species vetiver (*Vetiveria zizanoides*) showed great promise by yielding 1.19 to 1.73 t ha⁻¹ root biomass. Lemon grass (*Cymbopogon flexuosus*), Palmarosa (*Cymbopogon martinii*), Ocimum (*Ocimum sanctum*), Celery (*Apium graveolens*), Mint (*Mentha piperita*), and Metha (*Trigonella foenum-graecum*) were irrigated either with saline or canal water alone or with the alteration of canal and saline water. These species have tremendous potential for cultivation with irrigation of underground saline waters having salinity up to 10 dS m⁻¹. Among the aromatic grasses, vetiver was found the most promising, followed by palmarosa and lemongrass (Dagar,

2007). Irrigation water having water EC 10.0 and EC 16.0 dSm⁻¹ does not have adverse effect on herb, essential oil yield and quality of lemongrass and palmarosa, respectively (Patra *et al.*, 1992). German chamomile is also tolerant to irrigation water salinity up to 8 dSm⁻¹ and decreased the yield beyond that level of salinity (Patra and Singh, 1995). In *Cymbopogon martinii* the detrimental effect of 4 EC dSm⁻¹ was not observed in sandy soil, while same EC level was detrimental for palmarasa on sandy clay loam and sandy loam soils (Aishwath *et al.*, 1990 and Aishwath and Pal, 2004).

The irrigation water applied in sandy loam and sandy clay loam soil having SAR 9.1 leaves adverse effect on the yield of palmarosa without any effect on oil content. However, at same level of SAR no adverse effect observed on yield of palmarosa in sandy soil (Aishwath and Pal, 2004). Up to seven numbers of irrigations with sodic (RSC 10.0) water, there was no adverse effect on growth and yield of ambrette, palmarosa and lemongrass. At the initial stage of establishment of kewada has problem with this sodic water, but after 3-4 years of its establishment no adverse effect on growth was observed with sodic water (Aishwath, 2005a). However, some of the aromatic plant like ginger, lemongrass and jasmine are sensive for sodic water (Aishwath, 2005b). The adverse effect of high SAR (10.0) of irrigation water did not appear on growth and yield palmarosa. However beyond that level the growth, yield and uptake of the N, P, K, Ca and Mg reduced (Pal *et al.*, 1993).

3. Spices, medicinal and aromatic plants and their adaptability in Acid soils

Almost all the seed spices are fairly tolerable to mild soil acidity and alkalinity while coriander, fenugreek and celery could comes up well even up to 5.5 pH level (Fig. 4). It has also been reported that coriander comes well with the soil pH 6.5-8. Cumin can be grown with pH ranging from 6.8 to 8.3. While, fennel could be grown in heavy clay to sandy loams soils with pH of 6.5-8.0. In case fenugreek, it is cultivated on well drained loams or sandy soils and the acceptable pH rang is 6.0-8.0. The heavy loam or clay with pH 6.0-7.5 is most suitable. Celery thrives well in acid soil but not beyond pH 5.5 and rich in organic matter. Micro and secondary nutrient are essential otherwise plant shows deficiency symptoms as Boron- Cracked stem, Ca - Black heart and Mg-Chlorosis. Anise requires sandy loam to clay loam soil and pH 6.0-8.5 is well suited to this crop. For nigella, well drained sandy to heavy clay soil acid to alkaline pH is suitable for optimum plant growth. Caraway requires neutral to slightly alkaline soil having pH 6.5-7.5 is most suitable. It shows that these crops have wide adaptability with respect to soil reaction could be fitted into any adverse soil conditions (Aishwath *et al*, 2008).

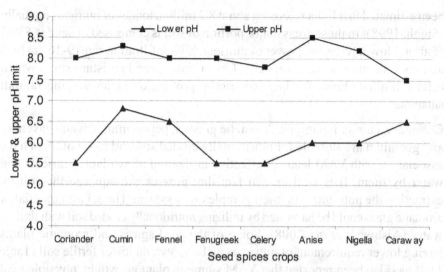

Fig. 4: Adaptability of seed spices crops with various pH ranges

4. Utilization of eroded lands through spices, medicinal and aromatic plants

Out of total land area, about 50 per cent is under degraded and cultivated land subjected to serious hazards of water, wind, shifting cultivation, water-logging, salinity and alkalinity and changing water courses. (Narayana and Babu, 1983). While seed spices crops and aromatic plants can combat against these problems of erosion on sloppy land. The cultivation of seed spices and aromatic plants like *Anethum sowa* and *Ocimum kilimandscharicum* recommended in degraded habitat orchards as inter-crop (Gupta *et al.*, 1998 and Narayana, 1998). Aromatic grasses on sloppy lands have wide potential for the extraction of oil. In *Cymbopogon species*, *Cymbopogan martinii* produced oil yield of 68.3 kg/ha/year followed by *C. winterianus*, 41.3kg/ha/year. At many a places, Cymbopogons are utilizing for dual purpose to control the gully erosion and oil extraction in Uttar Pradesh, Madhya Pradesh, Rajasthan and Gujarat (Verma and Chinnamani, 1998). Similarly, Eucalyptus globules and Eucalyptus robusta utilized in humid regions for dual purpose to control the soil erosion and stabilization of gully erosion as well as essential oil extraction (Ramamohan Rao, 1998). Lavender (*Lavandula officinale*) and rose (*Rosa domascena*) could also be grown in sloppy and erodes lands or terraces in hill areas (Shiva *et al.*, 2002).

5. Utilization of nutritionally eroded lands though spices, medicinal and aromatic plants

Fast mineralization of soil organic matter and release of nutrients in soil leads to losses of nutrients through water run off in arid and semi-arid areas. It has been estimated that India is loosing about 8.2 million tonnes of nutrients annually (Singh, 1998). In these areas major portion of soil is composed of sand (>85%) resulting lower retention power of nutrient, high infiltration rate (9-16 cm hr^{-1}) and poor moisture storage capacity. Lower fertility and moisture stress which offer a limited choice for increasing crop production under varying rainfall situation.

Coriander, cumin, fenugreek, Aromatic grasses, peppermint, davana, lavender and geranium are suitable for nutritionally eroded soil and some of them have association with VAM leads to effective utilization of available nutrients and water by them. It is well known fact that grasses can squeeze the soil by extracting the nutrients with their complex root system. Here, vast potential of aromatic grass could be harvested by utilizing nutritionally eroded soil with limited water (Aishwath *et al.*, 2008). Some of the seed spices and aromatic plants having lower requirement of nutrient can be grown on lower fertile soil (Table 1). It has also been reported that VAM simulate plant growth by physiological effects other than by enhancement of nutrient uptake or by reducing the severity of diseases caused by soil borne pathogens (Dehne, 1982). The responsiveness of VAM to the host specific seed spices and aromatic plant is given in table 2. Recently, possibilities have been explored for the cultivation of rose scented geranium (*Palargonium sps*) in semi-arid conditions by multiple cuttings/ harvesting (Kothari *et al.*, 2004).

Table 1: Nutrient removal by some spice and aromatic crops

Crop	N uptake (kg ha^{-1})	P uptake (kg ha^{-1})	K uptake (kg ha^{-1})
Menthol mint	109	28	129
Ctronella	181	33	255
Lemongrass	160	32	194
Palmarosa	430	62	339
Geranium	180	24	137
Coriander	30	4	36
Davana	179	26	171
Japanese mint	109	28	129
Pappermint	42	-	-

Source: Prakasa Rao, 1992 and 1993

Table 2: Per cent colonization of VAM fungi in some spices, medicinal and aromatic plants

Crop	Per cent colonization	Crop	Per cent colonization
Basil	63	*Mentha arvensis*	46
Citronella	75-91	*Mentha piperita*	100
Coriander	32	*Mentha spicata*	100
Cymbopogon citratus	80	*Mentha citrata*	40
Cymbopogon flexuosus	72	Musk	83
Cymbopogon jwarancusa	50	Palmarosa	41-98
Cymbopogon khasianus	76	Patchouli	43
Davana	86	Rose-scented geranium	22-59
Fenugreek	92	Sacred basil	77
Lavender	80	Vanilla	57
Lemongrass	21-75	Vetiver	42

(Aishwath and Tarafdar, 2007)

6. Utilization of arid and rain-fed areas through spices, medicinal and aromatic plants

The arid regions of the world occupy very extensive areas of the continental land masses and constitute nearly one – half of the land surface of the Earth. The major portion of the African continent and a great part of the Asian continent are arid.

Vegetation in the dry zones is very sparse. The general landscape is desolate and barren. Herbs appear only during a short period of the year when the conditions become favorable. The shrubs, herbs and trees have various means of storing water. These adaptations include increased ability to store water in their succulent stems or leaves; thickening of the leaf cuticle or reduction of leaf surface or the entire absence of leaves to lower the transpiration rate, and the ability to survive as a seed through many years of aridity. Drought – resisting plants may be divided into two groups: Succulents and Ephemerals. Seed spices are arid and semi-arid crops and come up with limited water conditions, while dill, ammi, Anise, fennel, papaver crops are fairly tolerant to moisture stress among the seed spices. The aromatic plant recommended for arid areas are ferula, artemisia, sage, lemongrass, guggal and Myrrh Recent research on *Ammi visnaga* and *A. majus*, *Pimpinella anisum*, *Foeniculum vulgare*, *Ferula asafetida*, *F. foetida*, Artemisia proved to be adaptive in dry areas. Some artemisias are of medicinal interest and a few are prized for their volatile oils. The chief source of santonin, in the U.S.S.R., is *A. cine* Berg. Growing in abundance in Russian Turkestan and Persia. Other species which have been found to contain santonin include *A. mexicana* Willd., A. neo – mexicana Woot, and *A. wrightii. A. grey*, found in America and *A. gallica* Willd. in East Germany,

France, England and Scotland, and *A. maritime* L., a fairly widely distributed species spread from England to as far east as Chinese Mongolia. The content of santonin in the American and English species is too low to warrant commercial extraction. *A. fragrans* Willd. and *A. parviflora* Roxb. occurring in Afghanistan have also been reported to contain santonin. *A. maritime* growing in certain areas of Kashmir (India) and Kurram (Pakistan) has been found to contain substantial amounts of santonin (1 – 2 per cent), which is commercially extracted for aromatic and medicinal purpose. Lemongrass or Indian Verbena (*Cymbopogon citrates, C. schoenanthus* and C. *nardus*) southern parts of the eastern Egyptian desert and which yields a volatile oil are adaptive to water stress. Myrrh is still used to some extent in pharmacology as an aromatic stimulant; fenugreek (*Trigonella foenumgraecum* L.) contains diosgenin content could serve for the demi-synthesis of cortisone derivatives and sex hormones are come up well under water stress (Chopra *et al.*, 2003).

7. Utilization of water logged areas through spices, medicinal & aromatic plants

There has been no accurate estimate of the area afflicted in India under water logging. Framji (1974) reported that nearly 3.5 million hectares areas become waterlogged with the rise in water table in the root zone. Recently, Sehgal *et al.*, (1998) reported about 20 million hectares area is under flooding causes water logging.

Most of the seed spices are susceptible for the water logging condition. However, water logging or submerged conditions can be utilized for the production of *Acorus calamus*, while, cardamom (*Elettaria cardamomum*) is best suited in swampy areas for its ideal production. Clove can withstand for a shorter period under water logging conditions. In case of kewada (*Pandanus species*), it is highly tolerable to water logging conditions for a longer period. The mint species *Mentha arvensis* is moderately tolerable to water logging for 8-10 dyas provided low atmospheric temperature. However, vetiver is capable of prolong water stagnation and also yielded with full potential (Shiva *et al.*, 2002).

8. Utilization of metal contaminated soils and environment protection through spices and aromatic plants

With our industrial progress, we are also progressing to contaminate our soil and water resource by various metals and heavy metals causes serious problem to the living ecosystem and their environment. Aromatic crops can provide economic return and metal-free final product, the essential oil. The study in Bulgaria using coriander, sage, dill, basil, hyssop, lemon balm, and chamomile grown at various distances from the smelter revealed that aromatic crops may not have significant phytoremediation potential, but growth of these crops in metal contaminated agricultural soils is a feasible alternative. Herbage essential oil yields of basil, chamomile, dill, and sage were reduced when they were grown closer to the contaminants. Metal removal takes place from the site with the harvestable plant parts was as high as 180 g ha^{-1} for Cd, 660 g ha^{-1} for Pb, 180 g ha^{-1} for Cu, 350 g ha^{-1} for Mn, and 205 g ha^{-1} for Zn. It also reported that high concentrations of heavy metals in soil or growth medium did not result in metal transfer into the essential oil. Of the tested metals, only Cu at high concentrations may reduce oil content (Valtcho *et al.*, 2008). No detectable amount of Cd, Cu, or Pb was found in the oils plant species Peppermint, basil, and dill. These can be grown in soils enriched with Cd, Pb, and Cu medium without risk for metal transfer into the oils, and without significant alteration of essential oil composition that may impair marketability (Valtcho *et al.*, 2006). Zheljazkov and Jekov (1995) investigated by Furnace Atomic Absorption (GFAA) and Inductively Plasma (ICP) techniques. The content of Cd, Pb, Cu, Mn, Zn and other trace elements determined in essential oils and plant extracts from the genera *Rosa, Lavandula, Mentha, Salvia, Ocimum, Foeniculum, Coriandrum, Anethum, Hyssopus* and *Rhus*. On the other hand the heavy metal content in oils from different regions were compared to heavy metal content in some oils obtained from species grown on heavily polluted soils and under severe atmospheric pollution. Results, obtained from the two apparatuses were very similar. They found that the concentration of the most hazardous heavy metals in all of the tested oils and plant extracts was very low, near the detection limits of the used apparatuses. It was concluded that most of the essential oil and medicinal crops could be successfully grown on heavy metal polluted soils and under atmospheric pollution as substitutes for some other edible crops.

References

Aishwath, O.P. 2005a. Appraisal of underground water in four consecutive years at NRC MAP, Boriavi (Gujarat) and impact of its irrigations on medicinal and aromatic plants. *Ecology, Environ. & Conserv.*, 11: 373-380.

Aishwath, O.P. 2005b. Coefficient of variation and correlation coefficient in underground water quality parameters in and adjoining municipal area of Boriavi, Gujarat. *Pollution Res.*, 24: 9-17.

Aishwath, O.P. and Pal, B. 2000. Performance of palmarosa (*Cymbopogon martinii* var Motia) under various soil texture and salinity levels. *Indian Perfumer*, 44: 285-290.

Aishwath, O.P. and Pal, B. 2004. Mineral composition of palmarosa (*Cymbopogon martinii*), soil EC and pH as influenced by saline water irrigation in texturally different soils. *J. Guj. Soc of Agron. Soil Sci.*, 4: 42-47.

Aishwath, O.P., Pal, B. and Singh, A.K. 1990. Effect of soil texture and water salinity on the growth, dry matter production, nutrients uptake and oil content of palmarosa (*Cymbopogon martinii*). *J. Agri. Sci. Res.*, 32: 45-47

Aishwath, O.P. and Tarafdar, J.C. 2007. Role of organic farming in medicinal and aromatic plants. In: *"Organic Farming and Sustainable Agriculture"* (Eds. Tarafdar *et al.*, 2007). Publisher - Scientific Publisher, Jodhpur. pp. 135- 185.

Aishwath, O.P., Vashishtha, B.B. and Malhotra, S.K. 2008. Scope of production of aromatic plants in arid and semi-arid regions. In: Proc. *National Workshop on Spices and Aromatic Plants*. Held on 6-7 February, 2008 at ARS (RAU) Mandor, Jodhpur, Rajasthan. P.10.

Anwar, M., Patra, D.D. and Singh, D.V. 1996. Influence of soil sodicity on growth oil yield and nutrient accumulation in vetiver (*Vetiveria zizanioides*). *Ann. Arid Zone*, 35: 49-52.

Chopra, L.C., Abrol, B.K. and Handa, K.L. (2003) Medicinal plants of the arid zone. Publisher Today & Tomorrow printers and Publishers. pp.11-55

Dagar, J.C. 2007. Alternative land uses in areas with water scarcity. In, On-farm land and water management (A Compilation of National Training Course). Pp. 154-165.

Dagar, J.C., Tomar, O.S., Kumar, Y. and Yadav, R.K. 2004. Growing three aromatic grasses in different alkali soils in semi-arid regions of northern India. *Land degradation and development*, 15: 143-151.

Dehne, H.W. 1982. Interaction between vesicular arbuscular mycorrhizal fungi and plant pathogens. *Phytopathol.*, 726: 1115-1119.

Farooqi, A.A. and Sreeramu, B.S. 2001. Ambrette, Chamomile, Dill, Lavender, and Palmarosa. In: *cultivation of medicinal and aromatic plants*. Pub. Univ. Press (India) Ltd. Hyderabad. Pp. 295-423.

Framji, K.K. 1974. International Commission on Irrigation and Drainage bull. Pp. 36-41.

Garg, V.K., Singh, P.K and Katiyar, R.S. 2000. Spices and aromatic plants. In: Proc. *Centennial conference on spices and aromatic plants* (20-23 September, 2000), Kalicut, Kerla, India. Pp. 133-138.

Gupta, R. and Gulati, B.C. 1965. Salt tolerance in *Cymbopogon nardus* (Lin) Rendle (Citronella cylon type) possibilities of commercial cultivation in *ushar* lands. Proc. Sem. *Sea salt and Plants*, held at Bhavnagar. Pp. 100-106.

Gupta, R.K., Bhardwaj, S.P. and Mittal, S.P. 1998. Management of degraded lands in the Himalayas. In, *Technologies for waste land development* (Eds Abrol I P and Dhurva Narayana, V.V.). Pub. Information Div. ICAR, KAB, New Delhi, pp. 196-208.

Gupta, S.K. and Gupta, I.C. 1997. Crop tolerance to saline conditions. In: *Management of saline soils and water*. Scientific Publishers, Jodhpur, India. Pp. 99-117.

Kelly, W.P. 1963. Use of saline irrigation water. *Soil Science*, 96: 305-391.

Kothari, S.K., Vijay Kumar, Y., Bhattacharya, A.K. and Ramesh, S. 2004. Harvest management in rose-scented geranium (*Pelargonium sps.*) for higher essential oil yield and superior quality under semi-arid tropics. *Archives of Agron. & Soil Sci.*, 50: 593-600.

Narayana, V.V. Dhurva, 1998. Land degradation due to water erosion. In, *Technologies for waste land development* (Eds Abrol, I.P. and Dhurva Narayana, V.V.). Pub. Information Div. ICAR, KAB, New Delhi. Pp. 190-213.

Narayana, V.V. Dhurva and Babu, R. 1983. Estimation of soil erosion in India. *J. Irri. Drain. Engng.* ASCE, 109: 419-434

Pal. B, Singh, A.K., Singh, P.K. and Singh, R.P. 1993. Effect of saline water on herb, oil yield and composition of palmarosa (*Cymbopogon martini* Var Motia). *Indian Perfumer*, 37: 267-269.

Pal, B., Singh, K.K., Jadaun, S.P.S. and Parmar, A.S. 1989. Effect of water salinity on herb and oil yield and composition of some *Cymbopogon species*. *Indian Perfumer*, 33: 196-199.

Patra, D.D., Anwar, M. and Singh, D.V. 1992. Herb yield essential, oil yield and nutrient accumulation by palmarosa under saline water irrigation. Abstract of technical paper presented at the conference on symposium on Environmental Soil Science, Canadian Society of Soil Science, August 8-13, Alberta, Canada, P. 36.

Patra, D.D. and Singh D.V. 1995. Utilization of salt affected soil and saline sodic irrigation water for cultivation of medicinal and aromatic palnts. *Current Res. on Med. and Arom. Pl.*, 17: 378-381.

Patra, D.D., Prasad, A., Anwar, M., Singh, D., Chand, S., Ram, B., Katiyar, R.S. and Kumar, S. 2002. Performance of lemongrass cultivars intercropped with chamomile under sodic soil with different levels of gypsum application. *Comm. Soil Sci. Pl. Analy.*, 33: 1707-1721.

Prakasa Rao, E.V.S. 1992. Fertilizer management in medicinal and aromatic crops. In: *Non traditional Sectors for fertilizer Use* (Eds Tandan, H.L.S). Fertilizer Development and Consultation organization, New Delhi, India. pp. 73-93.

Prakasha Rao, E.V.S. 1993. Nutrient management in some important tropical aromatic crops present and future. *Indian Perfumer*, 37: 35-39.

Prasad, A. and Chattopadhyay, A. 1999. Growth and yield of Japanese mint (*Mentha arvensis*) in acid sedentary and slightly alkaline soil. *Indian Perfumer*, 43: 9-14.

Prasad, A. and Singh, D.V. 1998. Effect of organic amendments on yield of palmarosa (*Cymbopogon martinii* Roxb. Wats) var Motia and chemical properties of sodic soil. *Indian perfumer*, 42: 21-26.

Ramamohan Rao, M.S. 1998. Approaches to management of denudated lands in peninsular, India. In, *Technologies for waste land development* (Eds Abrol, I.P. and Dhurva Narayana, V.V.). Pub. Information Div. ICAR, KAB, New Delhi. Pp. 232-244.

Rao, G.G., Nayak, A.K., Chinchmalatpure, A.R. and Khandelwal, M.K. 2001. Yield of dill (*Anethum graveolens*) on saline black soils of different unirrigated farm sites of Bhal area in Gujarat. *Indian Journal of Agricultural Sciences*, 71: 711-712.

Rao, G.G., Nayak, A.K., Chinchmalatpure, A.R. Singh, R. and Tyagi, N.K. 2001. Resource characterization and management options for salt affected black soils of agro-ecological regionV of Gujarat state. Pp. 48-53.

Sehgal, J.L., Saxena, R.K. and Pofali, R.M. 1998. Degraded soil-their mapping through soil surveys. In, *Technologies for waste land development* (Eds Abrol, I.P. and Dhurva Narayana, V.V.). Pub. Information Div. ICAR, KAB, New Delhi, pp. 1-20.

Sharma P.C., Mishra, B., Singh R.K., Singh, Y.P. 2001. Variability in the response of spinach, fenugreek and coriander to alkalinity and salinity stresses. *Indian Journal of Plant Physiology*, 6: 329-333.

Shiva, M.P. Lehri, A. and Shiva, A. 2002. Introduction, Calamus, Cardamom, Clove Kewada Mint and Vetiver. In. *Aromatic and Medicinal Plants*. Publisher, International Book Distributors, Dehra Dun, India. pp 67-307.

Singh, D.V. and Anwar, M. 1985. Effect of soil salinity on herb and oil yield and quality of some *Cymbopogons species. J. Indian Soc. Soil Sci.,* 33: 362-365.

Singh, R. 1999. Lavaniya jal aur mrda mai sounf ke kheti. Phal phool. Oct-Dec. pp. 30-35.

Singh, R., Bhargava, G.P., Singh, R. 1995. Response of safflower (*Carthamus tinctorius*) and dill (*Anethum graveolens*) to soil salinity. *Indian Journal of Agricultural Sciences*, 65: 442-444.

Singh, R.P. 1998. Land degradation problems and their management in semi-arid tropics. In, *Technologies for waste land development* (Eds Abrol, I.P. and Dhurva Narayana, V.V.). Pub. Information Div. ICAR, KAB, New Delhi. Pp. 125-136.

Valtcho, D. Z., Lyle E. C. and Baoshan, X. 2006. Effects of Cd, Pb, and Cu on growth and essential oil contents in dill, peppermint, and basil. *Environmental and Experimental Botany*, 58: 9-16.

Valtcho, D.Z., Lyle, E. C., Baoshan, X., Niels, E. N. and Andrew, W. 2008. Aromatic plant production on metal contaminated soils. *Science of the Total Environment*, 395: 51-62.

Verma, V. and Chinnamani, S. 1998. Rehabilitation of revine lands. In, *Technologies for waste land development* (Eds Abrol, I.P. and Dhurva Narayana, V.V.). Pub. Information Div. ICAR, KAB, New Delhi. Pp. 263-272.

Zheljazkov, V. and Jekov, D. 1995. Heavy metal content in some essential oils and plant extracts. In: *Proc. International Symposium on Medicinal and Aromatic held on 1 November 1995 at Amherst, USA, Published in Acta Horticulturae 2006*. Pp. 426.

30

Disease Management in Fruit Crops

S.K. Maheshwari, Hare Krishna and R.K. Sadh

Central Institute for Arid Horticulture, Bikaner, Rajasthan

Importance of fruits in human diet is well recognized since time immemorial. They are important part of a healthy diet as natural sources of energy, essential vitamins, minerals and antioxidants ever since the beginning of human civilization. Our country is bestowed with varied climatic conditions, which had made it possible to grow diverse group of fruits from its temperate to tropical, to subtropical and even to arid niches. The cultivation of fruit crops is, presently, one of the most important and profitable proposition. Growing fruit crops translate into realization of more yield/ income from a unit area of land than any of the agronomic crops. However, there is need to considerably increase the yield of fruit crops, particularly under changing environmental conditions as under such conditions diseases are getting more prevalent. Therefore by adoption of proper protection practices, yield of fruit crops can be improved.

Plant diseases caused by fungi, bacteria, viruses and phytoplasmas reduce crop yields (Khoury and Makkouk, 2010). Adoption and support for using participatory approaches help to farmers in disease management of fruit crops, reducing costs and improving production efficiency. Fungal, bacterial and viral diseases pose serious threat in the cultivation of fruit crops in India. To reduce the losses caused by plant disease, a complete knowledge of disease management strategies of fruit crops are essential which are as follows:

1. Aonla (*Emblica officinalis* Gaertn.)

i) Rust disease

Rust disease of *aonla* was first reported in Rajasthan (Tyagi, 1967). The disease caused considerable losses in major aonla growing areas of Uttar Pradesh (Rawal, 1993). The disease was prevalent in growing areas in Jobner (Jaipur), Ajmer, Jodhpur, Jhunjhunu and Nagaur districts as well as Bawal (Anon., 2010).

Symptoms

Reddish spots are formed on plant leaves from starting of August. Brown rust pustules are developed on the leaves. Black pustules develop on fruits, which later develop in a ring pattern. Many pustules coalesce together and cover a large fruit area in severe condition and results fruit dropping premature. The reddish brown uredo pustules on both surfaces of the leaflets can be seen.

Causal organism- *Revenelia emblicae*

Epidemiology

Disease development was more when relative humidity increased and the temperature decreased under Rajasthan conditions.

Disease management strategies

Cultural control

a)   Sanitation and clean growing of *aonla*.

b)   Pruning should be done properly in orchard for reducing humidity.

Chemical control

a)   Three sprays with dithane Z -78 (0.2%) at monthly interval during the months of July to September proves effective against this disease.

b)   This disease can also be controlled by wettable sulphur (0.25%) during July- September.

Host resistance

a)   Cultivar NA-6 was found to be resistant under Faizabad conditions (Anon. 2002).

C)   Cultivars Banarasi and Chakaiya are to be relatively free from disease.

ii) Anthracnose of *aonla*

Symptoms

It appears on leaflets and fruits in August-September. Initially, symptoms appear minute, circular, brown to gray spots with yellowish margin in leaflets. Severely infected leaves dried up. In fruits, pin lead like spots appear with dark brown to pink having red margin and yellow halo. On fruits, depressed lesions develop which later turn dark colour bearing dot like fruiting bodies. Infected fruits become shriveled and rot.

Causal organism- *Glomerella cingulata*

Epidemiology

The disease is favoured under hot and humid weather. This disease respond positively to increase in the temperature along with cloudy weather.

Disease management practices

Cultural control

(a)   Discard affected fruits and leaves at initial stage from the orchard.

(b)   Proper pruning should be done in orchard for air circulation.

(c)   Plant trees on a wide spacing and keep the surrounding area clear of vegetation.

Chemical control

Spraying of carbendazim (0.1%) or difolatan (0.2%) can reduce the disease (Bhardwaj and Sharma, 1999).

2. Ber : (*Ziziphus mauritiana* Lamk.)

Ber diseases like powdery mildew, leaf spots and fruit rots in arid and semi-arid region and post harvest diseases are economically important. Complete loss of ber fruits due to havoc nature of ber powdery mildew is being experienced in semi-arid and humid regions.

i) Powdery mildew

It is the most serious disease of ber and causes heavy losses in semi-arid and sub-tropical region but it is not serious in Bikaner and nearby regions from last few years due to unfavourable climatic conditions. Ber powdery mildew was first reported from Allahabad. Maximum disease incidence (38- 60%) was observed in local cultivars Kali, Bagwari whereas Seb was least susceptible.

Symptoms

Initial symptoms appear as white floury patches on young fruits at pea size and later cover the entire fruits. The infected fruits become misshapen, corky and finally are dropping. In severe incidence, floral parts, whole fruits, tender branches and leaves would appear with fungal conidia and less number of fruit setting and malformed fruits. *Ber* orchards are devastated completely when the powdery mildew occurs in severe form at flowering and fruit setting stages.

Causal organism- Oidium erysiphoides f. sp. *ziziphi*

Epidemiology

The cloudy, humid with moderate temperature favour its occurrence in ber orchards. In North India, infection appeared from October and reaching high by December. Maximum severity could occur at maximum temperature of 21.9°C, 59- 88 % relative humidity and 9.6 hours sunshine per day.

Disease management strategies

Cultural control

a)　Removal and destruction of infected leaves, twigs and collateral hosts from orchard.

b)　Ber orchards should be free from wild species of ber (*Ziziphus nummularia*).

c)　Summer ploughing in the orchard.

Chemical control

Karathane (Dinocap) at 0.1% dose resulted the best control of powdery mildew of ber (Singh *et al.*, 1995)

Biological control

a)　*Psendomonas fluorescens* CIAH- 196 (1%) was also moderately effective against this disease (Nallathambi, *et al.*, 2006).

Host resistance

Plantation of resistant varieties is the best method for reducing the disease. Use of resistant/tolerant varieties such as Sanur-2, Chinese, Jogia and Vikas must be grown.

ii) Black leaf spot

Disease incidence was noted during survey program of Rajasthan and Uttar Pradesh from 1.0- 8.0 per cent and 4.75 to 38.75%, respectively (Anon., 2010).

Symptoms

The disease is characterized by sooty tuft like circular to irregular black spots on lower surface of leaves. Leaves and twigs are dried rapidly under severe infection. Cloudy weather with moderate temperature in October- November is favourable for disease development.

Causal organism- Isariopsis indica var. *ziziphi*

Disease management strategies

Cultural

a) Removal and destruction of infected leaves and twigs from orchard.

b) To keep the ber orchard free from all weeds.

c) Pruning should be done properly in the orchard.

Chemical

Two fungicidal sprays of mancozeb or copper oxychloride @ 0.2% at 15 days interval were effective (Verma and Cheema, 1988).

Host resistance

Use of resistant varieties such as ZG-3, Seb, Bahadurgarhi and Safeda Rohtak.

iii. Alternaria *fruit rot: (C.o.-Alternaria* spp.*)*

Symptoms

It starts from pedicel and in bottom portion of fruits. Light brown to dark brown spots are produced on ber fruits. Mature fruits rot under severe conditions by brown blotching. Such fruits ultimately fall down resulting huge losses. Some fruits having concentric rings with a number of fungal spores. All the young fruits also fall down in severe case.

Epidemiology

Availability of high moisture, followed by warm weather conditions are most conducive for disease development. It is favoured between 20 to 30°C with an optimum temperature at 25°C.

Disease management strategies

Cultural control

a) Removal and destruction of infected leaves, twigs and weeds from orchard.

b) Pruning should

Chemical

a) Fungicidal sprays of mancozeb (0.2%) at 15 days interval was given to minimize the disease.

b) Mancozeb and copper oxychloride (0.2%) are also effective.

3. Date palm: *(Phoenix dactylifera L.)*

At Bikaner, survey of date palm diseases was carried out at farmers orchards like Bikaner, Bajju, Raiser, Khajuwala, Dantour, Nokha, Birmana and Fatehpur (Anon., 2010).

i) *Alternaria* leaf spot

It has been observed in severe form at Date Palm Research Centre, SKRAU, Bikaner and disease incidence was recorded from 14.16 to 72.50% depending upon cultivars as well as climatic conditions. Disease incidence was noticed in KVK, Fatehpur (Rajasthan) up to 45.50% (Anon., 2010).

Symptoms

The spots are most common on the lower leaves (pinnae) of the plants, whereas the spots are few and small size on upper leaves. The disease causes heavy losses to the date industries in both quality and quantity of production (Pal *et al.*, 2006).

Causal organism- Alternaria alternata

Epidemiology

It requires hot and humid weather conditions for disease initiation.

Disease management strategies

Cultural control

a) Removal and destruction of infected leaves, twigs and weeds from orchard.

b) Pruning should be done.

c) To disinfect all tools in pruning, etc. and cut surfaces.

Chemical control

a) Fungicidal sprays at 15 days interval (mancozeb @ 0.2%) was given to minimize the disease.

b)   Mancozeb and copper oxychloride (0.2%) are also useful for retarding the fungal growth.

Host resistant

Use of resistant varieties such as Medjool, Nagal Hilali, Sriganganagar and Medini (Pal *et al.*, 2006).

ii) Graphiola leaf spot

It is the most widely spread disease and occurs wherever the date palm is cultivated under humid conditions and partially irrigated areas.

Symptoms

This disease develops on both sides of the pinnae leaves, on rachis and leaf base. The numerous fruiting structures emerge as small-yellow/brown to black sori.

Causal organism- Graphiola phoenicis

Disease management strategies

Cultural control

a)   Removal and destruction of infected leaves, twigs and weeds from orchard.

b)   Pruning should be done in orchard.

c)   To disinfect all tools in pruning, etc. and cut surfaces.

Chemical control

Two sprays and soil drenching of copper oxychloride (0.2%)

Host resistant

i)   Grow Genetic tolerance varieties (Barhee, Rahman, Gizaz, Iteema, Khastawy and Tadala).

ii) Use of variety Hatemi was found to be resistant (Pundir, *et al.* 2006).

4. Pomegranate: (*Punica granatum*)

i) Bacterial leaf and fruit spot

This disease is prevalent in all the pomegranate growing states such as Maharastra, Karnataka, some parts of Gujarat of India.

Symptoms

Small and water soaked spots appeared on the leaves and became necrotic with dark brown. Many spots may coalesce and cover big area which are surrounded by chlorotic halo. Leaves are distorted and defoliated. Water soaked spots on fruits may turn dark brown. The spots are oily in appearance.

Causal organism- Xanthomonas axonopodis pv. *punicae*

Disease management strategies

Cultural control

a)   Removal of weed hosts and other infected plant parts.

b)   Pruning should be done in pomegranate orchard.

Chemical control

i)   Spraying of 200 ppm streptocycline can be controlled followed by pausamycin (0.05%) + copper oxychloride (0.2%) with 3 sprays at fortnightly intervals (Suriachandraselvan *et al.*, 1993).

Host resistant

Variety Jalore seedless was resistant to this disease (Nallathambi *et al*, 2006).

5. Bael: (*Aegle marmelos* Correa)

Alternaria Leaf spot

Alternaria leaf spot were reported by Madan and Gupta (1985). This disease was observed in Chomu area of Rajasthan (Anon., 2010).

Symptoms

The minute brown spots appear on the leaves which enlarge and become prominent with reddish brown lesions. These lesions later on coalesce into irregular patches. The affected leaves drop down finally.

Casual organism- Alternaia spp.

Disease management strategies

Cultural control

a)   Field sanitation

b)   Pruning of the plants in orchard.

Chemical control

Spraying of indofil M- 45 (0.2%)/ copper oxychloride (0.2%) at regular interval is very effective.

References

Anonymous. 2002. Epidemiological studies, In: Annual Report, All India Co-ordinated Research Project on Arid Zone Fruits. pp. 161.

Anonymous. 2010. Disease management, In: Annual Report, All India Co-ordinated Research Project on Arid Zone Fruits. pp. 135- 144.

Bhardwaj, S. S. and Sharma, I. M. 1999. Diseases of minor fruits. In: *Diseases of Horticultural Crops* (Eds. Verma, L. R. and Sharma, R. C.). Indus Publ. Comp., New Delhi, pp. 541-562.

Khoury, W. El. and Makkouk, K. 2010. Integrated plant disease management in developing countries. *J. Pl. Pathol.*, 92 (4): 35-42.

Madan, R. L. and Gupta, P. C. 1985. A new leaf spot of *bael* caused by *Alternaria alternata* (Fr.) Keissler. *Indian J. Plant Pathology*, 3: 239.

Nallathambi, P., Umamaheswari, C., Nagaraja, A. and Dhandar, D. G. 2006. Pomegranate diseases and management. Technical Bulletin, CIAH, Bikaner, pp.30.

Pal, V., Rathore, G. S. and Godara, S. L. 2006. Screening of genotypes of date palm against *Alternaria* leaf spot. *Indian Journal of Arid Horticulture*, 2: 62-63.

Pundir, J. P. S., Rathore, G. S., Naqvi, A. R. and Porwal, R. 2006. Date palm. In: *Advances in Arid Horticulture:* (Vol. II - Production technology of Arid and Semi- arid Fruits), International Book Distributing Co., Lukhnow.

Rawal, R. D. 1993. Fungal diseases of tropical fruits. In: *Advances in Horticulture* Vol. 3 *Fruit Crops*. (Eds. Chadha, K. L. and Pareek, O. P.), Malhotra Publ. House, New Delhi.

Singh, S. B., Maheshwari, S. K. and Singh, P. N. 1995. Field evaluation of fungitoxicants against powdery mildew of ber. *Annals of Plant Protection Sciences*, 3: 168- 169.

Suriachandraslevan, M., Jayasekhar, M. and Aubu, S. 1993. Chemical control of bacterial leaf spot and fruit spot of pomegranate. *South Indian Horticulture*, 41: 228- 229.

Tyagi, R. N. S. 1967. Morphological and taxonomical studies on the genus Revenelia Berk. Occurring in Rajasthan. Ph.D. Thesis, University of Rajasthan, Udaipur.

Verma, K. S. and Cheema, S. S. 1988. Chemical control trials against mouldy leaf spot of ber caused by *Isariopsis indica* var. *ziziphi. Plant Disease Research*, 3(1): 32-36.

31

Biointensive Integrated Pest Management for Horticultural Crops of Arid Region

Nisha Patel

ICAR-Central Arid Zone Research Institute, Jodhpur, Rajasthan–342003

Fruits with their delightful colours, flavours and tastes are nature's wonderful gift to mankind. Apart from being the most nutritious food in the form of easily digestible simple sugars, vitamins, minerals and many phyto-nutrients (Plant derived micronutrients), they provide unique health benefits due to their antioxidant properties and valuable fibre. Antioxidants help protect the human body from oxidant stress, diseases, and cancers, and also boost the immunity of the body thus developing the capacity to fight against various types of diseases. The insoluble dietary fiber in fruits is known to provide protection from conditions like chronic constipation, hemorrhoids, colon cancer, irritable bowel syndrome, and rectal fissures etc. Due to their unique nutrition-profile the importance of fruits in our diet cannot be over emphasised.

Insects are a major limiting factor in fruit production. The infestation of insects pests on fruit trees in orchards adversely affects the quality and yield of fruits. The size, shape, taste, attractiveness and palatability of fruit deterioriates, due to which the market price is reduced and the growers incur losses. Fruit growers resort to use of chemical pesticides under such conditions. However many reports suggest that although the global production and use of pesticides in agriculture is increasing every year, but the damage due to insects has not decreased rather it has also increased, indicating that the use of pesticides is not the panacea remedy for all insect related problems. Many scientists are of the opinion that over-reliance on the use of synthetic pesticides in crop protection programs around the world has resulted in disturbances to the environment, pest resurgence, pest resistance to pesticides, and lethal and sub-lethal effects on non target organisms especially beneficial flora and fauna, including humans. Farmers spend thousands of rupees on pesticides. The residues of pesticides in food and fruits is yet another cause of concern. Fruits which are considered to be wonderful for health are often laden with unacceptable levels of pesticide residues nulling their beneficial properties. Pesticides have been linked to many diseases especially cancer. All these issues related with use of pesticides has triggered rethinking on the methods of control.The recent examples of the results of indiscriminate pesticides use , consequent crop failures and suicide by farmers now calls for introspection in this matter.

However the scenario is not totally bleak and with a balanced, integrated and biointensive approach towards insect management, the population and damage due to harmful insects can be reduced and good yield and good quality of fruits can be obtained with less use of pesticides and without adversely affecting the the environment. Pest management is an ecological matter. The size of a pest population and the damage it cause reflects the planning and management of a particular agricultural ecosystem. Insect ecology means, the relationship between insects and their environment. The environment (including weather, food sources, and natural enemies) determines whether or not an insect population causes economic damage. Knowledge about ecology enables growers to make better decisions about managing pests. The management of insects needs to be done without affecting the environment and non target organisms. Integrated Pest Management (IPM) is a comprehensive approach for managing pests that uses multiple tactics to reduce the status of pests to tolerable levels while maintaining a quality environment. IPM is not a single pest control method but, rather, a series of pest management evaluations, decisions and controls.

The four important steps of IPM are for pest infestation includes setting action threshold, monitoring and identifying pests, prevention and control.

Set Action Thresholds : Before taking any pest control action, IPM first sets an action threshold, a point at which pest populations or environmental conditions indicate that pest control action must be taken. Sighting a single pest does not always mean control is needed. The level at which pests will either become an economic threat is critical to guide future pest control decisions.

Monitor and Identify Pests: Not all insects, weeds, and other living organisms require control. Many organisms are harmless, and some are even beneficial. IPM programs work to monitor for pests and identify them accurately, so that appropriate control decisions can be made in conjunction with action thresholds. This monitoring and identification removes the possibility of unnecessary use of that pesticides or use of wrong kind of pesticide.

Prevention

Using cultural methods, such as rotating between different crops, selecting pest-resistant varieties, and planting pest-free rootstock. These control methods can be very effective, cost-efficient, and present little to no risk to people or the environment.

Control: Once monitoring, identification, and action thresholds indicate that pest control is required, and preventive methods are no longer effective or available, IPM programs then evaluate the proper control method both for effectiveness and risk. Effective, less risky pest controls are chosen first, such as pheromones to disrupt pest mating, or mechanical control, such as trapping or weeding. If further monitoring, identifications and action thresholds indicate that less risky controls are not working, then additional pest control methods would be employed, such as targeted spraying of pesticides. Broadcast spraying of non-specific pesticides is a last resort.

Insect pests

Insects can cause damage to fruit plants or fruits in several ways. Insects can harm fruit plants by eating leaves i.e. defoliation, by sucking plant juices or by feeding inside the stem fruit, roots or leaves. However, not all insects on a plant are pests or all insect -feeding reduces yield. All plants have a certain amount of compensatory growth due to which plants can tolerate a lot of insect feeding because a plant compensates for the loss caused by an insect by producing more leaves and roots. In the arid region ber or Indian Jujube *Ziziphus mauritiana* Lam., Pomegaranate, citrus fruits and aonla Indian gosseberry) (*Emblica officinalis* G.) are the commonly grown fruits. Many species of insect have been recorded on these fruit crops. However, only few species have attained the pest status and caused substantial economic damage. Fruit flies, fruit borers, stem borers bark eaters, mealy bugs, caterpillars, whiteflies and aphids are the insect pests of major importance in fruit crops of the arid region.

Fruit flies Diptera, Tephritidae

The fruit flies such as *Carpomyia vesuviana* , *Bactrocera dorsalis* (Hendel) etc are one of the most destructive pests of fruits. These flies look like house flies but are smaller in size. The body color is variable from brown to black but usually there are prominant yellow and dark brown to black markings on the thorax. Generally, the abdomen has two horizontal black stripes and a longitudinal median which may form a T-shaped pattern. The wings are clear. The female has a pointed slender ovipositor to deposit eggs under the skin of host fruit. The adult is a strong flyer and can fly long distances. This ability allows the fly to infest new areas very quickly.

The female lays eggs under the rind of the fruits by puncturing. After hatching the caterpillars feed on the pulp. Fruit that has been attacked may be unfit to eat as larvae tunnel through the flesh as they feed. Fungi and bacteria enter through these holes, leading to the rotting of the fruit. This makes the fruit unfit for consumption.The greatest enemies of the ber/ jujube in India are fruit flies, *Carpomyia vesuviana*. The adult of this pest is small fly with black spots on the thorax and dark bands on the wings. Some cultivars are more susceptible than others. Larger sized and sweeter fruits are preferred more sometimes all fruits of such varieties may be damaged by fruit flies, while trees with smaller, less-sweet type fruits may have only negligible damage.

Life cycle: Females lay eggs 3 to 30 under the skin of host fruits; the female can lay more than 1,000 eggs in her lifetime. Time taken for development depends on the ambient temperature. Eggs are minute cylinders laid in batches. The maggots (larvae) are creamy-white, legless, and may attain a length of 10 mm inside host fruit. Maggots tunnel through the fruit feeding on the pulp, shed their skins twice, and emerge through exit holes in approximately 10 days. The larvae drop from the fruit and burrow two three cm into the soil to pupate. In 10 to 12 days, adults emerge from these puparia. The newly emerged adult females need eight to 12 days to mature sexually prior to egg laying. Breeding is continuous, with several annual generations. Adults live 90 days on the average and feed on honeydew, decaying fruit, plant nectar, bird dung and other substances.

Treatment: Fruit fly larvae pupate in the soil and it has been found that treatment of the ground beneath the tree helps reduce the problem. Raking of of the orchard soil during summer and rainy season destroys the hibernating pupae by exposing them to bright sunlight and birds and thus the extent of infestation is considerably reduced. Infested fruits having larvae should also be collected and destroyed.

In ber some varieties like Ilaichi, Mirchia, ZG-3 and 'Umran' are moderate to resistant to fruit fly damage. Early-maturing cultivars ('Gola', 'Jogia', 'Mundia' and 'Seb') attract higher fruit-fly infestation compared to those maturing late. The pest can be effectively controlled by two sprays of endosulfan or malathion. To prevent infestation, prophylactic sprays can be carried out with 0.03% oxydemeton or dimethoate starting from the stage when 70-80% fruits attain pea size and then repeating the spray at one month intervals. During the maturity of fruits, if necessary, spray with 0.5% Malathion+0.05% sugar solution at weekly intervals. Malathion has been observed to dissipate quickly in ber fruits decreasing well below the tolerance level of 3 ppm within two days.

Traps containing 1 ml methyl eugenol and 2 ml malathion/ litre of water can be used for the control of other fruit flies. Traps made of plastic bottles with holes big enough for flies to enter, are hung at 10 traps per hectare. Solution (100 ml) of methyl eugenol and malathion is put in each trap. Better control is obtained if all orchards of a particular area take up control measures collectively.

Bark-eating Caterpillar (*Inderbela* sp.): It damages stem and branches of grown up trees by eating bark and making tunnel into them. This insect is more common in neglected orchards. Several holes can be seen on the trunk and the trees lose productivity. Wood dust and faecal matter hanging in the form of a web around the affected portion is indication of the borer activity. Adult moths are active during May- June. Female moths deposit eggs under loose bark which hatch in 8-10 days. These larvae eat the bark and bore inside the tree. Larva remains hidden inside the tree during day time and becomes active at night. Severe infection reduces growth and productivity of the tree. The branches become weak and break from the point of attack by the larva.

Control: The webs around the affected portion should be cleaned. Cotton swab soaked in petrol or kerosene should be inserted in the holes and sealed with mud during September-October and February-March. Alternate sprays with Carbaryl (2.5 g/litre of water) or Quinalphos (2 ml/litre of water) or Methomyl (3.5 g /litre of water) is effective in controlling the pest.

Mealy bugs (Homoptera: Pseudococcidae)

Mealybugs (Homoptera: Pseudococcidae) are polyphagous insect pests which cause severe losses in fruits trees. Mealybugs are small oval, wingless, soft-bodied sucking insects which appear like cotton. Most females can move slowly and are covered with whitish, mealy or cottony wax due to which it is difficult to eradicate them. The bugs lay eggs in the soil . When their population increases they form colonies on stems and leaves developing into dense, waxy, white masses.

Damage

Due to mealybug infestation the productivity of fruit trees declines considerably. Adult mealybugs feed on leaves, stems and fruits. They suck a large amount of sap from the plants with the help of piercing/sucking mouth parts, depriving it of essential nutrients.The excess sap is excreted as honeydew which attracts ants and develops sooty mould inhibiting the plant's ability to manufacture food.

Infested fruits have uneven shapes, poor quality, and are susceptible to secondary infections by pathogens. High populations decrease plant growth and cause premature leaf or fruit drop.

Managemant

Management of adult mealy bugs is difficult because of its habit of hiding in crevices, and the waxy covering of its body. Unlike the adults, the crawlers are free from waxy coating and therefore the crawler stage is the more easily controlled by plant protection measures spraying pesticides. Successful control of mealy bug problem is possible with an integrated approach. For management of these pests, plants in the vicinity of the orchard serving as the alternate hosts for the mealy bugs should be destroyed. Ploughing of the orchard during November-December destroys the eggs in the soil. Flooding of orchard with water and raking of soil around tree trunk exposes the eggs to sun and natural enemies thereby destroying them.

Banding of tree trunk should be done with polythene sheets (400 gauge) 30 cm above ground level and just below the junction of branching to obstruct the ascent of the nymphs. Banding should be done well in advance before the hatching of eggs, i.e., around November - December. Pasting a grease band of 5cm width on the main stem prevents the crawlers from reaching the bunch. Early instar nymphs of the mealy bug can be controlled by spraying of 0.05 % carbaryl from January to March.Spraying of insecticides like Dichlorvos (0.02%), methyl demeton 25 EC 2 ml/l, chlorpyriphos 20 EC 2 ml/l, imidacloprid 200 SL 1ml/l or malathion 2.5ml/l of water at 15 days intervals has been found to control the insect population.

Application of entomopathogenous fungi viz., *Beauveria bassiana* (2g/L) or 5% NSKE around the tree trunk has also been found effective in reducing mealybug population. Encouraging biodiversity and beneficial insects in the orchards helps in increasing the natural predators such as ladybird populations in the orchard which helps in keeping the population of mealy bugs under control.

It is necessary to locate and destroy ant colonies with drenching of chlorpyriphos 20 EC @ 2.5 ml/l or apply 5% malathion dust @ 25 kg/ha As the ants provide the mealy bugs protection from parasitoids and predators and also help in spreading the crawlers to non-infested plants.

Recently papaya mealy bug has created havoc in Tamil Nadu especially in papaya orchard. Because of its damage, papaya production came down by 60 – 80 per cent. Chemical pesticides could provide only partial and temporary relief and farmers went for repeated sprayings of pesticides. The NBAIR, Bangalore and TNAU, Coimbatore worked with three exotic papaya mealy bug parasitoids, standardized the mass production techniques of these parasitoids, and distributed them to farmers for field release, free of cost. All these parasitoids proved highly successful in Tamil Nadu. These parasitoids especially *Acerophagus papayae*, established very well and brought significant control of papaya mealy bug not only on papaya, but also on other crops.

Fruit borers: Different types of caterpillars infest and destroy fruits. In Pomegranate the caterpillars of Pomegranate Butterfly (*Virachola isocrates*) is the most widespread, polyphagous and destructive pest of pomegranate. The adult female lays eggs on buds, flower or on young fruits. On hatching, the caterpillar bores into fruit and feed on the pulp. The infested fruit rots and drop off from the tree. The characteristic symptom is the bad smell and excreta of caterpillars coming out of the entry holes ultimately leading to fruit rot.

Management practices: The affected fruits should be collected and destroyed. Before maturity, the fruits should be bagged with butter paper. NSKE 5% or neem formulations 2 ml/1 should be sprayed at flowering stage. From the stage of flowering to fruit development, regular sprays of malathion 50 EC 0.1% or methomyl 40 SP @ 1.0 ml/l or azadirachtin 1500 ppm @ 3.0 ml/l should be done till harvesting subjected to the presence of fruit borer.

In ber, the caterpillars of *Meridarches scyrodes* feed on the ripe fruit and make it unfit for human consumption. The adult of this pest is small moth, dark brown in colour with fringed wings. Early instar larva is light yellowish and full-grown larva is red in colour. The larva bores into the fruits and feeds on the pulp around the seeds. The maggots bore into the ripe fruits and feed on the pulp. When full-grown, they come out of the fruit and drop to ground and pupate 6-15 cm below the soil. Adult emerges from the soil. For controlling the fruit borer effectively, the attacked fallen fruits should be collected regularly and destroyed. Wild ber trees present around or near to the orchard should be removed. Soil under the tree or near the trees should be raked to destroy the maggots and pupae present in the soil. Insecticides fenvelrate 0.4 per cent or chlorpyriphos 1.5 per cent dust should be mixed at 40 kg per hectare in the soil under the tree or near the trees to reduce the fruit borer incidence. Spray the crop three times, with endosulfan at 2 ml or malathion at 2 ml or fevelerate at 0.5 ml or carbaryl at 4 g /lit of water or neem seed kernel extract (NSKE) 5 per cent; first spray at marble stage, second spray at 15 days later and third spray at fruit ripening stage, by alternate use of insecticides. For the third spray add 10 g jaggery/ litre of spray mixture.

Whitefly

Adult females lay eggs on the lower surface of apical leaves. Eggs hatch after a week. The crawlers dig their mouth parts into the leaf tissue for sucking the sap and remain static as "scales" throughout the remaining part of their larval and pupal period. Serious damage is caused by the excretion of honeydew secreted by the by whitefly, which runs down to the fruit and the upper surface of leaves. Under moist conditions, sooty molds can develop on the honeydew, reducing photosynthesis and hindering respiration of plants. The damage by whitefly also leads to yellowing of leaves and stunted growth, in severe cases leading to shedding of leaves.

Management: White flies can be trapped by hanging bright yellow sticky traps coated with polybutene adhesive at the height of the crop canopy. Spraying water with high volume sprayer by focussing the nozzle towards the under surface of leaves helps in washing out the honeydew, eggs, larvae, pupae and adult whitefly. This should be followed by spraying Triazophos 40 EC (1.5 ml/ litre of water) or a mixture of 1.5 ml of Monocrotophos 36SL + 1.0 ml of Dichlorvos 76 EC per litre of water. The sprays are repeated at an interval of 8-10 days.

Aphids : Various species of aphids cause losses in fruit crops. Aphids have soft pear-shaped bodies with long legs and may be green, yellow, brown, red, or black depending on the species and the plants they feed on Aphids are have long slender mouthparts that they use to pierce stems, leaves, and other tender plant parts and suck out fluids. Most species have a pair of tubelike structures called cornicles projecting backward out of the hind end of their body.Almost every plant has one or more aphid species that occasionally feed on it. Many aphid species are difficult to distinguish from one another; however, management of most aphid species is similar.

Damage : They suck the cell sap from the lower surface of the leaves and devitalize the plant. They secrete sweet sticky substance, which attracts fungal growth. The affected leaves show chlorotic patches. High humidity favours the multiplication of aphids. Aphids may also transmit plant diseases. The viruses cause yellowing or , curling of leaves and stunt plant growth.

Management : Regular monitoring for aphids at least twice a week when plants are growing rapidly is important so that infestation can be caught early. Many aphid species prefer the underside of leaves, so turn leaves over when checking for aphids. On trees, clip off leaves from several areas of the tree. Also check for evidence of natural enemies such as lady beetles, lacewings, syrphid fly larvae, and the mummified skins of parasitized aphids. Look for disease-killed aphids as well; they may appear off color, bloated, flattened, or fuzzy. Substantial numbers of any of these natural control factors can mean the aphid population may be reduced rapidly without the need for treatment. Managing ants is a key component of aphid management Ants are often associated with aphid populations, especially on trees and shrubs, and frequently are a clue that an aphid infestation is present. If you see large numbers of ants climbing your tree trunks, check higher up the tree for aphids or other honeydew-producing insects that might be on limbs and leaves. To protect their food source, ants ward off many predators and parasites of aphids

Biological : Control Natural enemies can be very important for controlling aphids, especially in gardens not sprayed with broad-spectrum pesticides (e.g., organophosphates, carbamates, and pyrethroids) that kill natural enemy species as well as pests. Usually natural enemy populations don't appear in significant numbers until aphids begin to be numerous.Many predators also feed on aphids. The most well known are lady beetles adults and larvae, lacewing larvae, and syrphid flies larvae. Naturally occurring predators work best, especially in garden and landscape situations. Creating favourable conditions for these predators by enhancing plant diversity is an important tool in reducing population of this pest.

Chemical Control : Spraying with Dimethoate (0.03%) or Malathion (0.1%) at 15 days interval effectively controls the aphid population.

Scale insects : The scale insects can be identified by presence of small black swollen spots on the branches sand the fruits. Adults and pupa suck the cell sap from the fruit and tender shoots causing drying of branches. In case of severe infestation, the whole tree dries up. The insects secret honey dew like substance which attracts black sooty mould. As a result, all the leaves and the branches turn blackish affecting the growth of the plant.

Control : Removal and destruction of alternate hosts, which harbor the scale insects. Spraying the affected patches with Rogar (0.1%) or Quinalphos (0.06%) at 15 days interval helps to control the pest.

Beneficial insects

Contrary to popular perception that all insects are harmful and cause damage to our plants animals and to human beings, many of the insects are beneficial and aid in food production by way of pollination and controlling insects which are pests. All insects that are on the plant are not are "pests". Actually, some insects are needed to keep the natural enemy population alive. Natural enemies (predators, parasitoids, pathogens and nematodes) reduce pest insect populations. Natural enemies either can live naturally in and around the field, or be reared in large numbers to be released into the field. Rearing and releasing natural enemies is becoming an important IPM option in many crops.

Pollination

Pollinators are essential for fruit and seed production in many vegetables and fruits crops. For human nutrition, the benefits of pollination include not just abundance of fruits, nuts and seeds, but also their variety and quality.

Among insects bees (Hymenoptera: Apidae) are the most important and effective pollinators, but other insects such as wasps, flies, moths, butterflies, and beetles also have a major contribution as pollinating species. Vertebrate such as bats, birds, hummingbirds and rodents, squirrels birds are also important pollinators. In an agro- ecosystem with diversity of flowering plants, the there is abundance and diversity of pollinators too. At the organic farm in CAZRI there was a large diversity of pollinators which included different types of honeybees wasp, syrphids and many other dipterans and hymenopterans due to different types of flowering plants.

Natural enemies

The use of natural enemies to maintain pest populations below damaging levels is known as Biological control. Natural enemies of insects fall into three major categories: predators, parasitoids, and pathogens. Predators catch and eat their prey. Some common predatory arthropods include ladybird beetles, lacewings, syrphid flies, carabid (ground) beetles, big-eyed bugs, and spiders. Parasitoids (sometimes called parasites) do not usually eat their hosts directly. Adult parasitoids lay their eggs in, on, or near their host insect. When the eggs hatch, the immature parasitoids use the host as food. Pathogens are organisms that cause diseases in insects. The main groups of insect disease-causing organisms are insect-parasitic bacteria, fungi, protozoa, viruses, and nematodes. The bacterium *Bacillus thuringiensis* (Bt) is a well known microbial control agent that is available commercially. Several insect-pathogenic fungi are used as microbial control agents, including *Beauveria, Metarhizium,* and *Paecilomyces*

Enhancing biodiversity

The larval or young stages of many beneficial insects feed on other insects but the adults of most of the predators and parasitoids of pest insects require nectar and pollen at certain stages of their life for growth and reproduction. A diversified ecosystem with many types of trees, shrubs and small plants provides microhabitats, food sources (prey, nectar, pollen), alternative hosts and shelter for antural enemies and thus encourages colonization and population build up of natural enemies. The plant species chosen for this purpose should be such that they themselves do not attract insect pest of the crops chose for the farm and also the flowering should be in succession so that food is available for beneficial round the year.

Plants that have flowers with a good amount of nectar/ pollen, like Umbelliferae, *Compositae*, and *Brassicaceae* are especially attractive for insects. flowers which are attractive in colours, small shallow flowers extrafloral nectaries promote or encourage beneficial insects on the farm. Senna plant in the fields also supported parasitoid population as they have extrafloral nectaries which provided food to beneficial insect during off season.

Shrubs

Under arid conditions many plants like ber and henna in the organic farm provide pollen and ectar to variety of pollinators, predators and parasitoids besides providing shelter for insects like lacewing, many lacewing eggs were seen on henna and ber plants. Predatory wasps were observed in good numbers on ber flowers. which are encourage beneficial insects like coccinellids syrphids are *Calotropis* shrub was left on the farmand syrphids were observed on it. Moreover *Calotropis* and henna also supported a population of prey insects aphids and whiteflies which provided food to predators when the crops were not in field. *Acacia senegal* trees was also a good source of pollen and nectar and it attracted supported many types of pollinators and parasites, in the orchard.

Soil management

Soil management and plant health have a crucial role in prevention of insect damage. Healthier soils produce plants that are less damaged by pests. Practices that promote soil health help in keeping plants healthy and it can better express its inherent abilities to resist pests. Some soil-management practices boost plant-defense mechanisms, making plants more resistant and/or less attractive to pests. Healthier soils also harbor more diverse and active populations of the soil organisms that compete with, antagonize and ultimately reduce soil-borne pests. Application of nutrients only through organic source of fertilizers such as farm yard manure and oilseed cakes has long-term benefits in terms of building up of soil organic matter which favor multiplication of microorganisms besides improving the physical properties of soil. Oilseed cakes are a source of valuable major and micro-nutrients essential for optimal crop growth and yield and enrich the arid zone soils which are highly deficient in micronutrients. The utility of neem oil seed cake as a fertilizer as well as a pesticide for plants.

Precautions

The waiting period needs to be strictly observed so that the fruit is free from harmful residues.the chemical pesticides also affect the beneficial insects pollinators predators and prasitoids so spraying should be done in the morning and evening hours when these insects are not active.

32

Scientific Management and Feeding of Calf and Heifer

Subhash Kachhawaha, Dheeraj Singh and Basant Kumar Mathur

ICAR-Central Arid Zone Research Institute, KVK, Pali, Rajasthan–306401

Heifers are future cows of the herd. Production and reproduction performance of cows depends on the care and management during calf and heifer stage. Rearing period of heifers can be divided into two stages, viz. from weaning to first service and service to calving.

Care of calf before birth

During the last 2-3 months of gestation the fetus grows rapidly. Calf draws a significant quantity of nutrients from the mothers system. It has to be replenished by additional feeding of the cow. A cow should be fed well-balanced ration for maintenance.

- Cow, growth of foetus, production of colostrums with high nutritive value and sufficient body reserves of nutrients for ensuing lactation.

Care of calf at births

At the time of birth the cow does neat job of licking the calf dry and stimulate its circulation and respiration with hard licking. Some primiparous cows may be nervous and inexperienced. Under such circumstances, the husbandman should assist by removing the phlegm (mucus) from the nostril of the new born calf and wipe it dry with a clean towel. If respiration is not stimulated than light jolt may be given or by pouring cold water on the calf's head, which causes the grasping reflex in the calf. Examine the calf for injuries and birth defects.

Importance of colostrums

Colostrums contains large amount of gamma globulins which are nothing but anti-bodies gives passive immunity. It has 7 times more protein and twice the total solid of normal. Time of feeding colostrums is equally important to the quality. The Ig must be absorbed as such across its intestinal wall. Ig passes across the gut wall at the most rapid rate during the first 1-2 hours of life. So, it will be highly useful to feed colostrum in the first 15-30 minutes followed by a second dose in approximately 10-12 hours.

Newborn calves need

A. 2 litres in first 6 hours

B. 4 litres in first 24 hours

Inadequate colostrum intake in first 24 hours results in calves that are

*   9 times more likely to get sick

*   5 times more likely to die

Colostrum should be stored

*   In refrigerator for 7 days

*   Frozen for weeks/months

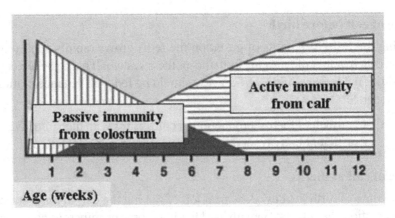

Fig. 1: Immunity level at different age in calf

*   In small containers (1-2 litres), Store colostrum for each cow separately.

Table 1: Relationship between chest size and weight of HF breed animals

S. No.	Chest size (cm)	Approx. Weight (kg)
1.	70	34
2.	80	49
3.	90	68
4.	95	85
5.	100	90
6.	110	117
7.	120	149
8.	130	186
9.	140	228
10.	150	275
11.	160	329
12.	170	389
13.	180	455
14.	190	528
15.	200	609

Table 2: Gamma globulin concentration in blood serum of calves in relation to the time of feeding colostrums after birth.

Estimate	Time of Colostrum feeding after birth (min.)	15 Min	2 Hrs	4 Hrs	3rd Day	30th Day
Gamma globulin as	15	13.65	17.13	17.04	20.92	25.20
% of total	120	16.78	16.39	15.66	32.52	17.75
	240	16.94	8.75	17.80	16.53	13.83

Table 3: Types of colostral immunoglubulins

S.No.	Ig class	Proportion of total (%)	Function
1.	IgG	85-90	Systemic immunity
2.	IgM	7	Early immunity and prevention of septicemia
3.	IgA	5	Not clear

Fig. 2: Determination of Colostrum Quality by instrument

- The colostrometer classifies colostrum as poor (red) with less than 22 mg Ig/mL,
- Moderate (yellow) with 22-50 mg Ig/mL
- Excellent (green) with greater than 50 mg Ig/mL.

Quantity of colostrum & milk

- Quantity of colostrum to be fed to calves depends on their body weight
- Indian recommendations- 1/10 of body weight for five days,
- Feeding is the most important factor in the management of young calf
- The calves are fed whole milk @ 1/10th of body weight along with ad libitum legume hay and calf starter
- The calves are weighed every week and the quantity of milk to be fed is adjusted accordingly
- Leguminous hay like Lucerne hay or berseem hay is made available ad libitum from the 4th day onward.

First dry feed to calves

- A good quality calf starter should be started from 2nd week of age
- Containing 75% TDN and 14-16% DCP
- Rumen development takes place rapidly between 4 and 8 weeks of age
- Starters are more important than forages as a source of fermentable carbohydrate for rumen development,
- Production of volatile fatty acids, particularly propionic and butyric acid in the rumen stimulates development of the rumen and the reticulorumen,
- Grain starter should be offered as early as four days after birth and should continue until about four months of age
- Feeding hay encourages early rumen development which will mean that the calf can be fed as a ruminant from an early stage

Fig. 3: Ruminal Papillae of a calf fed grain and grain starter

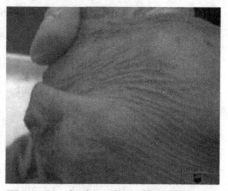

Fig. 4: Ruminal papillae of a calf fed all + calf starter

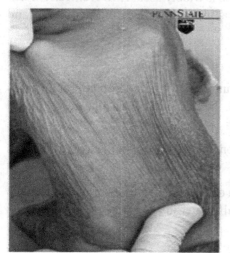

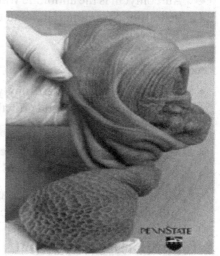

Fig. 5: Mialk and hay

Table 4: Schedule of feeding calves raised with calf starter

Age Days	Whole milk (kg)	Skim milk (kgs)	Calf starter(kg)
0-5	Colostrum feeding		
6-7	2	75	----
8-14	3	25	----
15-21	2.75	1.0	0.1
22-28	1.75	2.0	0.2
29-34	1.0	3.0	0.3
35-42	0.5	3.5	0.5
43-56	-	3.5	0.75
57-84	-	2.5	1.0
85-112	-	0.5	1.25
113-140	-	-	1.75
141-182	-	-	2.0

Advantages of Antibiotic feeding to calves in calf starter/milk replacer:

- Increase feed efficiency
- Decrease their vitamin B12 requirement
- Exert a protein sparing effect
- increase livability
- reduce incidence of calf scours and other disease,
- reduction in calf mortality rate
- more rapid growth by 15-20 percent
- Aureomycin is the antibiotic which is widely used. However, Terramycin Pencillin, Streptomycin, Bacitracin, Chloromycetin, Neomycin and Tetracycline etc.

Feeding milk and milk replacer

- Until the rumen can start supplying energy and microbial protein sufficient for maintenance and growth, the calf must have a high quality liquid milk or milk replacer diet.
- Milk Replacer with a lower fat content of 2 percent will have to be fed @ 2.5 – 3 liters in each feeding (5 – 6 liters / day).
- While feeding too little milk at an early stage depresses growth, too much milk for longtime will depress calf starter intake.

Adequate Housing is Important

- Keep calves in individual pens until they reach weaning age.
- Separate pens prevent the calves from suckling one another and reduce the spread of calf disease.
- Calf pens must be clean, dry, and properly ventilated.
- Ventilation should provide- fresh air at all times without drafts blowing directly on the calves.
- Calf housing should be bedded to keep the calves comfortable and dry.
- Sawdust or straw is most commonly used.
- Outdoor calf pens must be partially covered and walled to prevent excessive heat caused by the sun and to guard against cold winter rains and wind.

- Pens open to the east gain warmth from the morning sun and provide shade during the warmer parts of the day. Rain seldom falls from the east.

Heifer Management

A. Feeding of heifers

- Heifer is growing animals and so the requirements for growth is of higher order

- During early stage relatively more protein is required than energy

- Most young heifers grow well if excellent hay is given as much as they eat

- No grains need to be fed after the calf is 9 months

- Ad libitum green fodder should be fed so that the animal gets enough carotene

- If leguminous fodders are fed it gives enough calcium, and other minerals

- An extra amount of 1.25 to 1.75 kg concentrate may be provided to pregnant heifers to allow the growth of the foetus normally

- 6 weeks before calving 2-3 kgs of concentrates should be given

- Laxative -feeding should be given from two weeks before calving

B. Breeding of heifers

- The size and age of the heifers should be the basis for giving first service

- Poor conformation and poor growth are associated with poor fertility

- The mean age at first calving (month) in UK(26),USA(27) and Italy (28)

- At time of first service animals should have achieved 55-60 per cent of their mature body weight

- Target weight should be 350 kg - 380 kg at time of breeding

- Optimum growth rate should be between 0.6 kg - 0.8 kg/ day

- Inseminations are carried out within 12 hours of standing heat.

Table 5: Weight of indigenous and crossbred animals at different stages

Breed	Breedable weight (kg) and Age (months)	Weight (kg) and Age (months) at calving
Kankrej/Gir	230-250 (24-28)	300-350 (36-40)
Jersey cross	225-230 (15-18)	300-350 (26-28)
HF cross	270-280 (18-20)	350-400 (28-30)
Meshani/Surti	240-250 (24-30)	350-400 (36-40)

Table 6: Effect of body weight on calving rate (Friesian heifers)

Body weight (kg)	% calving to first serve	% not in calf
Less than 260	21	33
260-299	47	13
300-339	59	08
More than 340	57	07

Breeding Management

- The normal variation in the period of oestrus is 18-24 hours. The duration of heat is shorter in zebu.

- The heifers in heat should be identified and checked routinely.

- The heifers should be observed at least 3 times a day for signs of behavioural heat; one should observe specially early in the morning.

- The heifer in heat will be awake and on feet quite ahead of other animals.

- Standing heat is the best sign of heat in which the heifer will stand and allow animals to mount her.

- Producers should consider breeding heifers 15 to 20 days before the cow herd.

- This practice has several advantages.

- It permits more time and labour to be given to the heifers during the calving season.

- Heifers can be watched more closely and delivery assisted if necessary.

- Heifers should conceive earlier for their second calf, and conception rates should be higher in a short breeding season.

- Calves will be older and heavier at weaning.

- The disadvantages are that it lengthens the calving season and requires more feed after calving.

- Calving difficulty (dystocia) is of great concern to producers with first-calf heifers

- The major causes of dystocia in first calf heifers are an oversized calf or an undersized heifer

- A large calf and/or a heifer with a small birth canal (pelvic size) cause difficult deliveries

- Two practices can reduce calving difficulty

- One practice is to grow-out and develop heifers to be larger at calving.

- If heifers weigh 80% to 85% of their mature weight at calving, they should have less dystocia

- Pelvic area of heifers can also be measured, and heifers with small areas culled

- Since the heritability of pelvic area is about 50%, selecting heifers and bulls on pelvic size can be beneficial

- The second practice is to reduce calf size or birth weight

- Calf birth weight is the major cause of calving difficulty

- Since birth weight is 40% heritable, selecting bulls that sire calves with moderate birth weights can be effective in reducing difficulty

- Birth weight information on a bull and his sire can also be helpful

- Producers must be careful in selecting bulls if no prior calving information is known

- Bulls should be selected for moderate birth weights adjusted for dam age, and selected against large shoulders, bones, and head; thick muscling; and large hindquarters

- Along with better feeding, heifers should be protected from thermal stress

- By showering or splashing cold water on animals

- By providing cooled drinking water

- By protecting under shade during day and keeping in the open at night

- Housing heifer in loose house is found to be most efficient in all seasons

- Heifers should be provided 2.5 to 3 sq.m of covered floor space

- Introduction of heifers to the routines of the milking parlour is also important

- It may be a good idea to let pregnant heifers in later stage into the milking parlour and go through the routine

- This will give an opportunity for them to get adapted to the milking routine

33

Agri-Entrepreneurship
A New Avenue for Agriculturist

*Dheeraj Singh, Chandan Kumar, M.K. Chaudhary, M.L. Meena
and A.K. Shukla*

ICAR-Central Arid Zone Research Institute, KVK, Pali, Rajasthan–306401

Over 85% of the rural population in India is dependent on agriculture for their livelihood. About 50% of them being poor, most of their earnings are spent on meeting their basic needs, particularly food. With the increasing population over the last five decades, their per capita share of land and water resources has reduced substantially. As a result, rural people are faced with the problem of unemployment.

Inspite of agriculture being a major source of livelihood, the productivity as well as profitability in agriculture have been significantly low in the country. While the productivity can be attributed to illiteracy, lack of awareness, poor dissemination of technology, inadequate investment in agricultural inputs and poor communication and information services, lack of profitability is mainly due to inadequate and inefficient infrastructure required for forward and backward integration, poor post harvest and processing facilities and poor connectivity with market, resulting in exploitation by large number of middlemen. Traditionally, the farmers have been dependent on the Agricultural Extension Agencies of the State Government for information, input supply and marketing services. Over the years, these agencies are not being able to cope up with the growing responsibilities and specific needs of the farmers. To overcome this problem, Farmers' Cooperatives have been promoted to supply various agricultural inputs and organise the marketing of farm produce. Although, this was an excellent concept, most of these cooperatives could not carry out the work economically and efficiently due to lack of commitment from the elected leaders and unfair trade practices by the competing traders. The farmers, not realising the inability of the extension agencies, are still dependent on the Government and external agencies to help them in managing their agri-business.

Indeed, a major problem of Indian farmers is that, as agriculture is considered as a family tradition, a majority of the farmers continue to practise what their forefathers or their neighbours practised. There is a need for change. Agriculture must be considered as an enterprise, which should have a sound management back-up. As in any other enterprise, there should be proper planning about demand forecast, choice of technology, inventory of resources, need for external inputs, skill level of the available human resources and their training needs, infrastructure and services needed for carrying out various operations and marketing. This change in the mind set among the farmers and Agricultural Extension Agencies is the primary step for promoting successful entrepreneurship in agriculture.

Entrepreneurship in Agriculture

In the absence of local entrepreneurship, the opportunities in agriculture are high jacked by outsiders, particularly the urban businessmen and traders, leading to exploitation and deprivation of employment for the farmers. Considering the growing unemployment in rural areas and slow growth of the agricultural sector, it is necessary to tap the opportunities for promoting entrepreneurship in agriculture, which in turn can address the present problems related to agricultural production and profitability.

Types of Enterprises

While promoting entrepreneurship, we may consider different types of enterprises in agri-business.

1. **Farm Level Producers:** At the individual family level, each family is to be treated as an enterprise, to optimise the production by making best use of the technology, resources and demand in the market.

2. **Service Providers:** For optimising agriculture by every family enterprise, there are different types of services required at the village level. These include the input procurement and distribution, hiring of implements and equipment like tractors, seed drills, sprayers, harvesters, threshers, dryers and technical services such as installation of irrigation facilities, weed control, plant protection, harvesting, threshing, transportation, storage, etc. Similar opportunities exist in the livestock husbandry sector for providing breeding, vaccination, disease diagnostic and treatment services, apart from distribution of cattle feed, mineral mixture, forage seeds, etc.

3. **Input Producers:** There are many prosperous enterprises, which require critical inputs. Some such inputs which can be produced by the

local entrepreneurs at the village level are biofertilizers, biopesticides, vermicompost, soil amendments, plants of different species of fruits, vegetables, ornamentals, root media for raising plants in pots, agricultural tools, irrigation accessories, production of cattle feed concentrate, mineral mixture and complete feed. There are good opportunities to support sericulture, fishery and poultry as well, through promotion of critical service facilities in rural areas.

4. **Processing and Marketing of Farm Produce:** Efficient management of post-production operations requires higher scale of technology as well as investment. Such enterprises can be handled by People's Organisations, either in the form of cooperatives, service societies or joint stock companies. The most successful examples are the dairy cooperatives and fruit growers' cooperatives in many States. However, the success of such ventures is solely dependent on the integrity and competence of the leaders involved. Such ventures need good professional support for managing the activities as a competitive business and to compete well with other players in the market, particularly the retail traders and middlemen.

Problems of Entrepreneurship Development

Entrepreneurship in agriculture is not only an opportunity but also a necessity for improving the production and profitability. However, the rate of success is very low in India, because of the following reasons.

1. For most of the farmers, agriculture is mainly a means of survival. In the absence of adequate knowledge, resources, technology and connectivity with the market, it is difficult for the illiterate small holders to turn their agriculture into an enterprise.

2. Before promoting various services by self employed persons, there is a need to create awareness among the farmers, who are the users, about the benefits of these services.

3. For popularisation of services, the present practise of providing free service by the Government agencies should be discontinued. In fact, many farmers, particularly the politically connected leaders are of the impression that the government is responsible for providing extension and technical advisory services to the farmers. However, over the years, the credibility has eroded and the services of these agencies are not available to small farmers, particularly those living in remote areas. Nevertheless, the concept of free service makes the farmers reluctant to avail of paid services, offered by the local self-employed technicians.

4. The self-employed technicians need regular back up services in the form of technical and business information, contact with the marketing agencies, suppliers of critical inputs and equipment and research stations who are involved in the development of modern technologies.

5. There are several legal restrictions and obstacles, which come in the progress of agri-business, promoted by the People's Organisations and Cooperatives. Private traders engaged in such business tend to ignore these rules and disturb the fair trade environment.

6. People's Organisations often hesitate in taking the risk of making heavy investments and adoption of modern technologies, which in turn affect the profitability. With low profitability and outdated technologies, farmer members lose interest in their own enterprises as well as in that of their leaders.

Strategy for Promotion of Successful Enterprises

Considering the present problems faced by the entrepreneurs engaged in agri-business, it is necessary to create a congenial atmosphere in the field. Some of the important conditions necessary for successful agri-business are presented below:

1. There should be a unanimous option among government officials and farmers about the need and benefits of promoting self-employed youth or private entrepreneurs to facilitate the farmers to enhance agricultural production and profitability.

2. The Government should discontinue the practice of providing free services in those sectors where the work has been assigned to private entrepreneurs.

3. The technical skills and ability of the entrepreneurs should be evaluated to ensure high standards. There should be a monitoring agency to check the quality of the services and the charges collected from the farmers to avoid exploitation.

4. To popularise the services of the entrepreneurs, the Agricultural Extension Agencies and Farmers Organisations should give wider publicity about the services available to the farmers. Such publicity can enhance the credibility of the services provided by the entrepreneurs.

5. The Government should encourage the entrepreneurs by introducing various concessions and incentives.

6. Networks of entrepreneurs may be established to share their experiences. These networks can also establish a close link with Research Institutions and Universities to become acquainted with the latest research findings and seek solutions for their field problems.

Experience in Promoting Entrepreneurship in Agriculture

Carrot Pusa Rudhira Triggers Profitability and Entrepreneurship

Soodna village on the outskirts of newly carved Hapur district of western Uttar Pradesh is poised to become 'carrot village'. Thanks to *Pusa Rudhira*, an improved carrot variety developed and introduced by Indian Agricultural Research Institute (IARI), New Delhi. Soodna is one of the four villages adopted by IARI in the year 2010 for improving farm production and profitability in an integrated manner and to develop it into a 'model' village. Conventional cropping pattern of wheat-rice-sugarcane-vegetables was prevalent there. Farmers of this village were growing carrot among many other vegetables. However, carrot cultivation was not much remunerative before the introduction of *Pusa Rudhira*. Keeping in view the potential of carrot cultivation in the village, the IARI in 2011-12 introduced *Pusa Rudhira* on a small area of 1.75 acres of Sh. Charan Singh, a marginal farmer in the village.

Regular advisories were provided to optimise use of farm inputs in the crop. The farmer could harvest a bumper crop of *Pusa Rudhira* yielding 393.75 q/ha, which is about 10 q/ha higher than the prevailing variety. It provided a net return of Rs 2,64,286.00/ha, which was 37% higher than that of local variety. Superior quality traits of the newly introduced variety led to 18% higher price in the local market. *Pusa Rudhira* fetched higher price of Rs 928.00/q which was Rs 140.00/q higher than prevalent variety, due to its attractive long and deep red roots, red coloured core, uniformity in shape and size and more sweetness having TSS value of 9.5 oBrix.

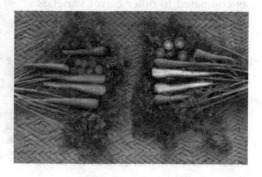

As seeing is believing, better performance of *Pusa Rudhira* in terms of higher productivity and profitability also interested other farmers of the village. During winters of 2012, another 20 farmers requested the seed of *Pusa Rudhira* and 200 kg seed was provided on cost basis. Consequently, *Pusa Rudhira* spread in about 60% area (90 acres) under carrot cultivation in the village within a year of its introduction. Enthused with the profits from *Pusa Rudhira,* carrot farmers adopted mechanized cleaning of carrots by procuring three cleaning machines on community sharing basis, which facilitated faster washing and also minimized damage to carrots. Higher productivity of *Pusa Rudhira* coupled with premium price in the next season provided farmers with an impressive net income of Rs 2,22,690.00/ha.Popularity of *Pusa Rudhira* has spread to different markets in Delhi and NCR. In rabi 2013-14, 120 acres (75%) of carrot area was under this variety. In years to come, *Pusa Rudhira* is expected to cover more and more area and remain dominant variety.

Consequently there is increasing demand for *Pusa Rudhira*. Sh Kamal Singh and Sh. Jaibhagwan Saini also marginal farmers, ventured to take up the seed production of the variety. The IARI scientists provided technical guidance and support. An extra income of Rs. 58000.00 is expected from sale of Rs 145 kg of carrot seed, which was produced in last year. No wonder these farmers are potential entrepreneurs of tomorrow.

Pusa Rudhira is also nutritionally rich as compared to other carrot varieties. The variety was tested to have higher levels of carotenoid (7.41mg) and phenols (45.15 mg) per 100 g. The primary benefit of these substances lies in their antioxidant property that guards against certain types of cancer, apparently by limiting the abnormal growth of cells. *Pusa Rudhira* is a boon to farmers and consumers as well.

(Source: CATAT, IARI, New Delhi)

Empowering Farm Women through Food Processing Entrepreneurship

Ujjain district has a predominance of Soybean based cropping system with only soybean during kharif in about 95 per cent area followed by chickpea, wheat, potato, garlic and onion in Rabi season. Earlier, farmers used to grow tomato with open pollinated varieties without giving due consideration to its production technology and hence got low yield of around 150 quintals per hectare. KVK, Ujjain after identifying the potential area promoted the cultivation of hybrid tomato in the year 2007 with all the appropriate technologies such as use of hybrid varieties, management of nursery, integrated nutrient and pest management, use of pro-trays for raising seedlings and special interventions such as use of boron to tackle physiological disorders in tomato. Today, tomato is cultivated in 3500 ha area in the district with an average productivity of 250 to 325 quintals per hectare.

Due to very limited food processing facilities in Madhya Pradesh, the farmers had no option except to sell their produce at very low rates sometimes as low as rupees 2 to 3 per kg. Keeping this in view, KVK took initiative to provide alternative intervention so that 35 per cent of the total produce is preserved in the form of home made products. For this a tomato grower group was randomly selected by the adopted cluster area of KVK namely village Dewarakhedi, Bhesoda and Kapeli. Various types of on and off training programmes, OFTs and FLDs were organized in the mentioned villages' especially on tomato preservation. Preservation of tomato products was demonstrated practically step wise to the farm women. For conducting the OFT and FLD purposively selected plants were tagged for preparation of 'ketchup'.The impact of conducted programmes was achieved in terms of formation of two SHGs viz. 'Ozen Self Help Group' and 'Jai Ma Durga' established in 2008 with a capital asset of over 1.2 and 1.5 lakh rupees collected from the members. Farm women learned the Vocational skill of preparing tomato ketchup at domestic level under the guidance of Home Scientist of KVK. Looking in to the potential of the SHG and the market avenues, KVK decided to extend its platform for marketing of the product in the brand name 'Raj Vijay Tomato Ketchup' which is an abbreviation of all the products marketed by Rajmata Vijayaraje Scindia Krishi Vishwa Vidyalaya (RVSKVV), Gwalior. After three years of frontline demonstrations the economics of tomato cultivation and that of preserved product was worked out which gives a clear picture and the farmers are able to earn 9.8 times more than what they use to earn during seasonal glut.

At present both the groups are capable of producing 2000 to 2500 kg of finished products. With market support the groups are planning to install a small processing unit with financial assistance from Bank of India in the next financial year.

Krishi Vigyan Kendra working at District level is bridging the gap between the farmers and research system by providing appropriate technology and platform for the farmers to strengthen their economic status.

Sustainable Farm Income from Integrated Farming in Arid Region

Shri *Nand Kishore Jaisalmeria* a farmer of Manaklao village of Jodhpur (25 km north) owns 3.1 ha agricultural land under rainfed conditions (average rainfall 365 mm), which was earlier a degraded land. His income from crops was very low during the year of drought. In the early 1980s, he came in contact with scientists of Central Arid Zone Research Institute (CAZRI) and planted 750 plants of improved  *(Source: KVK, Ujjain)* varieties of ber (Gola, Seb, Umran) provided by the institute, along with recommended package of practices. The piece of land was developed as an integrated farming system. Along with Ber orchard, he also maintained annual crops, intercropping, honey bee keeping and 10 + 1 goat unit.

Fencing of the farm was created by developing shelterbelts of multipurpose tree species like *Acacia tortillis, Prosopis juliflora*, Neem and other drought hardy tree species of economic value. The shelterbelt developed as fencing of the farm not only provided sustainable output in terms of fuel wood and fodder but also provided protection and conserved the soil and improved its fertility levels. For the first 20 years, the land was totally rainfed, and then a bore well was established whose water is saline. Ber is tolerant to saline water irrigation to some extent. The supplemental irrigation with saline water especially during fruit setting stage helped produce good yield. Even after 35 years, at present the farm and the orchard are well maintained and is a source of sustainable income.

The major revenue for the farm comes from sale of Ber fruit, rental charges for keeping Honey Bee Boxes, leaves of Ber and other trees, fuel wood from pruning and goats. Each plant of Ber produces about 30 kg fruits per annum. The farmer earns a total annual net income of about Rs 1,25,000 from 3 ha arid land giving a net income of about Rs.41,000/- per ha per annum, which is much higher as compared to traditional annual crops in the region. This perennials based farming system not only gave higher income but provided stability of production and income in the years of drought which is very common in this region. In addition to his own farm, Mr. Jaisalmeria also started a nursery of budded ber and sold lakhs of plants to the farmers, NGOs, Government departments from different states and contributed in spread of improved germplasm and package of practices of Ber cultivation.

ICAR Awardee Entrepreneur Invented Multipurpose Processing Unit

His life has seen a real rag to riches story. 46-year-old Dharamvir Kamboj, away from his village Dhamla of Yamnanagar in Haryana, used to pull a rickshaw till year 1987 on streets of Delhi. Besides little education, his conviction and innovative mind today has made him a successful entrepreneur with global reach.

Life has come full circle for Dharamvir as he got success at his doorstep after roaming on streets of metro city to make few hundred bucks. Inventor of the portable multipurpose processing unit, he is now providing employment to three dozen women of his village at his aloe vera and amla processing plant.The idea of inventing this machine came when he saw women in his village peeling herbals with their hands few years back. "They used to cuts their hands besides able to process very low quantity." Narrating his story, he said that once he used to pull a rickshaw for making his living. "The development of the machine has changed my life completely. Combination of grinder, processor and juicer, it has 200 kg per hour processing capacity for herbal products," he said, adding that besides selling more than 55 machines in India, he already exported few of

the machines to Kenya, making machines footprints on global map.Having only two acres of land at his native village, he started growing aloe vera and amla after the development of machine. "I am now producing juices, gel, sweets, shampo, face creams from Amla and Aloe vera and selling them in various parts of the country. The machine has made it possible to produce these products at large scale," he added. "Besides, more than 35 women are directly employed in my business and many in and around my village have started growing Amla and Aloe vera. This has made them prosperous too," he said. Awarded by various organizations including ICAR with Farmer scientist award in year 2010 by Honorable agriculture minister Shri Sharad Pawar, Dharamvir said that state horticulture department now provides subsidy on the machine developed by him, which is sold at price of Rs 1.35 lakh. His innovation received the fifth place at the National Competition of Grassroots Innovations and Traditional Knowledge organized by National Innovation Foundation in year 2009. Though, his condition could allow him to get quality education and study till 10th standard only, he has managed best of education for his children. Recently, he had displayed his machine at exhibition organized under NAIP Mass Media Project at Central Institute of Post Harvest Engineering and Technology, once again attracting attention of public and media. "I am earning around Rs 2 lakh per month," he said.

His story will continue to keep common man's hope alive that determination and hard work can change their destiny and rags to riches story not only exist in proverb.

(*Source:* NAIP Sub-Project on Mass Media Mobilization, DKMA with input from CIPHET, Ludhiana)

Processing using local solar tunnel drier

Shri Madan Lal Deora is a progressive farmer of Nimaz village of Pali, district of Rajasthan had established himself as a successful entrepreneur using self made tunnel solar drier after getting regular trainings. The solar tunnel drier is a poly house framed structure with UV-stabilized polythene sheet, where products on large scale could be dried under controlled environment. The enterprise consists of different aonla products including juice, squash, candy, powder, pickle, dry aonla, churan tablets, preserve and ladoos. Aonla pulp from moisture content of 80.95 % to safe moisture content 9.0 % can be successfully dried in 2 days at the prevailing conditions in the drier. Shri Madan Lal had become a successful entrepreneur and now he is motivating other farmers also to adopt his path. The enterprise is having a selling counter at the farmers home itself where the local people come and take various products. The farmer is earning Rs. 4.2 laks profits a year besides providing employment to many needy persons of his village.

Women as agri-enterpreneur

Srimati Surja Devi belongs to village Chimanpura and due to her poor financial conditions could study upto 5th standard only. She was very much interested in cooking and processing of different foods. As the result of KVK trainings, she learned the scientific methods of processing, preservation and value addition to different fruits and vegetables. Today Mrs. Surja Devi is a well known entrepreneur excelling in processed products and earns Rs. 400000 per annum by selling her homemade products. She concentrates primarily on the pickles prepared from raw mango, acid lime, green and red chillis, kair, sangria and aonla. In her range of fruit products includes aonla and karonda candy, fruit preserve prepared from aonla and bel and squashes prepared from orange and acid lime. One of the remarkable features of Surja Devi is sun dried products prepared from different fruits and vegetables which include kair, kumat, kachri, lasora, sangria, ber, chillis and aonla.

Conclusion

There is a need for efficient support organisations to monitor the activities of small enterprises. Prediction of the future demand, introduction of modern technologies, cost control and business expansions are the important areas, where entrepreneurs need regular support. Suitable legal support may also be required to protect the traders engaged in unfair trade practices.

It is better to promote agro-based enterprises in rural areas, as the local people have the required skills and most of the businesses help the entrepreneurs to ensure food security. The outputs of such business have ready demand even in rural areas and hence the market opportunities are better. With a strong agro-based programme, non-farm activities can also be initiated when the entrepreneurs are more experienced and capable of taking risk and can manage the programme better. Promotion of direct marketing by establishing close interaction between producers and consumers will further enhance the benefits, while encouraging a large number of unemployed rural youth to turn into micro-entrepreneurs and traders.

34

Organic Farming: Potential and Strategies in Arid Zone

Arun K.Sharma

ICAR-Central Arid Zone Research Institute, Jodhpur, Rajasthan–342003

In last few decades awareness about social and environmental issues has been increased. Agriculture is one of basic enterprise that has major role in these two issues . Organic agriculture is one of the agriculture production system that not only supportive to the environment but also sensitive to the social issues like employment, health, migration etc. Definitions given by two international organizations also verify this concept. These definitions are-

1. Organic agriculture is a holistic management system ,which enhances agro-ecosystem health, utilizing both traditional &scientific knowledge. Organic agriculture systems rely on ecosystem management rather than external agricultural inputs. (IFOAM, 2006)

2. Organic agriculture is an environmentally and socially Sensitive food supply system. The primary goal of organic agriculture is to optimize the health and productivity of independent communities of soil life, plants, animals and people.(FAO, 2002)

In simple words organic agriculture is the production system with the optimum utilization of local resources in such a way so that sustainability of production and wellness of the society & environment can be maintained for fairly long time.

Organic Agriculture Vs Conventional (Chemical) Agriculture-Ideological Differences

Although organic agriculture is seems to be just the exclusion of synthetic external inputs but it is the ideological differences with conventional agriculture (Sharma, 2001) that makes OA, friendly to society and environment. These differences are:

Organic Agriculture	Conventional (chemical) Agriculture
Holistic approach	Reductionist approach.
Decentralize production	Centralize production
Harmony with NATURE (Harness the *benefit*)	Domination on NATURE (Exploit for *profit*)
Diversity	Specialization
Input optimization(save more)	Output maximization(spend more)

Potential in Arid Zone

In India about 41% agriculture area receives rainfall less than 500 mm/year. These areas are mainly in Rajasthan, Haryana, Gujrat, Andhra Pradesh ,Karnataka, Maharashtra and Tamilnadu states. Low rainfall combines with high variability in rainfall, wide range of annual temperature range from 0°C to 49° C and light soils, makes a complex that makes agriculture is a challenge in these areas. Many a times crops either not able to complete life cycle or yield poorly due to this complex of climatic conditions. This variability has been further aggravated due to climate change effects as evidenced by trend of climatic data and experienced by the farmers in these areas in last decade. Agriculture systems with annuals (crops) and perennial (trees, shrubs and grasses) has been following traditionally in these areas to spread the risk of climatic uncertainty and to get food, fodder and fuel for sustenance of life. This is more or less subsistence agriculture only. In the last 4-5 decades several attempts have been made to improve productivity of this traditional system with the use of external inputs(synthetic) e.g. fertilizers, pesticides, weedicides but success limited to the good rainfall year as these inputs performs good with assured water availability, rather during below average rainfall years use of these synthetic inputs given negative results too.

In arid zone due to high risk associated with crop production, the use of agro chemicals is very less. Thus farmers are practicing organic agriculture since age by default. The system despite its low productivity compared to conventional system has established itself highly efficient in terms of resource recycling and providing better food and economic sustainability in the arid region (Sharma and Tewari, 2009). Some promising crops for production under organic system are cumin (spice), psyllium (medicine), sesame (oilseed), cluster bean (gum), moth bean (traditional confectionery items e.g. *papad, bhujia*), etc. Enhanced quality of these crops's products in the existing production system with better utilization of local resources could make a considerable contribution in strengthening the economic sustainability of this region (Sharma and Tewari, 2009). The improvement in this production system is possible by intervention in nutrient management and plant protection aspects. For nutrient management practice of ley agriculture, inclusion of legume in crop rotations and efficient recycling of biomass are important for the region. Even under prolonged dry spell it provided some produce to meet the farmer needs. For effective non-chemical plant protection some measures e.g. crop rotation, use of compost as soil application along with foliar spray of compost extract, application of oil cakes or residues of mustard with one summer irrigation and application of neem based pesticides have found promising. Thus organic cultivation of selective crops having higher export potential provides an opportunity to strengthen the economy of the region.

Productivity of traditional systems is low that is mainly due to inefficient methods of preparation and application of organic inputs. Therefore if these technologies can be improved and use integratedly there is every possibility of achieving sustainability combines with high productivity in these areas of second green revolution, as the support system for organic agriculture is already available.

No Constrain but Opportunity for Organic Agriculture

In arid zone where shortage of rainfall and light soils are constrains for intensive chemical input based conventional agriculture while these constrains become opportunities for organic agriculture. Hence organic agriculture not only suitable due to climatic uncertainties but also feasible due to availability of support system of the following favorable conditions(Sharma, 2011).

Low fertilizer use-early conversion : In these areas due to erratic pattern of rainfall, the rate of fertilizer application is very low (36.4kg/ha) as compared to national average of 76.8 kg/ha. This can be a good opportunity for early and easy conversion into organic agriculture (Sharma, 2001). Fortunately, according to the priority areas of National Programme for Organic Production (NPOP) of Ministry of Commerce and Ministry of Agriculture, this part of the country comes under the priority I. This may facilitate to get financial help for these areas too.

Diversified agriculture system :Traditional agriculture system in the region is highly diversified in nature that includes crops, trees, animals, grasses etc. This system is scientifically efficient in nutrient recycling and restoration of soil fertility. In these systems 10-30 trees/ha are available and 2-5 animals per reared by a farm family. This integrated agriculture system minimizes pest incidence as well as favors organic agriculture.

Employment opportunities : High density as well as high growth of human resource remains underutilized throughout the year due to erratic rainfall and limited irrigation facilities. Migration of human resources during drought imbalances the development of these areas. Since the organic agriculture is labour intensive and input preparation is made at local level, there is ample opportunity for round the year employment and proper utilization of human resource. Now under MNREGA scheme this work of water harvesting structure, input preparation will get good support. Therefore there are ample opportunities in arid areas for promotion of organic agriculture

Model Organic Farm (MOF) at CAZRI Jodhpur for Research on Low Rainfall Organic System

Considering the possibilities and to explore the potential through experimentation on organic production system for arid zone, a 2.0 ha model organic farm was established during 2008, within the central research farm CAZRI, Jodhpur. The farm registered for certification and got status of "Certified Organic Farm" in the year 2011. A rotation of four crops including cluster bean and sesame in rainy season (kharif) and cumin and psyllium in winter (rabi) was selected for the study. Yields were found to be increasing significantly with the increase in use of organic manures from 2.5 to 7.5 tons/ha in all the crops. Legume cultivation in kharif season on an average contributed to 25-30% increase in yield in the subsequent crops of cumin and psyllium.

Increase in soil moisture retention with the use of organic manure was observed that helped in better growth and yield of crops. Similarly increase in soil organic carbon from 0.23% to 0.29% was recorded after three year application of compost @ 5.0/ha. Crop resilience to climatic variability was enhanced with the use of organic manure.

There are general apprehensions that organic system is poor yielder. However, finding at MOF, CAZRI shows that, at initial developmental stage of organic system there may be reduction in yield but after 2-3 years, once the system developed the yield levels were comparable to the conventional (chemical input based) system. At third year yield of sesame 886.6kg/ha, cluster bean 630.2 kg/ha, cumin 516.9 kg/ha and psyllium 808.4 kg/ha was recorded. This was comparable to the average yield of conventional system (Sharma, 2013).

Recommendations for Organic System Management in Arid Zone

On the basis of experimentation done and experience gained at MOF following practice need to be integratedly used for successful organic agriculture in arid zone(Sharma, 2013)

- Soil incorporation of well decomposed compost @ 5.0 t/ha + 200 kg neem seed powder at the time of field preparation.

- Adoption of measures for rainwater harvesting and efficient utilization in high value crops

- Use of healthy organic seed, free from weed seed and treated with trichoderm@ 4.0g/kg seed, before sowing.

- Follow crop rotation essentially with the legumes.

- Installation of pheromone traps

- Regular visit to the field for crop health and timely protection measures

- Spray of Neem seed kernel extract @ 5.0% in the water extraction of neem + calotropis+ adhatoda leaves.

Promotion of Organic Agriculture:Need Integrated Efforts

Considering the export demand and contribution in the economy of this region; it is need of the hour to do integrated efforts for quality organic production . Policy plays major role in promotion of any programme. Policy in terms of supporting rules & regulations, subsidies, facilities, allocation of budget & personnel etc. can alone is sufficient if executed properly. The best example is Cuba(Latin America), where organic agriculture was made a national policy and now whole of the country is organic. Similarly some of the state like Uttranchal, NEH states etc. has declared organic state and they are taking lead. Although in India, organic movement was started in early 80s but it got momentum only after 2001 when govt. of India lunched National Programme on Organic Agriculture (NPOF). Later on most of govt. agencies have started to give priority to organic agriculture. However simply giving budget, subsidies etc. may not be sufficient to promote organic agriculture, as least development has been done in arid & semi arid areas even, they are kept at Priority I & II in NPOF. For better development of organic agriculture in arid zone additional measures need to be taken. Some of the proposed policy initiatives (Sharma, 2011) can be

1. Priority to organic agriculture in ongoing all rural development programmes e.g. watershed,SGSY,MNREGA. Food security mission, horticulture mission etc.

2. Popularization of organic agriculture without compulsion of certification and it should be promoted for improving soil fertility, reducing cost of production and other environmental advantages.

3. Dissemination of organic agriculture in holistic manner that includes of all these aspects which can make a sustainable organic agriculture in real term.

4. Coordination among supporting agencies for Integrated efforts Encouragement of decentralized input supply

5. Demonstrations, training, conferences, seminars, farmers fair etc. may be organized to make a general consensus about organic agriculture and good organic management.

6. Provision of subsidy may be made for organic inputs to make organic produce more competitive.

7. The demand of spices and medicinal plant is increasing when grown organically, so it must be promoted organically in the various agro-climatic regions of the drylands.

Table 1: Potential high-value crops for organic agriculture in different agro-climatic regions

Agro-climatic zone	Rainfall average (mm)	Suitable species and medicinal crop
Arid	100-250	Cumin, Senna, Guggul, Psyllium,
Transitional	250-350	Fenugreek, Cumin, , Aloe, Senna
Semi arid	350-500	Fennel, Asvagandha, Fenugreek, Coriander, Lucorice, Aloe, Aonla

Conclusion

Organic agriculture is a holistic production system run with the efficient use and recycling of locally available resources. Due to scarcity of water and light soils areas it is best suitable and applicable to low rainfall areas. Some monopoly high value crops of this region like seed spices are having great international demand if produce organically. Organic production in arid zone not only boosts the economy but also sustain the productivity of natural resources. The need is to do research on development of feasible & economic technologies, development of processing and marketing infrastructure, financial as well as technical support for quality organic production and strong policy support from government.

References

FAO. 2002. Organic Agriculture ,environment and food security. FAO, Rome, Italy, P.252.

IFOAM. 2006 . http://www.ifoam.org/growing_organic/definitions/doa/index.html

Sharma, A.K. and Tewari, J.C. 2009. Improvement in Traditional Agroforestry Systems with Organic Inputs in Arid Zone. In Conservation Farming (eds.), S. Bhan and R.L. Karle. Soil Conservation Society of India, New Delhi. pp. 292-295.

Sharma, A.K. 2001. A handbook of organic farming. Agrobios, Jodhpur, P.626.

Sharma, A.K. 2011.Action plan for organic seed spices production. In: Recent advances in seed spices (Eds.)Y.Rabindrababu,R.K.Jaiman and K.D.Patel. Daya Pub.House. New Delhi. Pp.85-93.

Sharma, A.K. 2013. Organic system management. *Indian Farming*, 63(4): 13-16.

35

Organic Certification Standards and Agencies

Arun K. Sharma

ICAR-Central Arid Zone Research Institute, Jodhpur Rajasthan–342003

Organic certification is the procedure by which the accredited certification Body by way of a Scope Certificate assures that the production or processing system of the operator has been methodically assessed and conforms to the specified requirements as envisaged in the National Programme for Organic Production (NPOP).

The NPOP provides for Standards for organic production, systems, criteria and procedure for accreditation of Certification Bodies, the National (India Organic) Logo and the regulations governing its use. The standards and procedures have been formulated in harmony with other International Standards regulating import and export of organic products.

Organic standards

Under organic certification certain standards have to be maintained. In organic crop production, management should cover a diverse planting scheme. For perennial crops, this should include plant-based ground cover crops. For annual crops, this should include diverse crop rotation practices, cover crops (green manures), intercropping or other diverse plant production methods.

1. Crop Production Plan

The producer seeking certificate under the NSOP is required to develop an organic crop production plan. This plan include: Description of the crops in the production cycle (main crop and intercrop) as per the agro climatic seasons.

  i. Description of practices and procedures to be performed and maintained.

  ii. List of inputs used in production along with their composition, frequency of usage, application rate and source of commercial availability.

iii.   Source of organic planting material (seeds and seedlings).

iv.   Descriptions of monitoring practices and procedures to be performed and maintained to verify that the plan is being implemented effectively.

v.   Description of the management practices and physical barriers established to prevent commingling and contamination of organic production unit from conventional farms, split operations and parallel operations.

vi.   Description of the record keeping system implemented to comply with the requirements.

2. Conversion Requirements

i.   The establishment of an organic management system and building of soil fertility requires an interim period, known as the conversion period. While the conversion period may not always be of sufficient duration to improve soil fertility and for re- establishing the balance of the ecosystem, it is the period in which all the actions required to reach these goals are started.

ii.   A farm may be converted through a clear plan of how to proceed with the conversion. This plan shall be updated by the producer, if necessary and shall cover all requirements to be met under these standards.

iii.   The requirements prescribed under these standards shall be met during the conversion period. All these requirements shall be applicable from the commencement of the conversion period till its conclusion.

iv.   The start of the conversion period may be calculated from the date first inspection of the operator by the Certification Body.

v.   A full conversion period shall not be required where de facto requirements prescribed under these standards have been met for several years and where the same can be verified on the basis of available documentation. In such cases inspection shall be carried out in reasonable time intervals, before the first harvest.

3. Duration of conversion period

i.   In case of annual and biennial crops, plant products produced can be certified organic when the requirements prescribed under these Standards have been met during the conversion period of at least two (2) years (organic Management) before sowing (the start of the production cycle).

ii.   In case of perennial plants other than grassland (excluding pastures and meadows), the first harvest may be certified as organic after at least thirty six (36)months of organic management according to the requirements prescribed under these Standards.

iv.   The accredited Certification Bodies shall decide in certain cases, for extension or reduction of conversion period depending on the past status/ use of the land and environmental condition.

v.   Twelve months reduction in conversion period could be considered for annuals as well as perennials provided, documentary proof has been available with the accredited Certification Body that the requirements prescribed under these Standards have been met for a period of minimum three years or more. This could include the land that been certified for minimum three years under the 'Participatory Guarantee System' implemented by the Ministry of Agriculture and wherein, the products approved for use in organic farming as listed in Annex 1 and 2 of this Appendix have been applied. The accredited Certification Bodies shall also consider such a reduction in conversion period, if it has satisfactory proof to demonstrate that for three years or more, the land has been idle and/or it has been treated with the products approved for use in organic farming.

vi.   Organic products in conversion shall be sold as "produce of organic agriculture in conversion" or of a similar description, when the requirements prescribed under these Standards have been met for at least twelve months.

4. Landscape

i.   Organic farming shall contribute beneficially to the ecosystem. The certification programme shall set standards/procedures for a minimum percentage of the farm area to facilitate biodiversity and nature conservation.

ii.   Areas which are managed organically shall facilitate biodiversity, *inter alia*, in the following manner:

iii.   Extensive grassland such as moorlands, reed land or dry land

iv.   In general all areas which are not under rotation and are not heavily manured.

v.   Extensive pastures, meadows, extensive grassland, extensive orchards, hedges, hedgerows, groups of trees and/or bushes and forest lines.

vi. Ecologically rich fallow land or arable land.

vii. Ecologically diversified (extensive) field margins.

viii. Waterways, pools, springs, ditches, wetlands and swamps and other water rich areas which are not used for intensive agriculture or aqua production.

5. Choice of Crops and Varieties

i. All seeds and plant material shall be certified organic. Species and varieties cultivated shall be adapted to the soil and climatic conditions and be resistant to pests and diseases. In the choice of varieties, genetic diversity shall be taken into consideration.

ii. When organic seed and plant materials are available, they shall be used.

iii. When certified organic seed and plant materials are not available, chemically untreated conventional seed and plant material shall be used.

iv. The use of genetically engineered seeds, transgenic plants or plant material is prohibited.

6. Diversity in Crop Production & Management Plan

i. The basis for crop production in organic farming shall take into consideration the structure and fertility of the soil and the surrounding ecosystem, with a view to minimize nutrient losses.

ii. Where appropriate, the organic farms shall be required to maintain sufficient diversity in a manner that takes into account pressure from insects, weeds, diseases and other pests, while maintaining or increasing soil, organic matter, fertility, microbial activity and general soil health. For non perennial crops, this is normal, but not exclusive, achieved by means of crop rotation preferably by leguminous crops.

iii. Soil fertility shall be maintained through, among other things, the cultivation of legumes or deep rooted plants and the use of green manures, along with the establishment of a programme of crop rotation several times a year and fertilization with organic inputs.

7. Nutrient Management

i. Sufficient quantities of biodegradable material of microbial, plant or animal origin produced on organic farms shall form the basis of the nutrient management programme to increase or at least maintain its fertility and the biological activity within it.

ii. Fertilization management should minimize nutrient losses. Accumulation of heavy metals and other pollutants shall be prevented.

iii. Non synthetic mineral fertilisers and brought-in bio fertilisers (biological origin) shall be regarded as supplementary and not as a replacement for nutrient recycling.

iv. Desired pH levels shall be maintained in the soil by the producer.

v. The certification programme shall set limitations to the total amount of biodegradable material of microbial, plant or animal origin brought onto the farm unit, taking into account local conditions and the specific nature of the crops.

vi. The certification programme shall set procedures which prevent animal runs from becoming over manuring where there is a risk of pollution.

vii. Permission for use shall only be given when other fertility management practices have been optimized

8. Pest, Disease and Weed Management

i. Organic farming systems shall be carried out in a way which ensures that losses from pests, diseases and weeds are minimized. Emphasis is placed on the use of a balanced fertilizing programme, use of crops and varieties well-adapted to the environment, fertile soils of high biological activity, adapted rotations, intercropping, green manures, etc. Growth and development shall take place in a natural manner.

ii. Weeds, pests and diseases shall be controlled through a number of pre-ventive cultural techniques which limit their development in a balanced nutrient management programme, e.g. suitable rotations, green manures, early and pre drilling seedbed preparations, mulching, mechanical control and the disturbance of pest development cycles. Accredited certification programmes shall ensure that measures are in place to prevent transmission of pests, parasites and infectious agents.

iii. Pest management shall be regulated by understanding and disrupting the ecological needs of the pests. The natural enemies of pests and diseases shall be protected and encouraged through proper habitat management of hedges, nesting sites etc. An ecological equilibrium shall be created to bring about a balance in the pest predator cycle.

iv. Products used for pest, disease and weed management, prepared at the farm from local plants, animals and microorganisms, shall be allowed. If the ecosystem or the quality of organic products might be jeopardized, the certification programme shall judge if the product is acceptable as per the procedure given to evaluate additional inputs to organic agriculture.

v.   Thermic weed control and physical methods for pest, disease and weed management shall be permitted.

vi. Thermic sterilization of soils to combat pests and diseases shall be restricted to circumstances where a proper rotation or renewal of soil cannot take place. The certification programme on a case-by-case basis may only give permission.

vii. All equipment from conventional farming systems shall be properly cleaned and free from residues before being used on organically managed areas.

9. Contamination Control

i.   All relevant measures shall be taken to minimize contamination from outside and within the farm.

ii. Buffer zones shall be maintained to prevent contamination from conventional farms. The buffer Zone should be sufficient in size to prevent the possibility of unintended contact of prohibited substances applied to adjacent conventional land areas/farms.

iii. In case of reasonable suspicion of contamination, the certification programme shall make sure that an analysis of the relevant products and possible sources of pollution (soil and water) shall take place to determine the level of contamination.

iv. Polyethylene and polypropylene or other polycarbonates coverings such as plastic mulches, fleeces, insect net and silage wrapping, only are allowed. These shall be removed from the soil after use and shall not be burnt on the farmland. The use of polychloride based products is prohibited.

10. Soil and Water Conservation

i.   Soil and water resources shall be handled in a sustainable manner. Relevant measures shall be taken to prevent erosion, salination of soil, excessive and improper use of water and the pollution of ground and surface water.

ii. Clearing of land through the means of burning organic matter, e.g. slash-and-burn, straw burning shall be restricted to the minimum. The clearing of primary forest is prohibited.

iii.   The certification programme shall require to check appropriate stocking rates which does not lead to land degradation and pollution of ground and surface water.

This is broad outline of standards for perennial and annual crop production under certified system. In the following tables list of permitted and restricted materials are given.

Products for Use in Fertilizing and Soil Conditioning

In organic agriculture the maintenance of soil fertility may be achieved through the recycling of organic material whose nutrients are made available to crops through the action of soil micro organisms. Many of these inputs are restricted for use in organic production. In this annex "restricted" means that the conditions and the procedure for use shall be subjected to condition. Factors such as contamination, risk of nutritional imbalances and depletion of natural resources shall be taken into consideration.

Table 1: Products to be used as fertilizes in organic farming

Inputs	Condition for use
Matter Produced on an Organic Farm Unit	
Farmyard & poultry manure, slurry, cow urine	Permitted
Crop residues and green manure	Permitted
Straw and other mulches	Permitted
Matter Produced Outside the Organic Farm Unit	
Blood meal, meat meal, bone meal and feather meal without Preservatives	Restricted
Compost made from any carbon based residues(animal excrement including poultry)	Restricted
Farmyard manure, slurry, cow urine (preferably after control fermentation and/or appropriate dilution) "factory" farmings ources not permitted Fish and fish products without preservatives	Restricted Restricted
Guano	Restricted
Human excrement	Prohibited
By-products from the food and textile industries of biodegradablematerial of microbial, plant or animal origin without anysynthetic additives	Restricted Prohibited for
Peat without synthetic additives	soil conditioning
Sawdust, wood shavings, wood provided it comes from untreated wood	Permitted
Seaweed and seaweed products obtained by physical processesextraction with water or aqueous acid and/or alkaline solution	Restricted
Sewage sludge and urban composts from separated sources whichare monitored for contamination	Restricted
Straw	Restricted
Vermicasts	Restricted
Animal charcoal	Restricted
Compost and spent mushroom and vermiculate substances	Restricted

Contd.

Compost from organic household reference	Restricted
Compost from plant residues	Permitted
By products from oil palm, coconut and cocoa (including empty fruit bunch, palm oil mill effluent (pome), cocoa peat and empty cocoa pods)	Restricted
By products of industries processing ingredients from organicagriculture	Restricted
Minerals	
Basic slag	Restricted
Calcareous and magnesium rock	Restricted
Calcified seaweed	Permitted
Calcium chloride	Permitted
Calcium carbonate of natural origin (chalk, limestone, gypsumand phosphate chalk)	Permitted
Mineral potassium with low chlorine content (e.g. sulphate ofpotash, kainite, sylvinite, patenkali)	Restricted
Natural phosphates (e.g. Rock phosphates)	Restricted
Pulverised rock	Restricted
Sodium chloride	Permitted
Trace elements (Boron, Ferrous, Manganese, Molybdenum, Zinc)	Restricted
Wood ash from untreated wood	Restricted
Potassium sulphate	Restricted
Magnesium sulphate (Epson salt)	Permitted
Gypsum (Calcium sulphate)	Permitted
Silage and silage extract kainite, sylvinite, patenkali)	Permitted excluding
Aluminum calcium phosphate	Restricted
Sulphur	Restricted
Stone meal	Restricted
Clay ((bentonite, perlite, zeolite)	Permitted
Microbiological Preparations	
Bacterial preparations (biofertilizers)	Permitted
Biodynamic preparations	Permitted
Plant preparations and botanical extracts	Permitted
Vermiculate	Permitted
Peat	Permitted

"Factory" farming refers to industrial management systems that are heavily reliant on veterinary and feed inputs not permitted in organic agriculture.

Products for Plant Pest and Disease Control

Certain products are allowed for use in organic agriculture for the control of pests and diseases in plant production. Such products should only be used when absolutely necessary and should be chosen taking the environmental impact into consideration.

Many of these products are restricted for use in organic production. In this annex "restricted" means that the conditions and the procedure for use shall be subjected to conditions.

Table 2: Product to be used/unused for plant protection in organic farming

Inputs	Condition for use
Substances from plant and animal origin	
Azadirachta indica (neem preparations)	Permitted
Neem oil	Restricted
Preparation of rotenone from *Derris elliptica Lonchocarpus, Thephrosia spp*	Restricted
Gelatine	Permitted
Propolis	Restricted
Plant based extracts – garlic, pongamia etc.	Permitted
Preparation on basis of pyrethrins extracted from *Chrysanthemum cinerariaefolium*, containing possibly a synergist *Pyrethrum cinerafolium*	Restricted
Preparation from *Quassia amara*	Restricted
Release of parasite predators of insect pests	Restricted
Preparation from *Ryania species*	Restricted
Tobacco tea	Prohibited
Lecithin	Restricted
Casein	Permitted
Sea weeds, sea weed meal, sea weed extracts, sea salt and salty water	Restricted
Extract from mushroom (Shitake fungus)	Permitted
Extract from Chlorella	Permitted
Fermented product from Aspergillus	Restricted
Natural acids (vinegar)	Restricted
Minerals	
Chloride of lime/soda	Restricted
Clay (e.g. bentonite, perlite, vermiculite, zeolite)	Permitted
Copper salts / inorganic salts (Bordeaux mix, copper hydroxide, copper oxychloride) used as a fungicide depending upon the crop and under the supervision of accredited Certification Body	Restricted
Mineral powders eg : stone meal	Prohibited
Diatomaceous earth	Restricted
Light mineral oils	Restricted
Permanganate of potash	Restricted
Lime sulphur (calcium polysulphide)	Restricted
Silicates, clay (Bentonite)	Restricted
Sodium bicarbonate	Restricted
Inputs	Condition for use
Sulphur (as a fungicide, acaricide, repellant)	Restricted
Microorganism used for biological pest control	
Viral preparation (eg. Granulosis virus, Nuclear Polyhedrosis Virus	Permitted
Fungal preparations (*Trichoderma spp.*)	Permitted
Bacterial preparations (*Bacillus spp)*	Permitted
Parasites, Predators and sterilized insects	Permitted
Others	
Carbon dioxide and nitrogen gas	Restricted
Soft soap (potassium soap)	Permitted
Ethyl alcohol	Prohibited
Homeopathic and Ayurvedic preparations	Permitted

Herbal and biodynamic preparations	Permitted
Traps	
Physical methods (Chromatic traps, Mechanical traps, sticky trapsand Pheromones	Permitted

List of certification agencies approved by APEDA, Govt. of India

1. Bureau Veritas Certification India (BVCI) Pvt. Ltd.Andheri (East), Mumbai-400 072

2. ECOCERT India Pvt. Ltd.,Gurgaon – 122018, Haryana, India

3. IMO Control Pvt. Ltd., Bangalore-560 008.

4. Indian Organic Certification Agency, Cochin (Kerala)

5. Lacon Quality Certification Pvt. Ltd., Thiruvalla - 689 101 (Kerala)

6. OneCert Asia Agri Certification (P) Ltd., Jaipur-302020, Rajasthan

7. SGS India Pvt. Ltd., Gurgaon-122016, Haryana

8. Control Union Certifications, Navi, Mumbai - 400709

9. Uttarakhand State Organic Certification Agency, Dehradun, Uttarakhand

10. APOF Organic Certification Agency, Bangalore-560 004 (Karnataka)

11. Rajasthan Organic Certification Agency (ROCA), Jaipur-302005 Rajasthan

12. Vedic Organic Certification Agency, Hyderabad-500050

13. ISCOP Coimbatore – 641 001

14. Food Cert India, Secunderabad - 500 003, Telangana

15. Aditi Organic Certifications Pvt. Ltd, Bangalore - 560010

16. Chhattisgarh Certification Society, India (CGCERT), Raipur- 493 111

17. Tamil Nadu Organic Certification, Coimbatore

18. Intertek India Pvt. Ltd., New Delhi - 110 044

19. Madhya Pradesh State Organic Certification Agency, Bhopal - 462 023

20. Biocert India Pvt. Ltd, Pune - 411041

21. Odisha State Organic Certification Agency (OSOCA), Bhubaneswar-751003

22. Natural Organic Certification Agro Pvt. Ltd., Pune - 411058

23. Fair Cert Certification Services Pvt. Ltd., Khargone-451001